Advanced in Remote Sensing Approaches for Agricultural Monitoring at Field and Regional Scale

Advanced in Remote Sensing Approaches for Agricultural Monitoring at Field and Regional Scale

Guest Editors

Jose Antonio Dominguez-Gómez
Jitka Kumhálová

MDPI

Basel • Beijing • Wuhan • Barcelona • Belgrade • Novi Sad • Cluj • Manchester

Guest Editors

Jose Antonio Dominguez-Gómez
Ingeniería Topográfica y Cartografía
Universidad Politécnica de Madrid
Madrid
Spain

Jitka Kumhálová
Department of Vehicles and Ground Transport
Czech University of Life Sciences Prague
Prague
Czech Republic

Editorial Office

MDPI AG
Grosspeteranlage 5
4052 Basel, Switzerland

This is a reprint of the Special Issue, published open access by the journal *Remote Sensing* (ISSN 2072-4292), freely accessible at: https://www.mdpi.com/journal/remotesensing/special_issues/RAVD7YUXWI.

For citation purposes, cite each article independently as indicated on the article page online and as indicated below:

Lastname, A.A.; Lastname, B.B. Article Title. *Journal Name* **Year**, *Volume Number*, Page Range.

ISBN 978-3-7258-5843-9 (Hbk)
ISBN 978-3-7258-5844-6 (PDF)
https://doi.org/10.3390/books978-3-7258-5844-6

Contents

remote sensing

MDPI

Article

Field-Level Classification of Winter Catch Crops Using Sentinel-2 Time Series: Model Comparison and Transferability

Kato Vanpoucke [1,2,*], Stien Heremans [1,2], Emily Buls [3] and Ben Somers [2,4]

1 Research Institute for Nature and Forest (INBO), Havenlaan 88, 1000 Brussels, Belgium
2 Division of Forest, Nature and Landscape (FNL), Department of Earth and Environmental Sciences, KU Leuven, Celestijnenlaan 200E, 3001 Leuven, Belgium
3 Vlaamse Landmaatschappij (VLM), Koning Albert II-laan 15, 1210 Brussels, Belgium
4 KU Leuven Plant Institute (LPI), Kasteelpark Arenberg 31, 3001 Leuven, Belgium
* Correspondence: kato.vanpoucke@inbo.be

Abstract: Winter catch crops are promoted in the European Union under the Common Agricultural Policy to improve soil health and reduce nitrate leaching from agricultural fields. Currently, Member States often monitor farmers' adoption through on-site inspections for a limited subset of parcels. Because of its potential for region-wide coverage, this study investigates the potential of Sentinel-2 satellite time series to classify catch crops at the field level in Flanders (Belgium). The first objective was to classify catch crops and identify the optimal model and time series input for this task. The second objective was to apply these findings in a real-world scenario, aiming to provide reliable early-season predictions in a separate target year, testing early-season performance and temporal transferability. The following three models were compared: Random Forest (RF), Time Series Forest (TSF), and a One-Dimensional Convolutional Neural Network (1D-CNN). The results showed that, with a limited field-based training dataset, RF produced the most robust results across different time series inputs, achieving a median F1-score of >88% on the best dataset. Additionally, the early-season performance of the models was delayed in the target year, reaching the F1-score threshold of 85% at least one month later in the season compared to the training years, with large timing differences between the models.

Keywords: catch crop; early classification; temporal transferability; machine learning; satellite image time series; Sentinel-2

1. Introduction

When mineral nitrogen remains in the soil after harvest of the main crop, nitrate leaching can lead to elevated nitrate levels in surface and groundwater, adversely affecting the environment and human health [1]. For this reason, the Nitrates Directive in the European Union aims to reduce water pollution from agricultural sources (91/676/EEC) and thereby requires Member States to monitor their water quality and implement good agricultural practices. An example of such practice is the cultivation of catch crops, which are also supported by the Common Agricultural Policy (CAP) in the context of soil health and are often referred to as cover crops in this context [2]. Catch crops are grown from late summer to early spring, bridging the gap between main cropping seasons [3], with the goal of reducing nitrate leaching. Replacing fallow with catch crops reduces nitrate leaching through multiple mechanisms, including active uptake of both nitrogen and water [4,5]. Their effectiveness is influenced by factors such as (i) biomass, (ii) species, and (iii) sowing and termination dates, with a recent meta-analysis showing an average reduction in nitrate leaching by 68% [4,5].

Monitoring the adoption of catch crops by farmers currently relies on on-site field visits by authorities [3]. Field visits are, however, very labor- and time-consuming, and therefore, only a small fraction of all fields can be visited. To get a more complete overview of the extent to which catch crops are adopted across a region, satellite remote sensing can be used to map catch crop presence and prioritize field visits to areas of higher uncertainty [3]. Satellite remote sensing offers large-scale coverage and high spatio-temporal resolution and has already been successfully used in crop monitoring and policy support for the CAP [6].

Catch crop monitoring with remote sensing has not yet been researched extensively, but the field has recently been growing due to the mainstreaming of these cropping practices combined with the recent rise in the availability of satellite data with sufficiently high spatial, spectral, and temporal resolution [7]. As outlined by [8], past studies have focused on three main aspects of catch crops: (a) their presence [3], (b) their emergence and termination date [9], and (c) their biomass [8,10,11]. These studies applied a wide range of methods, including phenology through optical satellite time series, to monitor catch crops [3,12,13]. The use of phenological profiles can help differentiate catch crops from other crops and vegetation because the spectral differences between vegetation types may be small at any single moment in time, while differences in timing may be more pronounced [14]. Especially for crop type classification, the different growth patterns and differences in the timing of the growth stages can be used as discriminating features [15,16]. Distinguishing between catch crop types can help estimate their effect on N-leaching, as different types have different nitrogen uptake capacities. Both the type of catch crop and its biomass accumulation influence the amount of nitrogen retained, making catch crop type an important variable to consider when modeling N-uptake by catch crops across a region [11].

Simultaneously, with the increased focus on time series data, data-driven machine learning models have become prominent in (catch) crop monitoring [3,17,18]. For example, for the classification of catch crops, ref. [3] used satellite time series and Random Forest (RF), a model that is also popular in main crop classification [18,19]. Some advantages of RFs are their easy-to-tune parameters and their robustness to overfitting [17,20,21]. However, RFs do not explicitly take into account the temporal dependencies between the observations, as they handle each time step as an independent variable. This drawback has been overcome by the introduction of a new branch of RFs specifically designed for time series classification and prediction. Time Series Forest (TSF) is an example of such an RF where temporal features—mean, standard deviation, and slope—are calculated over predefined (randomly chosen) temporal intervals, and these deduced features are subsequently used for classification [22]. The model, therefore, incorporates crop phenology by explicitly using the temporal dimension.

Over the last few years, deep learning algorithms have also appeared in crop monitoring research [15,23]. Different kinds of deep learning architectures have been compared for the task of crop classification, such as RNNs, MS-ResNet, and TempCNN [23]. Notably, TempCNN was introduced by [17], exploring the use of One-Dimensional CNNs (1D-CNN) for time series-based crop type classification, a lightweight model compared to other deep learning architectures. The use of deep neural networks comes with two major challenges: (1) the expertise required for optimizing both the model architecture and the hyperparameter values, shifting the difficulty from feature engineering to architecture engineering, and (2) the large number of parameters that have to be estimated, which makes deep learning methods prone to overfitting and thus reliant on a large set of labeled training data [17].

Even though deep learning models have shown promising results in main crop type mapping, catch crop monitoring has predominantly relied on machine learning models to date [3,24,25]. However, for catch crop mapping, the use of a Multilayer Perceptron network (MLP) has demonstrated improvements over the use of machine learning methods [26]. To our knowledge, this study is one of the few to compare traditional machine learning with deep learning methods for the end-of-season classification of catch crops while using optical time series input data. In contrast with traditional machine learning algorithms, deep learning models can learn feature representations from raw input (time series) without feature engineering [15,23,27], but in real-world applications with limited training samples, optimizing the input dataset in both the time and spectral dimensions may still be necessary [14].

Moreover, deep learning models have shown promising results in early-season classification [19], which is useful when applying the models in the operational scenario of guiding in-season fieldwork to areas where the catch crop presence is most uncertain. In this application, the models also need to be generalizable in time, i.e., transferable between years. However, as most crop classification studies use training and testing data from the same year, the generalizability of models across years has not been well studied [28]. The temporal generalization of models, also named temporal transferability, is a crucial aspect of early-season classification in an operational setting when only training labels from previous years are available [29]. Specifically in the context of early-season mapping, this topic has received increasing attention recently [19,28–31].

The main objective of this paper is to test the potential of different machine learning and deep learning methods that differ in the way they handle the temporal dimension of the data for classifying catch crops from a time series of medium-resolution satellite images. We will also assess the impact of (1) the temporal resolution of the input time series and (2) the spectral resolution of the input features in order to further optimize the performance of the classification models. In the second objective, the optimized models are applied in a real-world scenario, where we aim to test the potential of these models for early-season catch crop classification. In this case, the target year is different from the training year(s), so this is an assessment of temporal model transferability. These research objectives give rise to four research questions:

(1) What is the (relative) accuracy of different machine learning and deep learning classifiers for winter crop classification from optical satellite time series?
(2) To what extent do the temporal and spectral resolution of the input data affect the models' classification accuracy?
(3) How early in the season can the presence of catch crops reliably be detected to help prioritize field visits in areas of higher uncertainty?
(4) How well do the trained models maintain their accuracy when applied to a target year not included in the training dataset?

2. Materials and Methods

This section will first describe the data that were used, then the models and their training process, and lastly, the experimental setup. An overview of the workflow is given in Figure 1.

2.1. Study Area

The study area covers the region of Flanders, which is located in the North of Belgium and is characterized by a temperate climate with warm summers and the absence of a dry season according to the Köppen–Geiger classification (Figure 2) [32]. Agriculture covers a total of 45% of the region's total land area, of which 55% is used for fodder production and 35% for arable farming [33]. In Flanders, the cultivation of catch crops in winter has been introduced as a measure to decrease nitrate leaching from agricultural fields. Reference data for modeling was collected throughout the region, with a focus on the municipalities with the highest nitrogen concentrations in their surface and groundwater. These municipalities are spread over Flanders and are located in the main agricultural clusters of the region.

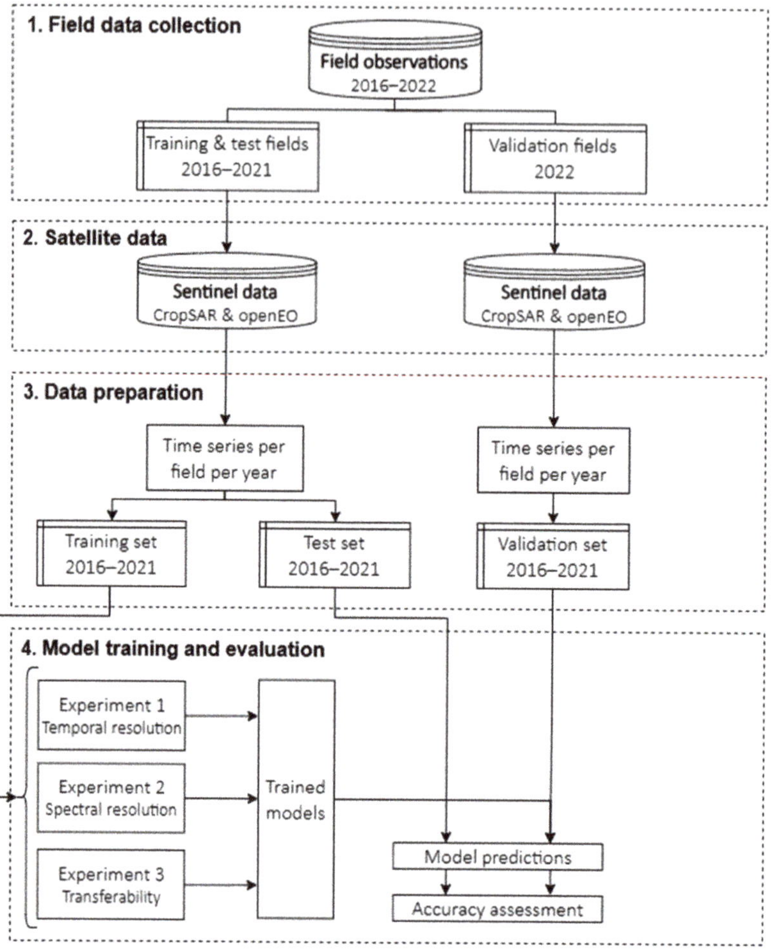

Figure 1. Highlights of the applied workflow. Three experiments were conducted: The first and second experiments aimed to study the influence of temporal resolution and input features on the classification, respectively on the classification. The third experiment evaluated the models' performance early in the winter season and their temporal transferability.

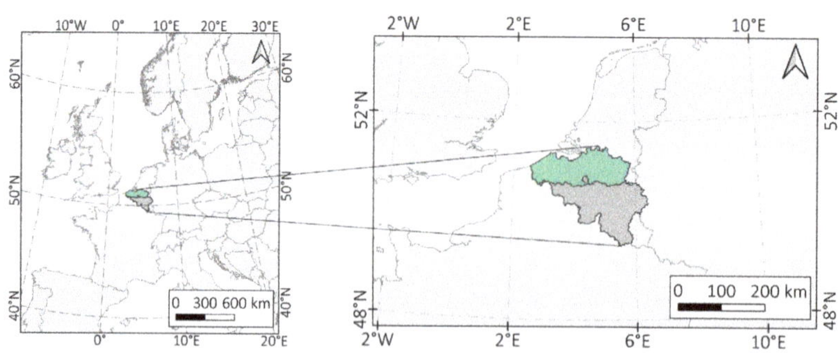

Figure 2. The location of the study region in Europe; Flanders is colored in cyan. Administrative borders: European Environment Agency (www.eea.europa.eu/datahub (accessed on 22 November 2024)).

2.2. Data

2.2.1. Reference Data

Field observations (2016–2021) were used as ground truth labels and were gathered by the Agency for Agriculture and Fisheries in the context of monitoring farmers' compliance and, more broadly, EU greening measures. Because this dataset was collected in a broader context, it contained unnecessary observations and, therefore, needed preprocessing, which is described in Appendix A. The ground truth labels obtained for this dataset consisted of the (catch) crop species present on a field during the winter period, from approximately October to January.

In addition to this dataset, extra field observations were collected during two dedicated field campaigns in collaboration with the Flemish Land Agency (VLM), the government agency responsible for monitoring farmers' compliance with the catch crop regulation. The two field campaigns were carried out in the winter seasons (October to January) of 2021 (i.e., winter 2021–2022) and 2022 (i.e., winter 2022–2023) and were focused on clusters of municipalities with the lowest water quality, which, due to regulations, had a high presence of catch crops. Data from the first field campaign were added to the training dataset, while those from the second field campaign were added to an independent validation set used for testing the temporal transferability of the trained models. Between field campaigns, the field selection process differed slightly. In the first field campaign (winter 2021–2022), fields were selected based on an annual farmer survey and a stratified random sample proportional to the declared catch crops, ensuring an even distribution across all municipalities. The second field campaign (winter 2022–2023) comprised two subsets: (i) fields selected based on label occurrence in the training dataset with priority for underrepresented labels (minority classes), and (ii) fields visited by a government agency targeting fields with previous absence of catch crops. In total, the dataset consisted of 6881 fields, of which 5921 fields were visited in observation years ranging from 2016 to 2021 (i.e., winter 2016–2017, 2017–2018, etc.). These observations comprise both the data collected by the Agency for Agriculture and Fisheries and the data collected during the first field campaign. These observations were used for training and testing the models. An additional 970 fields were observed during the second field campaign in 2022 (i.e., winter 2022–2023), which were used as the independent validation set (Figure 3).

2.2.2. Classification Groups

Because of the similarity in temporal profiles of (catch) crop species [3], species with similar spectral and temporal characteristics were grouped together. The catch crop species are listed in Table 1, and other crops cultivated during winter are referred to as winter crops. The catch crop species were subdivided into three groups based on their plant morphology and phenology (Table 1) [34,35]. The first group consists of leafy plant species; two example species in this group are white mustard and phacelia. This group is characterized by broad leaves and high aboveground biomass, properties that cause a high nitrogen uptake [34]. Plants in this group are very sensitive to frost damage and often do not survive winter. The second group consists of grass-like species with more narrow and upright leaves; two commonly used species are grass and oats (Table 1) [34,35]. As opposed to the first group, grass-like crops are winter-hardy and have a slower nitrogen uptake throughout the season. Their aboveground growth is slower, but their root system is extensive, which contributes to reduced soil slaking and an increased build-up of soil organic carbon. Mixtures of any of the allowed species can be considered as a third group, where mixtures can either consist of only leafy species, only grass-like species, or a combination of species from both groups.

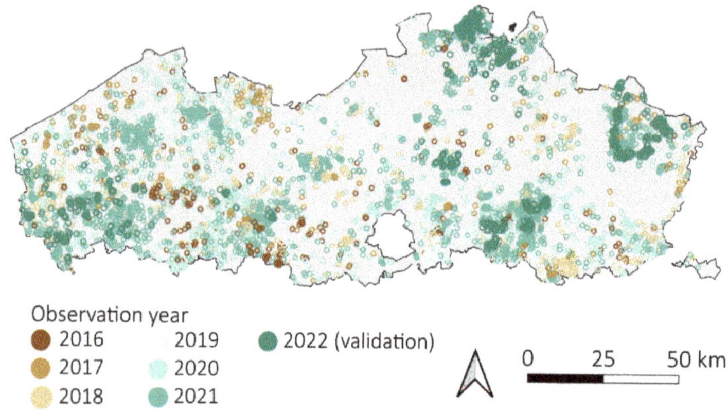

(**A**) Location field observations colored by year

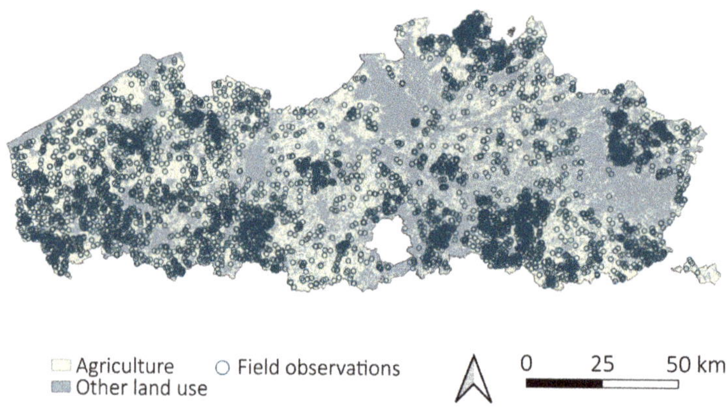

(**B**) Location field observations and agricultural land use

Figure 3. Locations of the 6881 field observations in the study area: (**A**) colored by year of observation where the color indicates the year in which the winter year started (e.g., 2022 are observations in the winter year 2022–2023) and (**B**) the field observations located on an agricultural land use map. Land use: Department of Environment and Spatial Development (www.vlaanderen.be/datavindplaats (accessed on 22 November 2024)).

The non-catch crop winter crop species were also grouped into three groups: (1) legumes, (2) winter cereals, and (3) other winter crops. Legumes are used as an agricultural practice to increase soil nitrogen and, therefore, are not used as a catch crop [34]. The winter cereals, winter barley (*Hordeum vulgare* L.) and winter wheat (*Triticum aestivum* L.), were put into a separate group because farmers can use these crops as an 'equivalent measure' instead of sowing catch crops. The difference with sowing catch crops is that the winter cereals sown in the current year remain on the field as the main crop in the following year, so they are not removed from the field after the required period, as is the case with catch crops. Examples of the third group, other winter crops, are cash crops sown during the study period, for example, leek (*Allium ampeloprasum* L.) and spinach (*Spinacia oleracea* L.).

Table 1. Overview of all allowed catch crop species with their common name, scientific name [36], and group.

Common Name	Scientific Name	Classification Group
Bristle oat	*Avena strigosa*	Grass group
Festulolium	*Lolium* × *Festuca* spp.	Grass group
Grasses Grass herbs Grass sods Undersowing grass	e.g., *Lolium multiflorum* *Lolium multiflorum* var. *westerwoldicum*	Grass group
Oat (spring)	*Avena sativa*	Grass group
Oat (winter)	*Avena sativa*	Grass group
Rye	*Secale cereale*	Grass group
Sudan grass	*Sorghum* × *drummondii*	Grass group
English ryegrass and white clover	*Trifolium repens* + *Lolium perenne*	Grass group
Arugula	*Eruca vesicaria* subsp. *Sativa*	Leafy group
Buckwheat	*Fagopyrum esculentum*	Leafy group
Cow cabbage	*Brassica oleracea convar. acephala*	Leafy group
Leaf mustard	*Brassica juncea*	Leafy group
Phacelia	*Phacelia tanacetifolia*	Leafy group
White mustard	*Sinapis alba*	Leafy group
Borage	*Borago officinalis*	Leafy group [1]
Nyger	*Guizotia abyssinica*	Leafy group [1]
Oil radish	*Raphanus sativus* supsp. *Oleiferus*	Leafy group [1]
Oilseed turnip	*Brassica rapa* subsp. *oleifera*	Leafy group [1]
Tagetes	*Tagetes* spp.	Leafy group [1]
Mixtures of non-legume cover species		Mixtures

[1] Interpretation based on plant morphology.

To classify these six different groups, the approach of a classification hierarchy was taken, dividing the complex classification problem into progressive classifications for improved accuracy in the case of complex classes with a lot of intra-class variability [37–39]. A limited two-level hierarchy limits process complexity and training time, as separate classifiers were trained for each level (Figure 4A). On the first level, fields were classified on winter crop presence or absence, separating bare soil and crop residues from fields occupied by a winter crop. In the second level, sown fields were further classified into six groups: three catch crop groups (leafy, grass-like, and mixtures) and three non-catch crop groups (legumes, winter cereals, and other winter crops). The resulting label distribution shows that (1) the overall majority of the dataset consists of fields with catch or winter crops present, (2) most observations were collected in the winter years of 2020–2022, and (3) there is a label imbalance within the six crop groups (Figure 4B).

2.2.3. Satellite Time Series

For every labeled parcel, the spectral signal was aggregated by taking the mean value over the field. These field boundaries are made publicly available by the government and updated annually, ensuring up-to-date field boundaries. They are also referred to as the Land Parcel Information System (LPIS). The object-based approach reduces noise and heterogeneity within a field and is useful in a decision-making context, as decisions are generally made at the parcel level [25]. To further reduce noise, we applied a 10 m inward buffer at each parcel, a technique that removes the spectrally mixed pixels at the edges of the parcels, thereby reducing the edge effects [10]. Input time series were downloaded for the period of August–April, chosen based on the crop calendars for catch crops and the other prominent winter crops in Flanders. This, therefore, delineates our study period.

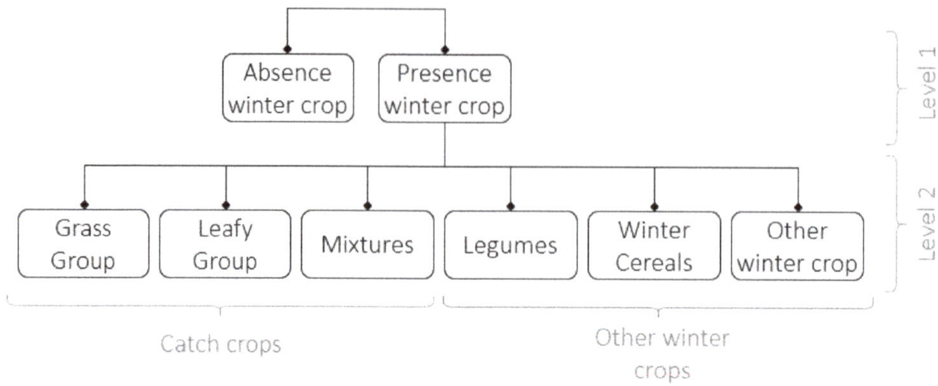

(A) Classification hierarchy

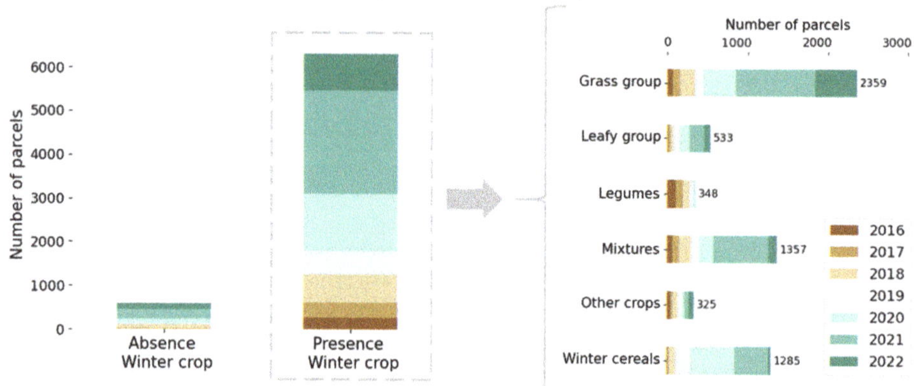

(B) Label distribution

Figure 4. Overview of (**A**) the classification hierarchy and (**B**) the label distribution on both the first (left) and second level (right) of the hierarchy.

Time series data were calculated for every field using the Sentinel-2 optical satellite constellation, part of the Copernicus program. The mission consists of a two-satellite constellation in a sun-synchronous orbit with a 180° phase difference [40]. The first satellite was launched in 2015, and the second in 2017. Every 5 days, the same area is visited under equal viewing conditions, but due to overlap in swaths of different orbits, there is a revisit time of about 2–3 days over Belgium. Sentinel-2 uses optical remote sensing in the range of 400 nm to 2500 nm, which includes the Visible (VIS), Near Infrared (NIR), and Short-Wave Infrared (SWIR) regions. The multispectral sensor, MultiSpectral Instrument (MSI), measures 13 spectral bands at different spatial resolutions: four bands at 10 m, six bands at 20 m, and three bands at 60 m. To obtain data from this mission, two different services were used, CropSAR (version 1.0) and openEO (version 0.22.0).

CropSAR is a service that monitors crop growth in Belgium, and it is provided by VITO, the Flemish Institute for Technological Research [41]. The CropSAR service provides daily time series by applying a deep learning network trained on cloud-free images of both Sentinel-1 and Sentinel-2 to interpolate between the gaps of optical acquisitions, thereby combining optical and radar data [42]. CropSAR was used to download preprocessed, ready-to-use daily time series for three different indices: the Normalized Difference Vegetation Index (NDVI), the fraction of Absorbed Photosynthetically Active Radiation (fAPAR), and the Fraction of green vegetation Cover (FCover) [43]. While the NDVI is an estimation of the greenness of the land surface, fAPAR and FCover are both biophysical variables of vegetation and can also be measured on the land surface. VITO derived fAPAR and FCover from satellite observations by using a neural network to estimate these biophysical variables from Sentinel-2.

OpenEO provides an API that can be used to acquire and process large Earth observation datasets from different providers in the cloud [44]. The API was used to access Terrascope Level 2A Sentinel-2 data and subsequently transform the raw series of individual images into an equidistant time series. Processing consisted of first masking clouds and cloud shadows with the Scene Classification Layer (SCL) through a kernel process provided by openEO (standard deviation 1.5, window size 11). Then, gaps in the time series were filled by linear interpolation: the time series were first divided into equidistant time intervals, using the median value over a specified period, and then interpolation was applied to fill gaps in the time series. To fill gaps at the start or end of the time series, an extra month was added at the start (July) and end (May). Where gaps remained, the value of the next valid period was duplicated, and if gaps remained, the value from the previous valid period was duplicated. Time series were downloaded with dekadal time intervals for bands B02-04 (VIS), B05-07 (vegetation red-edge), B08 (NIR), and B11-12 (SWIR) of Sentinel-2. Dekadal time intervals consist of three intervals per month, with the first two being 10 days each and the third comprising the remaining 8–11 days, depending on the length of the month. In addition, time series were also downloaded with monthly time intervals for the same bands of Sentinel-2.

2.3. Methods

We compared three machine learning models for classifying time series into the winter crop classes: Random Forest (RF), Time Series Forest (TSF), and a One-Dimensional Convolutional Neural Network (1D-CNN). The latter two models explicitly account for the temporal order of the data, while the RF model is added as a baseline method. To compare the model accuracies, each unique model–dataset combination was trained five times with different random seeds, using the same five seeds for each combination to maximize result comparability. All model runs were trained on 75% of the reference dataset and evaluated on the 25% remaining data. The train–test split was stratified by (catch) crop species to maximize the comparability of the training and test sets. Hyperparameters were tuned to maximize the Cohen kappa coefficient, in accordance with [23], and determined at the first hierarchy level, then applied to the second level to reduce selection complexity. In what follows, the models and their training process will be described in more detail. The entire overview of the selected hyperparameters per dataset is given in Appendix B.

2.3.1. Random Forest

RF is a non-parametric classifier known for its high accuracy and fast processing, using multiple decision trees created through a bagging approach [20,45]. It assigns class labels based on the majority vote from these trees. The RF model was implemented using the sklearn (v1.3.2) library in Python (v3.11.6) [46,47]. We optimized only the number of estimators to ensure maximal comparability between the two forest-based models (RF and TSF) and set the maximum number of features to zero for the same reason (Table 2). Hyperparameters were tuned using the RandomizedSearchCV function from sklearn, randomly sampling 10 hyperparameter value combinations from the corresponding search space and a 5-fold cross-validation [46]. For the RF model, the search space was too small to sample ten settings, so a maximum of four samples was used, i.e., the entire search space for this model (Table 2).

Table 2. Hyperparameter search spaces for each model.

Model	Hyperparameters	Search Space
RF	n_estimators	200, 300, 400, 500
TSF	n_estimators min_interval	200, 300, 400, 500 3, 5, 7, 9
CNN	batch_size learning rate num_epochs dropout kernel_size stride	8, 16, 32, 64 0.001–0.1 15, 20, 25, 30 0.1, 0.2, 0.3, 0.4 3, 5, 7, 9 1, 2, 3, 5

2.3.2. Time Series Forest

A Time Series Forest (TSF) model is a modified Random Forest model specifically adapted to the classification of time series [22]. TSF is a feature-based classification where the feature's mean, standard deviation, and slope are derived from randomly chosen time intervals in the time series. These features are calculated over a number of intervals and then used for classification. The model was implemented using the sktime library in Python (v0.21.0), using both the TimeSeriesForestClassifier and ColumnEnsembleClassifier for univariate time series input and multivariate input time series, respectively [48,49]. Two hyperparameters were optimized again using the RandomizedSearchCV function: the number of estimators (trees) and the minimum interval length (Table 2).

2.3.3. One-Dimensional Convolutional Neural Network

CNNs have been used in several vegetation remote sensing applications, including main crop classification and tree crown delineation [27,50]. CNNs make use of convolutions, which can be seen as sliding a filter over input data to detect local patterns, with the weights of this filter optimized during training to identify discriminative features [51,52]. A key advantage of CNNs is weight sharing across all time stamps, reducing the number of parameters to be tuned [17,51].

The 1D-CNN was implemented using the PyTorch (v2.0.1) library in Python [53]. The model architecture consists of three sequential convolutional blocks, each consisting of a one-dimensional convolutional layer, a batch normalization, and a ReLU activation function (Figure 5). These three convolutional blocks are followed by two fully connected blocks, each consisting of a linear layer, a batch normalization, and a ReLU activation function (Figure 5). The batch normalizations standardize the inputs of each input batch and thereby help regularize the network and reduce overfitting [17].

For training the network, we used the minibatch stochastic gradient descent approach and implemented it using the SGD optimizer in Pytorch [54]. For the loss function, we used binary cross-entropy at the first level of the hierarchy and cross-entropy at the second level of the hierarchy because binary and multiclass classifications require different loss functions [54]. Using the Ray Tune library in Python, we optimized the following hyperparameters for the network: window size and stride for the convolutional layers, and batch size, learning rate, number of epochs, and dropout for training the network (Table 2) [55]. For each input data set, the same 50 hyperparameter combinations were tested out, and bad trials were terminated using the median stopping rule. Instead of a 5-fold cross-validation, the training data (75% of the dataset) was split once in an 80–20 split, stratified by the classification group. This means that, in total, 60% of the entire reference dataset was used for training the models on various hyperparameter sets and 15% for evaluating. The remaining 25% was allocated to the validation set.

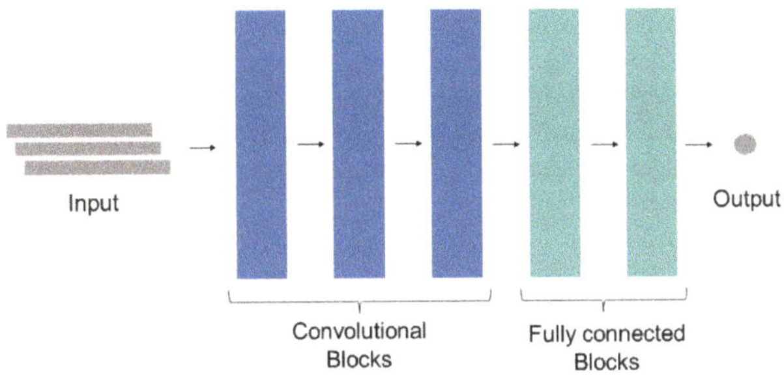

Figure 5. Network architecture of the One-Dimensional Convolutional Neural Network (1D-CNN). The convolutional blocks each comprise a convolutional layer (number of filters, 8, 16, 32), a batch normalization, and a Rectified Linear Unit (ReLU) activation function. The fully connected blocks each consist of a linear layer, a batch normalization, and ReLU activation function.

2.3.4. Experimental Setup

The first research objective aimed to compare models and inputs for catch crop classification, which was addressed through two experiments, evaluating multiple factors affecting classification accuracies for all three models. The first experiment aimed to study the influence of temporal resolution on the classification by using time series with three different temporal intervals: daily, dekadal, and monthly intervals. To achieve this, we used the daily NDVI time series downloaded from CropSAR and resampled it to dekadal and monthly intervals by taking the mean value in every interval (Figure 6). The experiment used the full time series length, ranging from August to April.

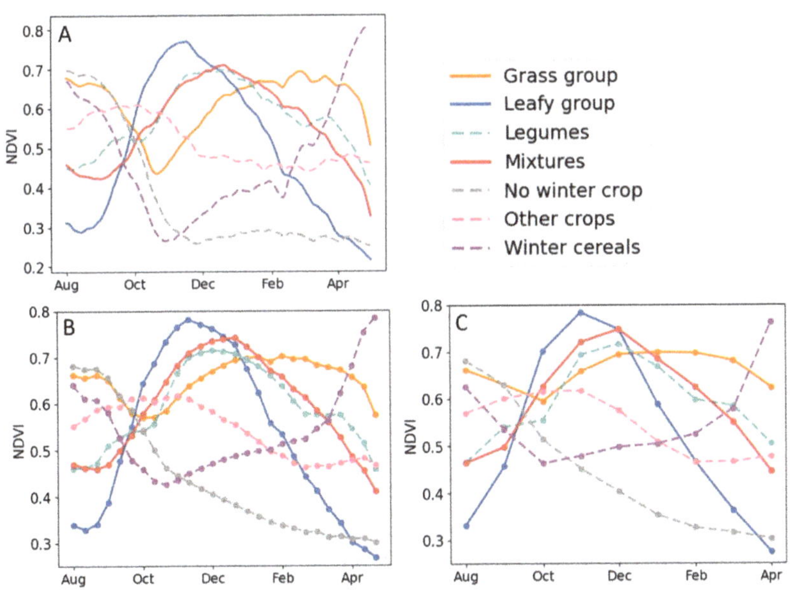

Figure 6. Average NDVI time series at (**A**) daily, (**B**) dekadal, and (**C**) monthly temporal resolution for each of the classes. Linear interpolation was used to create continuous time series.

The second experiment evaluated how input features affect classification accuracy, aiming to determine if additional information beyond NDVI could improve accuracy. Part A of this experiment compared daily NDVI time series with multivariate time series (NDVI, fAPAR, and FCover) to assess the added value of biophysical variables. In Part B, we used the dekadal Sentinel-2 dataset to construct three input feature sets: (1) dekadal NDVI time series, (2) dekadal time series with the raw reflectance values from each band of Sentinel-2 and (3) dekadal time series with normalized reflectance values from each band of Sentinel-2. This tested the impact of both additional features and normalization on classification accuracy. Z-normalization was applied on each timestamp–band combination separately and was calculated as follows:

$$z = \frac{x - u}{s} \quad (1)$$

where x is the reflectance value, u is the mean, and s is the standard deviation, calculated using the StandardScaler in sklearn [46]. Again, we used the full time series length, ranging from August to April, for this experiment. These first two experiments combined are used to answer the first two research questions.

In the second research objective, we focused on applying the models in a real-world scenario for early-season classification on a year excluded from training and validation. Therefore, the third experiment evaluated model performance throughout the season by varying time series lengths, always starting in August and extending month by month to April. Testing the temporal transferability of the models was done for the years the models were trained upon and on the dataset for the winter 2022–2023, which was excluded from training. This experiment used the input feature set with the highest accuracy, which was identified in the previous two experiments, and aimed to answer the third and fourth research questions.

2.3.5. Accuracy Assessment

To evaluate the accuracy of the models, the results from the classification hierarchy were first combined into two coarse-labeled groups: catch crops and no-catch crops. The catch crop group consists of the grass group, leafy group, and mixtures, while all other classes are grouped in no catch crops. Based on the classification of these two groups, the models were compared with a macro F1-score as a measure to compare the overall classification. The F1-score was chosen over the overall accuracy because of the role of chance agreement in an unbalanced dataset, where the overall accuracy favors the dominant class [29]. The calculation of the F1-score is given by:

$$F1 = 2 * \frac{Precision * recall}{Precision + recall} \quad (2)$$

where precision is the proportion of positive (negative) predictions that are actually positive (negative), and recall is the proportion of positives (negatives) that are actually predicted as positive (negative). Because the models were run five times on each unique input dataset, each time with a different random state, all the F1-scores were calculated five times.

3. Results

3.1. End-of-Season Classification

The first experiment examined the effect of temporal resolution on the classification performance of all models. All models achieved the highest F1-scores on the daily temporal resolution for the coarse-label classification, with a median F1-score of 86.5% for the RF model, 86.5% for the TSF model, and 85.9% for the CNN model (Figure 7). Every model type reached the lowest F1-score on the monthly time resolution, and here, the median values for the F1-scores were 85.1% for RF, 81.8% for TSF, and 83.6% for CNN. Further, each model type has a different rate in the decline of the F1-scores, where the TSF model has the largest decline in performance, while RF has the lowest decline.

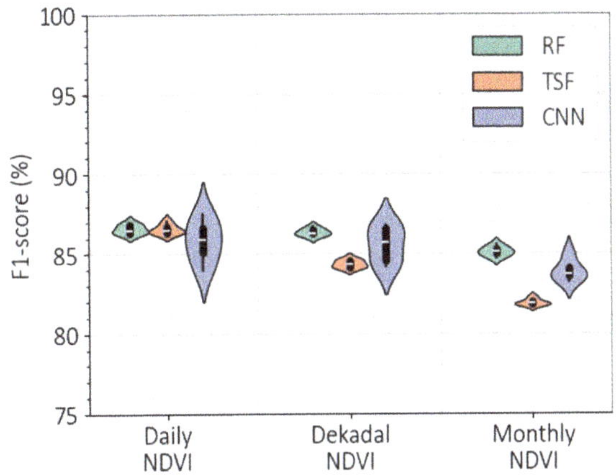

Figure 7. Violin plots for the F1-scores of the first experiment given for three temporal resolutions and three different models (Random Forest or RF, Time Series Forest or TSF, and a One-Dimensional Convolutional Neural Network or 1D-CNN). Within the violin plots, the boxplots are depicted in black.

In part A of the second experiment, we then looked at the influence of biophysical variables on the classification performance. At both daily and dekadal temporal resolutions, the dataset, including the biophysical variables (fAPAR and FCover), attained higher F1-scores for every model type compared to the NDVI-only dataset (Figure 8A).

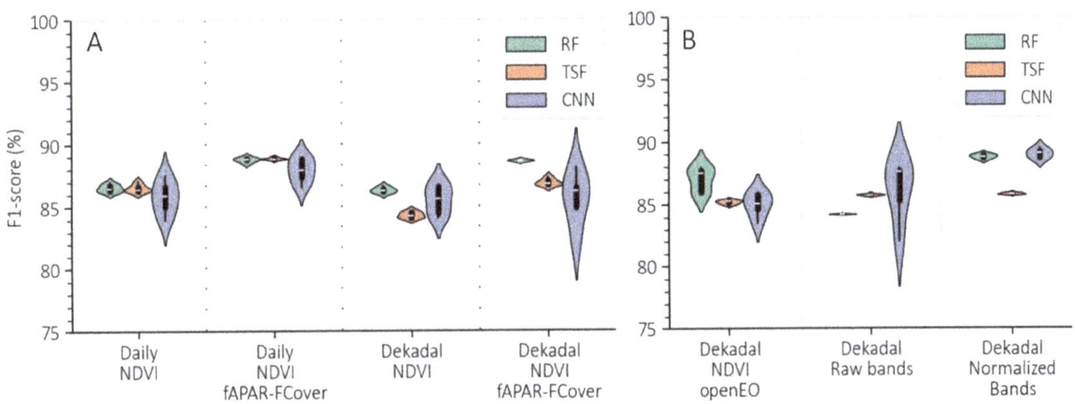

Figure 8. Violin plots for the F1-scores of the second experiment given for three temporal resolutions and three different models (Random Forest or RF, Time Series Forest or TSF, and One-Dimensional Convolutional Neural Network or 1D-CNN). (**A**) Part A of the second experiment and (**B**) part B of the second experiment. Within the violin plots, boxplots are depicted in black.

Part B of this second experiment looked at the classification performance of the models on the time series, purely derived from Sentinel-2 data. Three sets of input features were compared at a dekadal temporal resolution: NDVI, the raw bands, and normalized bands. When using NDVI as an input feature, the median F1-scores were in the range of 85.1–87.5% (Figure 8B). However, the CNN resulted in a larger range of F1-scores (83.5–86.0%) compared to the forest-based models. This is especially the case for the F1-scores with raw bands as an input dataset, where a wide range of values was obtained (F1-scores 82.0–87.9%). However, the CNN model showed less variability in model performance when the Sentinel-2 bands were normalized (F1-scores 88.5–89.5%). In contrast, the TSF model resulted in a constant performance with an F1-score of around 85.3–85.8%. The median F1-score for the RF model showed a decrease from 87.5% to 84.2% when using the raw bands of Sentinel-2 compared to that using NDVI only. When normalizing the bands of Sentinel-2, performance increases again to a median F1-score of 88.8%.

Looking more closely at the fine-label classification of the three models, it was shown that all models confused the three catch crop groups (leafy, grass-like, and mixture) with each other (Figure 9). The leafy group was mostly confused with the mixtures, and the mixtures were mostly confused with the grass group. There were also some misclassifications between catch crops and winter cereals, mostly between the grass-like and mixture groups. Small groups like legumes, other winter crops, and the no-crop group mostly ended up in the three biggest groups: the grass group, mixtures, and winter cereals.

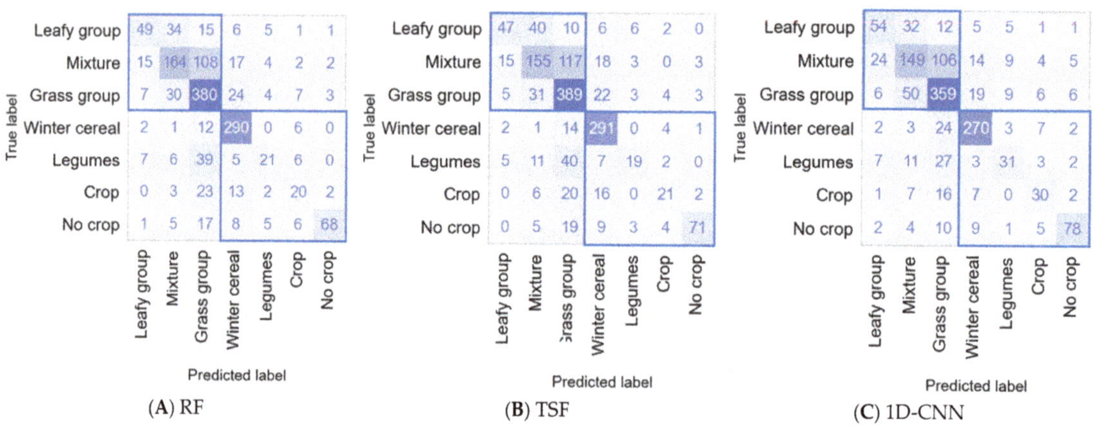

Figure 9. Confusion matrices for the fine-label, end-of-season classification for three different models: (A) Random Forest (RF), (B) Time Series Forest (TSF), and (C) One-Dimensional Convolutional Neural Network (1D-CNN). The blue frames indicate the coarse-label classification groups.

3.2. Early-Season Classification

From the first two experiments, the input feature set that achieved the highest overall F1-scores was selected, which was the daily NDVI, fAPAR, and FCover time series. The third experiment focused on assessing the early-season performance of the models and their temporal transferability using this 'best' input feature combination. When training and evaluating the same years, all models showed an increase in F1-scores when including extra time steps throughout the first three months (September, October, November) (Figure 10). However, adding extra time steps only marginally improves model performance from December onwards, as indicated by a stagnation in the F1-scores. The inclusion of March and April into the time series again resulted in an increase in F1-scores.

When evaluating the early-season performance of the models separately, the forest-based models (RF and TSF) attain a median F1-score surpassing 85% in November, whereas the CNN model only reaches this threshold in March (Figure 10). By April, all models reach their highest median F1-score. However, the CNN model resulted in lower and more variable F1-scores, particularly earlier in the season.

When applying the trained models to the time series of a target (winter) year to assess transferability, the increase in F1-score with time steps remained for the forest-based models (Figure 10). In contrast, the CNN model showed a very unstable performance that lacked this general upward trend and produced a larger range of F1-scores compared to the other two models. All models reached maximum model accuracy in April with median F1-scores of 88.8% for RF, 88.3% for TSF, and 81.5% for the CNN model. Therefore, the forest-based models reached very similar median F1-scores when training and evaluating in the same years in April (RF 88.7% and TSF 88.9%), while the CNN model reached a lower median F1-score in comparison (87.9%).

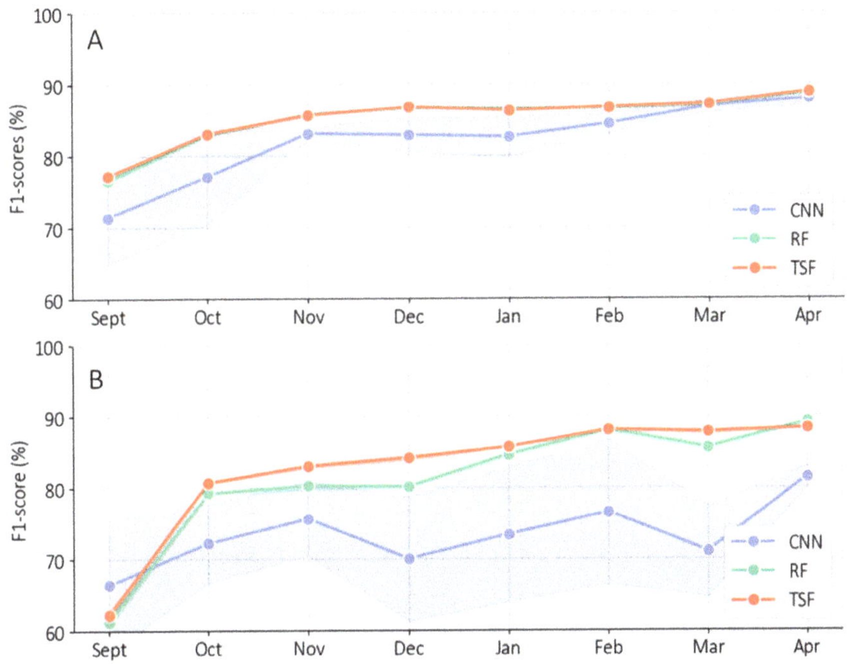

Figure 10. F1-scores for the third experiment, which was completed on daily time series with features NDVI-fAPAR-FCover for three models (RF, TSF, and 1D-CNN). (**A**) Results when trained and tested on the same years; (**B**) results when the models are applied on a target year. Month on the x-axis shows the end point of the time series: e.g., Dec represents a time series starting in August and ending in December. Shadowing indicates the standard deviation on the results from the diferent model runs. Note that the results for RF and TSF largely overlap in panel (**A**).

Revisiting the early-season performance of each model, TSF achieves a median F1-score of 85% in December, followed by RF in February, whereas the CNN model fails to reach a median F1-score of 85%. Compared to the same year training and testing, the models attain this threshold later in the season during the separate validation year. Maximum F1bscores were again all reached in April. The CNN model reached a score of 83%, while RF and TSF reached F1-scores of 89% and 88%, respectively. These classification results were illustrated on a map using the RF model (Figure 11).

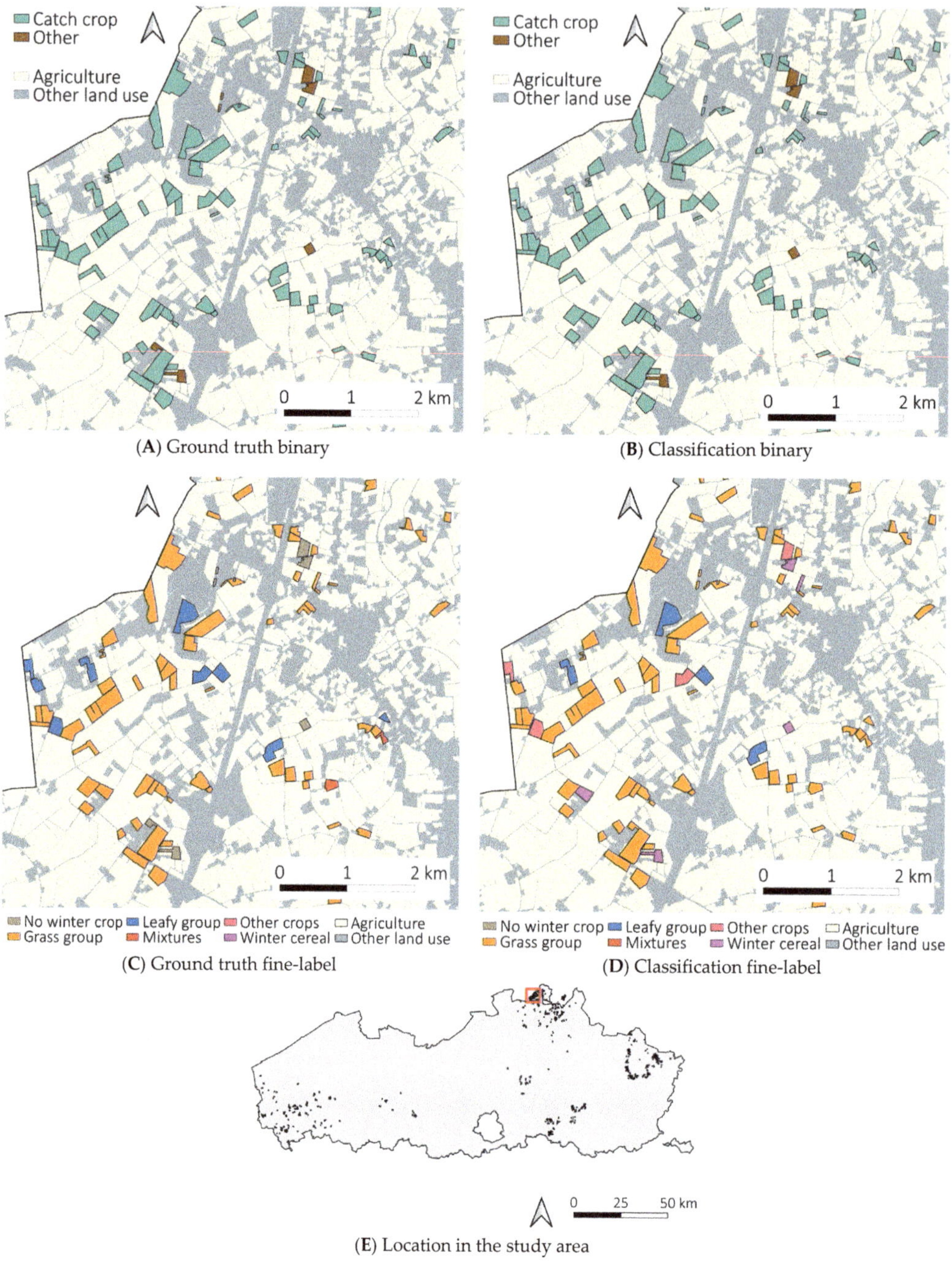

Figure 11. Maps to illustrate the results of the Random Forest (RF) model on time series until April. Binary (**A**) ground truth and (**B**) classification results, fine-label (**C**) ground truth and (**D**) classification results. (**E**) Location of maps (**A–D**) in the study area. Land use: Department of Environment and Spatial Development (www.vlaanderen.be/datavindplaats (accessed on 22 November 2024)).

4. Discussion

4.1. Model Comparison

Our results show marked differences in the performance of three tested classification models (RF, TSF, and 1D-CNN). While the RF algorithm demonstrated its versatility by delivering consistent results on different datasets, the TSF model was more dependent on the input dataset than RF, and the 1D-CNN model often resulted in lower accuracies than the other models and displayed a high degree of variability between model runs. Given that RF has been used in a wide range of remote sensing classification tasks, as reviewed by [20], our finding that it performs robustly across datasets is consistent with previous research. In contrast, the results of our time series-specific algorithms differ from previous research. For example, in their study comparing time series classification algorithms on univariate time series, TSF outperformed both baseline classifiers, with RF showing similar performance to the baseline classifiers [56]. However, this study did not include remotely sensed time series, and their datasets mostly contain time series with <1000 training instances that are longer than 50 time steps. In contrast, our datasets contain >5900 training instances with 9, 27, or 273 time steps, depending on the temporal resolution (respectively monthly, dekadal, and daily), which may also explain the different results.

Further, in time series-based crop classification, several studies have demonstrated enhanced classification results with the use of a deep learning classifier in comparison with shallow machine learning techniques. For example, refs. [15,17] reported improved crop classification results using 1D-CNNs compared to RFs. However, in our application, 1D-CNNs often did not outperform a baseline RF, with the exception of the normalized dekadal band dataset in experiment 2B. A very likely explanation for these contrasting results is the limited number of training samples in our field-visit and object-based approach, resulting in approximately 6000 training samples over multiple years and ranging from 325 to 2359 samples per class (Figure 4B). A recent study on time series-based crop classification using 1D-CNNs used 4000 training samples per class, a number chosen after an analysis of the amount of training data and the mapping accuracy [57]. These results suggest that 28,000 training samples, spread evenly across classes, would be required for deep learning to be effective in our catch crop classification. Often, many more training samples are used. References [15,17] used labels available from a single-year crop survey in combination with a pixel-based approach, resulting in approximately 1 million and 2 million training samples, respectively, with samples per class ranging from 9000 to 292,000. Due to its data limitation, the catch-crop classification could not fully benefit from the advantages that deep learning models provide. Therefore, the large number of labeled data needed for deep learning models proved to be a limitation for improved catch crop-type classification with these models. Switching to a pixel-wise classification would be a quick win for creating more training data without increasing the field efforts. Other options for handling the large data requirement of deep learning models are data augmentation and transfer learning; the latter will be discussed below.

4.2. Fine-Label Classification

Across all models, it proved difficult to make a fine-label classification of winter crops based on their temporal profile alone (Figure 9). The fine-label classification resulted in confusion between different groups, notably between the grass group and the mixtures, which was expected because mixtures can also be a mix of two grass-like catch crops. The same can be said about confusion between leafy species and mixtures, as mixtures can consist of multiple leafy species. Theoretically, mixtures between grass-like and leafy groups could result in a unique canopy structure and, thus, a distinguishing feature from the other catch crop groups. However, in reality, competition between species makes the proportion of a single catch crop in the canopy unpredictable, resulting in canopy heterogeneity [58]. Factors influencing the canopy composition are management, soil type, and weather conditions rather than the ratio in the seed mix. The heterogeneity of the mixture group limits the separability of the catch crop groups.

Our results are similar to those found by [3], who also had difficulty distinguishing between catch crop species based on a time series-based classification with RF. Further, [59] similarly performed a time series-based classification with RF to distinguish two cover crop species: cereal rye and red clover. In particular, they noted that rye was often misclassified as fallow, which they attributed to (1) the highly variable biomass production depending on the planting date, causing fields with low biomass to resemble fallow fields with weeds, and (2) the interseeding of rye with crop residue, resulting in fields with low biomass production to appear similar to fields with maize stubble. These observations of high biomass variability and interseeding with crop residue were also noted in our fieldwork for the grass group (including rye), which also resulted in variability in spectral signals within this group for a single winter cropping year (Appendix C). The high variability within the grass group could, therefore, explain the high confusion of this group with the other classes.

4.3. Effect of Temporal Resolution and Input Variables

The first experiment revealed that a higher temporal resolution of the input time series resulted in a more accurate classification. This is consistent with the results of crop classification studies by [19,60], who also reported improved accuracy with higher temporal resolution, although with marginal improvements in [60]. The highest accuracies were achieved using multivariate daily time series consisting of NDVI, fAPAR, and FCover inputs. This suggests that incorporating a wide range of features can indeed be beneficial, contrary to the expected limitations of the Hughes phenomenon, where the addition of too many features beyond a certain point does not improve classification accuracy [61]. This added value of using more input features is observed at both daily and dekadal temporal resolution.

In addition to the use of other engineered features, the use of (normalized) reflectance bands improved the performance of most of the RF and CNN models over using NDVI alone. Notably, the combination of CNN with normalized input features resulted in the best performance for the CNN model. This finding indicates a potential loss of information when overly relying on feature engineering in the case of CNN models [50]. Normalization of input features proved crucial when using reflectance band inputs in combination with the CNN model: the normalization not only increased the F-scores but also yielded more consistent predictions between model runs. This improvement can be attributed to the normalization process ensuring that the variation in data is not overly dominated by features with a larger dynamic range, thus balancing the influence of each feature on the model's output [17]. Experiment 2B showed that both extensively preprocessed datasets for higher temporal resolution and raw input time series can achieve similar accuracies when used with the appropriate model type.

The experiments demonstrated that additional information could be extracted from the raw input by using biophysical variables in addition to the NDVI or by using a suitable model type in combination with the raw input itself. Future work could explore additional vegetation indices, similar to the approach taken by [59] in their catch-crop-type classification, especially since a number of alternative indices are aimed at addressing specific NDVI limitations. In addition, our study focused on the use of Sentinel-2, and although the daily NDVI, fAPAR, and FCover time series were derived from combining Sentinel-1 and Sentinel-2, a winter crop classification might benefit from a direct integration of radar-based (Sentinel-1) input features [8,62,63].

4.4. Early-Season Classification Performance for Separate Target Year

The third experiment showed the trends in classification accuracy throughout the winter season. The F1-scores increased until November, then stagnated, and increased again throughout March and April. Overall, the models improved as the time series lengthened, which is an expected result and consistent with previous research [14]. The stagnation could potentially be explained by the limited number of cloud-free images that are available in these parts of the season, which could restrict the early-season classification for winter crops compared to that of main crops. Peak F1-scores were reached by the end of April, which is likely due to the distinct temporal profile of winter cereals when April is included (Figure 6). Because winter cereals, unlike catch crops, remain on the field after winter, their time series show a continuous increase in the vegetation index values around March and April, which contrasts with the downward trend of catch crops over the same period (Figure 6). This discriminating characteristic is probably also the reason that the fine-label classification of winter cereals performed relatively well compared to other classes (Figure 9).

When comparing the models for their early-season classification capabilities on the validation year, we see that the forest-based models reach the threshold F1-score of 85% later in the season than for the years included in the training. A slight drop in performance on a separate target year was expected, as temporal (and spatial) transferability is generally a challenge in crop classification, even when multiple years have been used for training [28]. This is due to different weather conditions between years, which affect crop development and result in different temporal profiles for the same crop [28,29]. Reference [19] even reports that the variation in spectral profiles between years is the main factor in reducing the accuracy of the early-season mapping of main crops. Interannual differences could, therefore, offer an explanation for the slightly lower accuracies of the forest-based models earlier in the season (Appendix C). The differing weather conditions during the summer and autumn months every year impact the timing of growth stages of the catch crops, resulting in variations in crop development earlier in the season. For example, a large part of the training data come from the year 2021 (winter year 2021–2022), with an exceptionally wet summer and a rather dry autumn, while 2022 (winter year 2022–2023) had an exceptionally dry summer followed by a warm autumn, which is likely to have affected the growth stages of the catch crops differently compared to the previous year [64,65].

However, interannual differences do not entirely explain the lack of generalization to the target year for the 1D-CNN model; its results most likely point to an overfitting problem. Although various regularization methods, such as dropout and batch normalization, were used to prevent overfitting, the model type itself also plays an important role: given the large number of parameters that need to be fitted for a deep learning model such as a CNN, the limited number of samples is again the most likely explanation. Even though the potential of deep learning approaches, and more specifically, 1D-CNN, have been reported for early-season mapping [19], our application does not seem to benefit from deep learning models over machine learning models. RF is the better option for early-season classification, given our limited dataset.

This issue of temporal model generalizability, particularly in the context of early-season mapping, has gained attention in recent crop mapping research [28–30]. These studies have focused on early-season time series classification with transformer-based models, which represent the current state-of-the-art in deep learning. Transformers can exploit temporal dependencies and handle variable time series lengths as inputs [66], making them highly suitable for early-season mapping as a single model can be used for progressive mapping throughout the season [28,29]. Ref. [29] demonstrated the effectiveness of a transformer-based deep learning architecture for early-season crop mapping, highlighting its temporal (and spatial) generalizability. However, their approach requires a substantial amount of labeled data to train this transformer-based model, an architecture that often requires larger datasets in comparison to CNNs to achieve better performance [67]. Our study already faced the constraint of limited sample availability with CNNs, so it can be questioned how useful this specific model would be for our catch crop classification.

Fortunately, recent research also addresses this aspect of limited labeled data availability. Here, transfer learning, a machine learning technique that transfers knowledge between related tasks, emerged as a strategy to alleviate the need for labeled data in environmental remote sensing [67]. For example, ref. [28] demonstrated pre-trained models, providing a strong starting point for early-season crop mapping in a separate target year for which no data are yet available, even when the vegetation trend differs between years. Reference [30] similarly demonstrated the effectiveness of transfer learning for the transferability of crop mapping models on Sentinel-1 data. Therefore, there is a need for the development of transferable models or even foundation models in remote sensing, which are large pre-trained and often transformer-based models that can be applied to a variety of tasks downstream through transfer learning [67]. As such, future work can explore transfer learning for early-season catch crop classification—for example, with a recently developed pre-trained transformer model for remote sensing time series by [68]—as this approach seems to generalize well even with limited training data.

5. Conclusions

This study aimed to assess the feasibility of end-of-season catch crop classification with Sentinel-2 time series, with specific attention given to the optimization of the model type and the input features. The best model–input combination was then tested in a real-world scenario for early-season catch crop classification in a separate target year, thereby testing its temporal transferability. In summary, RF displayed the highest adaptability to input features and often outperformed time series-specific classification methods, TSF, and 1D-CNN. This study demonstrated the benefits of using high temporal resolution time series, as well as a diverse range of input features for the catch crop classification task. When the models were applied for early-season classification in a separate target year, the models reached the F1-score threshold of 85% at least one month later than for the training years. Overall, the advantages of deep learning approaches (1D-CNN) could not be demonstrated, probably due to the limited amount of reference data available to train the models. However, very recent research into early-season crop classification and temporal transferability shows opportunities for the real-world application of deep learning classification models for catch crop classification.

Author Contributions: Conceptualization, K.V, S.H., E.B., and B.S.; methodology, K.V., S.H., and E.B.; software, K.V. and E.B.; validation, K.V.; formal analysis, K.V.; investigation, K.V.; data curation, K.V. and E.B.; writing—original draft preparation, K.V.; writing—review and editing, K.V., S.H., E.B., and B.S.; visualization, K.V.; supervision, S.H. and B.S.; project administration, S.H. and B.S.; funding acquisition, S.H. and B.S. All authors have read and agreed to the published version of the manuscript.

Funding: This research was funded by Research Foundation Flanders (FWO), grant number S006421N.

Data Availability Statement: The data presented in this study are available on request from the corresponding author. The data are not publicly available due to privacy restrictions.

Acknowledgments: A special thanks to the Vlaamse Landmaatschappij (Flemish Land Agency) and the Agency for Agriculture and Fisheries for their help regarding data collection. Also, thanks to Zifu Wang for providing useful feedback regarding deep learning.

Conflicts of Interest: The authors declare no conflicts of interest. The funders had no role in the design of the study; in the collection, analyses, or interpretation of data; in the writing of the manuscript; or in the decision to publish the results.

Appendix A

For preprocessing the dataset from the department of Agriculture and Fisheries, we took a two-step approach. First, the useful observations were selected based on the label and the date of the observation. Second, the selected labels were checked with their corresponding time series to filter out any discrepancies. This was done because the field observations are taken at a single point in time, while fields are subject to change. For example, it could be that a field visit in October results in the label 'bare soil', while visiting the same field a few weeks later results in the presence of vegetation.

Two factors were taken into account for the selection of observations: (1) the observation date and (2) the relative occurrences of the land cover types in the reference dataset. First, most observations occurred in the period of October–January, which is also the period that most catch crops are required as a cover, and thus, the period that is under study here. Observations outside this period were compared with the crop calendar of the observed crop to determine if this crop can be on the field at the time of the field visit. Second, the observation could not be used if a winter crop type was only observed once. For model training and validation, reference data are split into separate training and validation sets stratified by crop species. This split requires a minimum of two observations per crop species. Also, one winter crop type, 'mixture of catch crops', was about ten times more abundant than the second most represented type. However, as a heavily unbalanced dataset is not beneficial for model training, only a subset of the parcels of this type was retained in the final reference dataset. Parcel selection within the most prevalent crop type depended on two principles: (1) parcels with a larger area were favored because it facilitates the next step of downloading the time series for these parcels, as it results in more usable pixels per field, and (2) only parcels with an agreement between the declared crop and observed crop (both were 'mixture of catch crops') were retained to maximize the reliability of the remaining training data. Observations of permanent land cover types, like pastures and greenhouses, were removed as no winter crops were found on these fields.

In addition to this, the field observation labels were compared with the corresponding daily NDVI time series. Because field observations are only a snapshot at a certain point in time and we worked with time series as input to the model, there was sometimes a disparity between the label and the time series. For example, if a field observation in October indicated that there was no catch crop present, the time series could still indicate a green-up of the field from November onwards. Therefore, we decided to trust the ground truth label unless the daily NDVI time series clearly disproved it. A crop observation was not used if, for the entire period of October–December, the NDVI value surpassed 0.35. Similarly, a bare soil observation was not used if, for the entire period of October–December, the NDVI value was higher than 0.35. We chose December as the endpoint for this because, in our application, there should still be time to visit the field after making a model prediction. Therefore, we trained the model on the examples that we wanted to detect, i.e., the field that contained a catch crop before the end of December, so that field visits can still be made in January.

Appendix B

Table A1. Selected hyperparameters for all models per dataset.

Model	Hyperparameters	Search Space	Selected Values (per Dataset)			
			S1 + 2 fAPAR + FCover + NDVI Daily	**S1 + 2 NDVI Daily**	**S1 + 2 fAPAR + FCover + NDVI Dekadal**	**S1 + 2 NDVI Dekadal**
RF	n_estimators	200, 300, 400, 500	500	400	400	500
TSF	n_estimators min_interval	200, 300, 400, 500 3, 5, 7, 9	300-400-200 3-3-3	300 3	400-200-300 9-3-5	500 5
CNN	batch_size learning rate num_epochs dropout kernel_size stride	8, 16, 32, 64 0.001–0.1 15, 20, 25, 30 0.1, 0.2, 0.3, 0.4 3, 5, 7, 9 1, 2, 3, 5	32 0.042 20 0.1 5 2	32 0.042 20 0.1 5 2	32 0.062 25 0.1 9 3	64 0.022 15 0.1 7 1
			S1 + 2 NDVI Monthly	**S2 NDVI Dekadal**	**S2 9 Bands Dekadal**	**S2 9 Norm Dekadal**
RF	n_estimators	200, 300, 400, 500	200	400	300	500
TSF	n_estimators min_interval	200, 300, 400, 500 3, 5, 7, 9	500 3 *	400 3	200-300-500- 200-400-200- 200-500-200 3-3-3 * 3-3-3 3-3-3	400-500-200- 200-500-300- 500-400-400 3-3-3 * 3-3-3 3-3-3
CNN	batch_size learning rate num_epochs dropout kernel_size stride	8, 16, 32, 64 0.001–0.1 15, 20, 25, 30 0.1, 0.2, 0.3, 0.4 3, 5, 7, 9 1, 2, 3, 5	64 0.022 15 0.1 7 1	64 0.022 15 0.1 7 1	64 0.022 15 0.1 7 1	64 0.022 15 0.1 7 1

* In the case of the dataset with nine bands from Sentinel-2, only the number of estimators was optimized because the search space was too big when the minimum interval length was included (monthly min interval was also not optimized—too short).

Appendix C

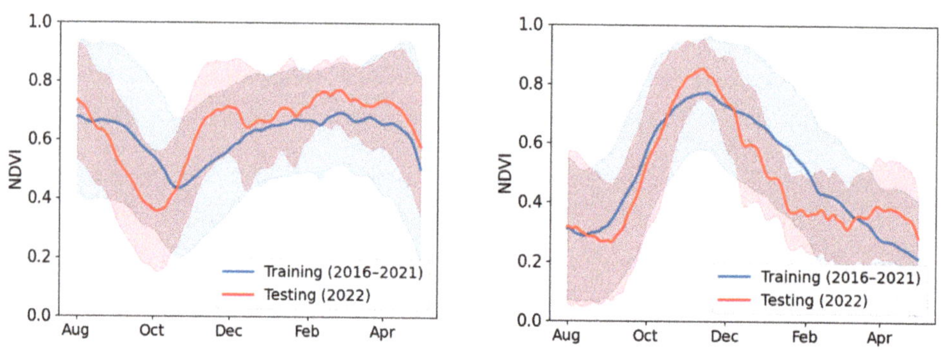

(**A**) Grass group (**B**) Leafy group

Figure A1. *Cont.*

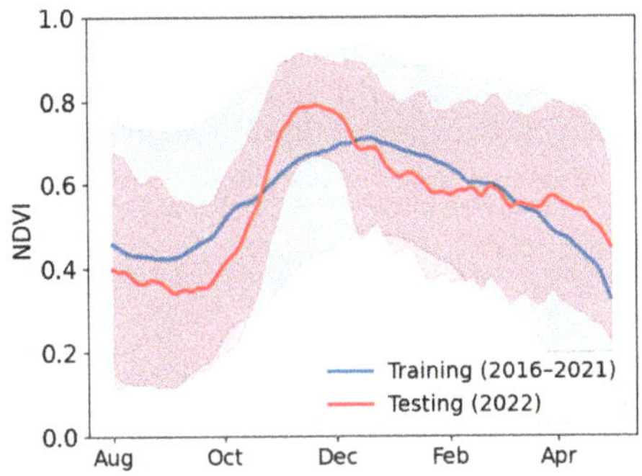

(**C**) Mixtures

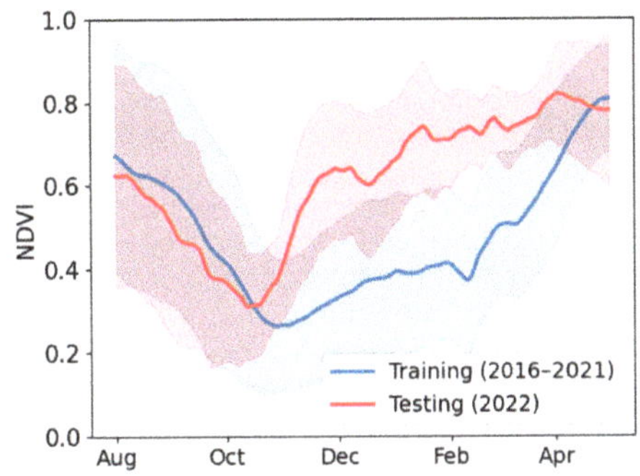

(**D**) Winter cereals

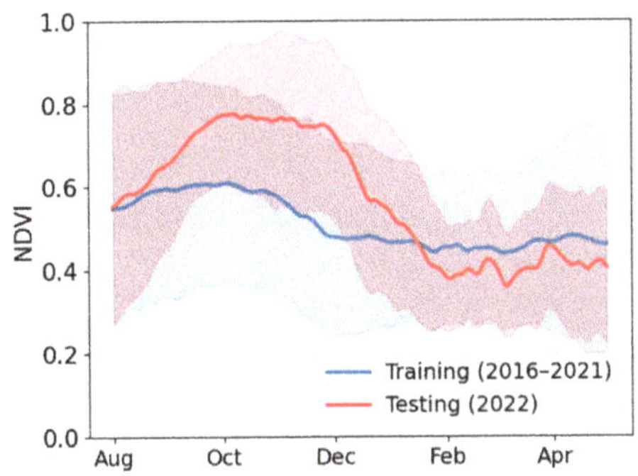

(**E**) Other crops

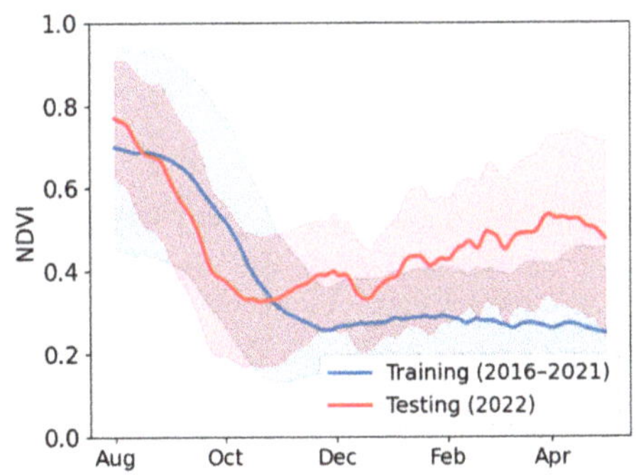

(**F**) Absence of winter crop

Figure A1. Comparison of the temporal profiles (NDVI, 1-day interval) of the different classification groups between the training data (2016–2021) and the validation data (2022). Shadowing indicates the standard deviation on each averaged profile. The group 'Legumes' is not visualized because of its absence in the validation dataset (2022).

References

1. Cameron, K.C.; Di, H.J.; Moir, J.L. Nitrogen Losses from the Soil/Plant System: A Review. *Ann. Appl. Biol.* **2013**, *162*, 145–173. [CrossRef]
2. Fendrich, A.N.; Matthews, F.; Van Eynde, E.; Carozzi, M.; Li, Z.; d'Andrimont, R.; Lugato, E.; Martin, P.; Ciais, P.; Panagos, P. From Regional to Parcel Scale: A High-Resolution Map of Cover Crops across Europe Combining Satellite Data with Statistical Surveys. *Sci. Total Environ.* **2023**, *873*, 162300. [CrossRef] [PubMed]
3. Schulz, C.; Holtgrave, A.-K.; Kleinschmit, B. Large-Scale Winter Catch Crop Monitoring with Sentinel-2 Time Series and Machine Learning–An Alternative to on-Site Controls? *Comput. Electron. Agric.* **2021**, *186*, 106173. [CrossRef]
4. Nouri, A.; Lukas, S.; Singh, S.; Singh, S.; Machado, S. When Do Cover Crops Reduce Nitrate Leaching? A Global Meta-Analysis. *Glob. Chang. Biol.* **2022**, *28*, 4736–4749. [CrossRef] [PubMed]
5. Blanco-Canqui, H. Cover Crops and Water Quality. *Agron. J.* **2018**, *110*, 1633–1647. [CrossRef]
6. Sitokonstantinou, V.; Papoutsis, I.; Kontoes, C.; Lafarga Arnal, A.; Armesto Andrés, A.P.; Garraza Zurbano, J.A. Scalable Parcel-Based Crop Identification Scheme Using Sentinel-2 Data Time-Series for the Monitoring of the Common Agricultural Policy. *Remote Sens.* **2018**, *10*, 911. [CrossRef]
7. Barnes, M.L.; Yoder, L.; Khodaee, M. Detecting Winter Cover Crops and Crop Residues in the Midwest US Using Machine Learning Classification of Thermal and Optical Imagery. *Remote Sens.* **2021**, *13*, 1998. [CrossRef]
8. Jennewein, J.S.; Lamb, B.T.; Hively, W.D.; Thieme, A.; Thapa, R.; Goldsmith, A.; Mirsky, S.B. Integration of Satellite-Based Optical and Synthetic Aperture Radar Imagery to Estimate Winter Cover Crop Performance in Cereal Grasses. *Remote Sens.* **2022**, *14*, 2077. [CrossRef]

9. Fan, X.; Vrieling, A.; Muller, B.; Nelson, A. Winter Cover Crops in Dutch Maize Fields: Variability in Quality and Its Drivers Assessed from Multi-Temporal Sentinel-2 Imagery. *Int. J. Appl. Earth Obs. Geoinf.* **2020**, *91*, 102139. [CrossRef]
10. Thieme, A.; Yadav, S.; Oddo, P.C.; Fitz, J.M.; McCartney, S.; King, L.; Keppler, J.; McCarty, G.W.; Hively, W.D. Using NASA Earth Observations and Google Earth Engine to Map Winter Cover Crop Conservation Performance in the Chesapeake Bay Watershed. *Remote Sens. Environ.* **2020**, *248*, 111943. [CrossRef]
11. Goffart, D.; Curnel, Y.; Planchon, V.; Goffart, J.-P.; Defourny, P. Field-Scale Assessment of Belgian Winter Cover Crops Biomass Based on Sentinel-2 Data. *Eur. J. Agron.* **2021**, *126*, 126278. [CrossRef]
12. Kc, K.; Zhao, K.; Romanko, M.; Khanal, S. Assessment of the Spatial and Temporal Patterns of Cover Crops Using Remote Sensing. *Remote Sens.* **2021**, *13*, 2689. [CrossRef]
13. Pan, L.; Xia, H.; Zhao, X.; Guo, Y.; Qin, Y. Mapping Winter Crops Using a Phenology Algorithm, Time-Series Sentinel-2 and Landsat-7/8 Images, and Google Earth Engine. *Remote Sens.* **2021**, *13*, 2510. [CrossRef]
14. Cai, Y.; Guan, K.; Peng, J.; Wang, S.; Seifert, C.; Wardlow, B.; Li, Z. A High-Performance and in-Season Classification System of Field-Level Crop Types Using Time-Series Landsat Data and a Machine Learning Approach. *Remote Sens. Environ.* **2018**, *210*, 35–47. [CrossRef]
15. Zhong, L.; Hu, L.; Zhou, H. Deep Learning Based Multi-Temporal Crop Classification. *Remote Sens. Environ.* **2019**, *221*, 430–443. [CrossRef]
16. Rußwurm, M.; Körner, M. Multi-Temporal Land Cover Classification with Sequential Recurrent Encoders. *ISPRS Int. J. Geo-Inf.* **2018**, *7*, 129. [CrossRef]
17. Pelletier, C.; Webb, G.I.; Petitjean, F. Temporal Convolutional Neural Network for the Classification of Satellite Image Time Series. *Remote Sens.* **2019**, *11*, 523. [CrossRef]
18. Van Tricht, K.; Gobin, A.; Gilliams, S.; Piccard, I. Synergistic Use of Radar Sentinel-1 and Optical Sentinel-2 Imagery for Crop Mapping: A Case Study for Belgium. *Remote Sens.* **2018**, *10*, 1642. [CrossRef]
19. Yi, Z.; Jia, L.; Chen, Q.; Jiang, M.; Zhou, D.; Zeng, Y. Early-Season Crop Identification in the Shiyang River Basin Using a Deep Learning Algorithm and Time-Series Sentinel-2 Data. *Remote Sens.* **2022**, *14*, 5625. [CrossRef]
20. Belgiu, M.; Drăguţ, L. Random Forest in Remote Sensing: A Review of Applications and Future Directions. *ISPRS J. Photogramm. Remote Sens.* **2016**, *114*, 24–31. [CrossRef]
21. Mellor, A.; Boukir, S.; Haywood, A.; Jones, S. Exploring Issues of Training Data Imbalance and Mislabelling on Random Forest Performance for Large Area Land Cover Classification Using the Ensemble Margin. *ISPRS J. Photogramm. Remote Sens.* **2015**, *105*, 155–168. [CrossRef]
22. Deng, H.; Runger, G.; Tuv, E.; Vladimir, M. A Time Series Forest for Classification and Feature Extraction. *Inf. Sci.* **2013**, *239*, 142–153. [CrossRef]
23. Rußwurm, M.; Körner, M. Self-Attention for Raw Optical Satellite Time Series Classification. *ISPRS J. Photogramm. Remote Sens.* **2020**, *169*, 421–435. [CrossRef]
24. Ahmed, Z.; Nalley, L.; Brye, K.; Steven Green, V.; Popp, M.; Shew, A.M.; Connor, L. Winter-Time Cover Crop Identification: A Remote Sensing-Based Methodological Framework for New and Rapid Data Generation. *Int. J. Appl. Earth Obs. Geoinf.* **2023**, *125*, 103564. [CrossRef]
25. Henits, L.; Szerletics, Á.; Szokol, D.; Szlovák, G.; Gojdár, E.; Zlinszky, A. Sentinel-2 Enables Nationwide Monitoring of Single Area Payment Scheme and Greening Agricultural Subsidies in Hungary. *Remote Sens.* **2022**, *14*, 3917. [CrossRef]
26. Tao, Y.; You, F. Prediction of Cover Crop Adoption through Machine Learning Models Using Satellite-Derived Data. *IFAC-PapersOnLine* **2019**, *52*, 137–142. [CrossRef]
27. Simon Sanchez, A.-M.; Gonzalez-Piqueras, J.; de la Ossa, L.; Calera, A. Convolutional Neural Networks for Agricultural Land Use Classification from Sentinel-2 Image Time Series. *Remote Sens.* **2022**, *14*, 5373. [CrossRef]
28. Račič, M.; Oštir, K.; Zupanc, A.; Čehovin Zajc, L. Multi-Year Time Series Transfer Learning: Application of Early Crop Classification. *Remote Sens.* **2024**, *16*, 270. [CrossRef]
29. Weilandt, F.; Behling, R.; Goncalves, R.; Madadi, A.; Richter, L.; Sanona, T.; Spengler, D.; Welsch, J. Early Crop Classification via Multi-Modal Satellite Data Fusion and Temporal Attention. *Remote Sens.* **2023**, *15*, 799. [CrossRef]
30. Pandžić, M.; Pavlović, D.; Matavulj, P.; Brdar, S.; Marko, O.; Crnojević, V.; Kilibarda, M. Interseasonal Transfer Learning for Crop Mapping Using Sentinel-1 Data. *Int. J. Appl. Earth Obs. Geoinf.* **2024**, *128*, 103718. [CrossRef]
31. Lin, C.; Zhong, L.; Song, X.-P.; Dong, J.; Lobell, D.B.; Jin, Z. Early- and in-Season Crop Type Mapping without Current-Year Ground Truth: Generating Labels from Historical Information via a Topology-Based Approach. *Remote Sens. Environ.* **2022**, *274*, 112994. [CrossRef]
32. Beck, H.E.; Zimmermann, N.E.; McVicar, T.R.; Vergopolan, N.; Berg, A.; Wood, E.F. Present and Future Köppen-Geiger Climate Classification Maps at 1-Km Resolution. *Sci. Data* **2018**, *5*, 180214. [CrossRef] [PubMed]
33. Statistics Flanders Landbouwareaal. Available online: https://www.vlaanderen.be/statistiek-vlaanderen/landbouw-en-visserij/landbouwareaal (accessed on 18 July 2024).
34. Hermans, I.; Elsen, A.; Bries, J. *Groenbemesters En Nitraatresidu*; Bodemkundige Dienst van België vzw: Heverlee, Belgium, 2010; p. 44.

35. Vandendriessche, H.; Vanongeval, L.; Smeets, E.; Geypens, M. Monitoring of N-Uptake by Green Manures and of the Influence of N-Release on N-Availability, Production and Quality of Sugar Beet. In *Progress in Nitrogen Cycling Studies: Proceedings of the 8th Nitrogen Workshop Held at the University of Ghent, 5–8 September 1994*; Van Cleemput, O., Hofman, G., Vermoesen, A., Eds.; Developments in Plant and Soil Sciences; Springer: Dordrecht, The Netherlands, 1996; pp. 147–151; ISBN 978-94-011-5450-5.
36. EPPO. EPPO Global Database. 2024. Available online: https://gd.eppo.int/ (accessed on 18 July 2024).
37. Waśniewski, A.; Hościło, A.; Chmielewska, M. Can a Hierarchical Classification of Sentinel-2 Data Improve Land Cover Mapping? *Remote Sens.* **2022**, *14*, 989. [CrossRef]
38. Xie, G.; Niculescu, S. Mapping Crop Types Using Sentinel-2 Data Machine Learning and Monitoring Crop Phenology with Sentinel-1 Backscatter Time Series in Pays de Brest, Brittany, France. *Remote Sens.* **2022**, *14*, 4437. [CrossRef]
39. Turkoglu, M.O.; D'Aronco, S.; Perich, G.; Liebisch, F.; Streit, C.; Schindler, K.; Wegner, J.D. Crop Mapping from Image Time Series: Deep Learning with Multi-Scale Label Hierarchies. *Remote Sens. Environ.* **2021**, *264*, 112603. [CrossRef]
40. European Space Agency. S2 Mission. Available online: https://sentiwiki.copernicus.eu/web/s2-mission (accessed on 18 July 2024).
41. Biesemans, J. Supporting Land and Infrastructure Management with CropSAR. Available online: https://vito.be/en/news/supporting-land-and-infrastructure-management-cropsar (accessed on 18 July 2024).
42. Van Tricht, K. Looking Through the Clouds to Improve Crop Monitoring! Available online: https://blog.vito.be/remotesensing/cropsar2019 (accessed on 18 July 2024).
43. Piccard, I.; Swinnen, E.; Keukelaere, L.D.; Kerchove, R.V.D.; Eerens, H. Terrascope Sentinel-2 Algorithm Theoretical Base Document S2—NDVI & BIOPAR—V102. 2020. Available online: https://tinyurl.com/4mru98y5 (accessed on 12 April 2023).
44. Schramm, M.; Pebesma, E.; Milenković, M.; Foresta, L.; Dries, J.; Jacob, A.; Wagner, W.; Mohr, M.; Neteler, M.; Kadunc, M.; et al. The openEO API–Harmonising the Use of Earth Observation Cloud Services Using Virtual Data Cube Functionalities. *Remote Sens.* **2021**, *13*, 1125. [CrossRef]
45. Breiman, L. Random Forests. *Mach. Learn.* **2001**, *45*, 5–32. [CrossRef]
46. Pedregosa, F.; Varoquaux, G.; Gramfort, A.; Michel, V.; Thirion, B.; Grisel, O.; Blondel, M.; Prettenhofer, P.; Weiss, R.; Dubourg, V.; et al. Scikit-Learn: Machine Learning in Python. *J. Mach. Learn. Res.* **2011**, *12*, 2825–2830.
47. van Rossum, G. *Python Reference Manual*; Centrum voor Wiskunde en Informatica: Amsterdam, The Netherlands, 1995.
48. Király, F.; Löning, M.; Bagnall, T.; Middlehurst, M.; Ray, A.; Ganesh, S.; Walter, M.; Oastler, G.; Lines, J.; ViktorKaz; et al. sktime/sktime. 2020. Available online: https://zenodo.org/records/14337818 (accessed on 18 July 2024).
49. Löning, M.; Bagnall, A.; Ganesh, S.; Kazakov, V. Sktime: A Unified Interface for Machine Learning with Time Series. *arXiv* **2019**, arXiv:1909.07872.
50. Kattenborn, T.; Leitloff, J.; Schiefer, F.; Hinz, S. Review on Convolutional Neural Networks (CNN) in Vegetation Remote Sensing. *ISPRS J. Photogramm. Remote Sens.* **2021**, *173*, 24–49. [CrossRef]
51. Ismail Fawaz, H.; Forestier, G.; Weber, J.; Idoumghar, L.; Muller, P.-A. Deep Learning for Time Series Classification: A Review. *Data Min. Knowl. Discov.* **2019**, *33*, 917–963. [CrossRef]
52. LeCun, Y.; Bengio, Y.; Hinton, G. Deep Learning. *Nature.* **2015**, *521*, 436–444. [CrossRef] [PubMed]
53. Paszke, A.; Gross, S.; Massa, F.; Lerer, A.; Bradbury, J.; Chanan, G.; Killeen, T.; Lin, Z.; Gimelshein, N.; Antiga, L.; et al. PyTorch: An Imperative Style, High-Performance Deep Learning Library. *Adv. Neural Inf. Process. Syst.* **2019**, *32*, 8024–8035.
54. Chollet, F. *Deep Learning with Python*; Manning Publications Co.: Shelter Island, NY, USA, 2018; ISBN 978-1-61729-443-3.
55. Liaw, R.; Liang, E.; Nishihara, R.; Moritz, P.; Gonzalez, J.E.; Stoica, I. Tune: A Research Platform for Distributed Model Selection and Training. *arXiv* **2018**, arXiv:1807.05118.
56. Bagnall, A.; Lines, J.; Bostrom, A.; Large, J.; Keogh, E. The Great Time Series Classification Bake off: A Review and Experimental Evaluation of Recent Algorithmic Advances. *Data Min. Knowl. Discov.* **2017**, *31*, 606–660. [CrossRef]
57. Pham, V.-D.; Tetteh, G.; Thiel, F.; Erasmi, S.; Schwieder, M.; Frantz, D.; Van Der Linden, S. Temporally Transferable Crop Mapping with Temporal Encoding and Deep Learning Augmentations. *Int. J. Appl. Earth Obs. Geoinf.* **2024**, *129*, 103867. [CrossRef]
58. Holzhauser, K.; Räbiger, T.; Rose, T.; Kage, H.; Kühling, I. Estimation of Biomass and N Uptake in Different Winter Cover Crops from UAV-Based Multispectral Canopy Reflectance Data. *Remote Sens.* **2022**, *14*, 4525. [CrossRef]
59. Wang, X.; Blesh, J.; Rao, P.; Paliwal, A.; Umashaanker, M.; Jain, M. Mapping Cover Crop Species in Southeastern Michigan Using Sentinel-2 Satellite Data and Google Earth Engine. *Front. Artif. Intell.* **2023**, *6*, 1035502. [CrossRef]
60. Huang, X.; Huang, J.; Li, X.; Shen, Q.; Chen, Z. Early Mapping of Winter Wheat in Henan Province of China Using Time Series of Sentinel-2 Data. *GISci. Remote Sens.* **2022**, *59*, 1534–1549. [CrossRef]
61. Hughes, G. On the Mean Accuracy of Statistical Pattern Recognizers. *IEEE Trans. Inf. Theory* **1968**, *14*, 55–63. [CrossRef]
62. Orynbaikyzy, A.; Gessner, U.; Conrad, C. Crop Type Classification Using a Combination of Optical and Radar Remote Sensing Data: A Review. *Int. J. Remote Sens.* **2019**, *40*, 6553–6595. [CrossRef]
63. Selvaraj, S.; Bargiel, D.; Htitiou, A.; Gerighausen, H. Analyzing Temporal Characteristics of Winter Catch Crops Using Sentinel-1 Time Series. *Remote Sens.* **2024**, *16*, 3737. [CrossRef]
64. Royal Meteorological Institute. Klimatologisch Jaaroverzicht 2021. Available online: https://www.meteo.be/resources/climatology/pdf/klimatologisch_jaaroverzicht_2021.pdf (accessed on 17 July 2024).
65. Royal Meteorological Institute. Klimatologisch Jaaroverzicht 2022. Available online: https://www.meteo.be/resources/climatology/pdf/klimatologisch_jaaroverzicht_2022.pdf (accessed on 17 July 2024).

66. Vaswani, A.; Shazeer, N.; Parmar, N.; Uszkoreit, J.; Jones, L.; Gomez, A.N.; Kaiser, Ł.U.; Polosukhin, I. Attention Is All You Need. In *Advances in Neural Information Processing Systems*; Curran Associates, Inc.: Vancouver, BC, Canada, 2017; Volume 30.
67. Ma, Y.; Chen, S.; Ermon, S.; Lobell, D.B. Transfer Learning in Environmental Remote Sensing. *Remote Sens. Environ.* **2024**, *301*, 113924. [CrossRef]
68. Tseng, G.; Cartuyvels, R.; Zvonkov, I.; Purohit, M.; Rolnick, D.; Kerner, H. Lightweight, Pre-Trained Transformers for Remote Sensing Timeseries. *arXiv* **2024**, arXiv:2304.14065.

remote sensing

MDPI

Article

Cotton Yield Prediction via UAV-Based Cotton Boll Image Segmentation Using YOLO Model and Segment Anything Model (SAM)

Janvita Reddy [1], Haoyu Niu [1,2,*], Jose L. Landivar Scott [3], Mahendra Bhandari [3], Juan A. Landivar [3], Craig W. Bednarz [4] and Nick Duffield [1,2]

1 Department of Electrical & Computer Engineering, Texas A&M University, College Station, TX 77843, USA
2 Texas A&M Institute of Data Science, Texas A&M University, College Station, TX 77843, USA
3 Texas A&M AgriLife Research and Extension Center, Texas A&M University, Corpus Christi, TX 78406, USA
4 Department of Agricultural Sciences, West Texas A&M University, Canyon, TX 79016, USA
* Correspondence: hniu@tamu.edu

Abstract: Accurate cotton yield prediction is essential for optimizing agricultural practices, improving storage management, and efficiently utilizing resources like fertilizers and water, ultimately benefiting farmers economically. Traditional yield estimation methods, such as field sampling and cotton weighing, are time-consuming and labor intensive. Emerging technologies provide a solution by offering farmers advanced forecasting tools that can significantly enhance production efficiency. In this study, the authors employ segmentation techniques on cotton crops collected using unmanned aerial vehicles (UAVs) to predict yield. The authors apply Segment Anything Model (SAM) for semantic segmentation, combined with You Only Look Once (YOLO) object detection, to enhance the cotton yield prediction model performance. By correlating segmentation outputs with yield data, we implement a linear regression model to predict yield, achieving an R^2 value of 0.913, indicating the model's reliability. This approach offers a robust framework for cotton yield prediction, significantly improving accuracy and supporting more informed decision-making in agriculture.

Keywords: cotton; semantic segmentation; unmanned aerial vehicles; segment anything; yield estimation; object detection

1. Introduction

Crop yield prediction is increasingly critical due to concerns about food security and the need to estimate food availability for a growing population [1]. Accurate early predictions are essential for reducing famine risks and improving resource management, which benefits farmers through better decisions on crop insurance, harvest planning, and budgeting [2]. Traditional methods, such as labor-intensive manual surveys and modeling approaches, are often costly, subjective, and inefficient, especially for large-scale operations [3]. Remote sensing methods use vegetation index, such as the normalized difference vegetation index (NDVI), to estimate crop yields [4]. For instance, Anastasiou et al. demonstrated that proximal sensing, particularly green normalized difference vegetation (GNDVI) during veraison, provided stronger correlations with grape yield compared to satellite-based methods [5]. Additionally, Kogan et al. used the advanced very high resolution radiometer (AVHRR)-based vegetation condition index (VCI) to estimate vegetation state and productivity, showing a strong correlation with crop density [6]. However, index-based approaches are limited by their reliance on calibration, manual interpretation, and difficulties in handling environmental variability and high-dimensional data. Unmanned aerial vehicle (UAV) technology has emerged as a valuable tool in ecological research for monitoring vegetation and ecosystems, offering a cost-effective and user-friendly alternative for agricultural remote sensing applications [7–9].

Machine learning and computer vision techniques have been developed for crop yield prediction. For example, Veenadhari et al. developed a classifier using the decision tree approach that helps to understand how different climatic factors influence the results [10]. Ramesh et al. developed statistical models, such as multiple linear regression and a density-based model, to estimate future year yield, taking into account variables such as snow area, fertilizers, rainfall, and production [11]. Khaki et al. trained deep neural network models that incorporate genotype, weather, and soil conditions to predict and check yield [12]. In addition, feature selection was performed to optimize the input space, without significant reduction in accuracy. Aggarwal et al. applied histogram equalization and k-means clustering to separate the crop from the background in images, increasing accuracy and computational efficiency [13]. You et al. effectively integrated spatial and temporal features by combining long short-term memory (LSTM) networks with Gaussian processes for accurate yield prediction of soybean crop [14]. Computer vision tools are powerful for automating the inspection tasks in agriculture. Image segmentation is a crucial step in computer vision tasks where the goal is to partition an image into meaningful segments, typically by classifying each pixel into a predefined category. In the context of agriculture, segmentation plays an essential role in extracting valuable information from images of crops. For example, Wang et al. developed an automated system for apple yield estimation by capturing the nighttime images of trees and segmented the apples based on color cues [15]. Morphological operations were applied to locate and count the apples. In [16], Sarkate et al. performed color segmentation using thresholding and histogram analysis in the hue, saturation, and value (HSV) color space to detect gerbera flowers.

Deep learning-based segmentation has gained significant prominence due to its robustness and adaptability. Unlike traditional methods, which are often sensitive to noise and require carefully tuned thresholding values, deep learning models excel in handling variability within agricultural images [17,18]. For instance, in [19], Palacios et al. focused on predicting the number of berries and the overall yield of grapevines by employing a SegNet-based convolutional neural network (CNN) to detect berries and canopy features in grapevine images, which were then used to train support vector regression (SVR) models for predicting berry count and yield. This approach allowed accurate yield forecasting up to 60 days before harvest. In [10], Veenadhari et al. created an ensemble framework based on the bagging strategy and the UNet network. It utilizes both RGB and HSV color spaces to improve the segmentation of maize crop. In [20], Yu et al. modified UNet by reducing the downsampling rate, and adding attention as well as feature extract blocks to help model distinguish oranges from complex backgrounds in orchard environments. CNNs have become highly significant in segmentation tasks due to their impressive accuracy and effectiveness. However, their ability to capture information is inherently limited to local regions [21], and they struggle to model long-range dependencies across an image. This limitation arises from the localized nature of convolutional operations and the fixed receptive field of convolutional layers. Following the introduction of the self-attention mechanism in [22], there has been a significant surge in the application of Transformer models. The integration of CNNs and Transformers is particularly promising for complex vision tasks that require both fine-grained feature extraction and an understanding of broader spatial relationships. For instance, in [23], Chen et al. proposed TransUnet, which used transformer to encode image tokens from a CNN feature map to extract global contexts, while the decoder upscales these encoded features and merges them with low-level feature maps to attain precise location. In [24], Silva et al. leveraged a hybrid approach by adding a resnet block after the transformer encoder to combine the advantages of both CNN and vision transformer (ViT) for soybean weed detection. In [21], Coro et al. employed an encoder CNN combined with a spatial and channel reconstruction unit to preserve essential spatial information, while developing a decoder transformer with multiple attention mechanisms to focus on the local features of crops.

More recently, the Segment Anything Model (SAM) was proposed, and gained a lot of attention because of its capability to accurately segment the images based on user prompts [25]. This foundational model was trained on a large dataset of 1 million images and over 1 billion masks, which makes it highly adaptable and able to transfer its knowledge to any new distribution of images. For instance, in [26], Zhang et al. evaluated the performance between SAM automatic mode and SAM box prompt for segmenting clinical radiotherapy images. In [27], Zhang et al. developed SAMed for medical image segmentation, which infused a low-rank adaptation fine-tuning strategy, also incorporating a warm-up strategy and the Adam optimizer for better convergence and lower loss. Though SAM has excellent segmentation capabilities, it is seldom used to segment on agriculture-specific domains. To address this issue, researchers are working on fine-tuning millions of parameter-trained models. In [28], Li et al. incorporated SAM adaptors between the decoder layers, keeping all other parameters constant, to better suit the segmentation task for agricultural images.

Despite the strong zero-shot segmentation capabilities of SAM, it struggles with multiple objects and domain-specific images, such as those from agricultural settings. To overcome these limitations, in this article, the authors explore the innovative use of vision foundational models for cotton yield prediction, which leverages trained prompts for segmenting various objects, and present a comparative analysis showcasing its superior performance against other models on cotton yield prediction. Cotton plays a vital role in the global economy and is one of the most widely used natural fibers worldwide. Its versatility makes it essential in various industries, from textiles to food production. In the United States (US), cotton is a major cash crop that contributes significantly to the agricultural sector. The United States is one of the largest producers and exporters of cotton globally, accounting for about 35% of global cotton production [29]. Among US states, Texas is the largest producer, responsible for approximately 40% of the nation's cotton production in recent years [30]. This makes cotton essential not only for the agricultural economy but also for various industries that rely on its diverse applications.

The objective of this study is to develop and evaluate deep learning models, specifically YOLO and SAM, for cotton boll image segmentation and yield prediction using UAV RGB images. The major contributions are (1) the application of YOLO for high-accuracy cotton boll detection; (2) the integration of the SAM model for semantic segmentation in cotton yield prediction; and (3) the provision of the model framework, code, and methodology to support future research and applications in agriculture. The rest of the manuscript is organized as follows: Section 2 outlines in detail the materials and methods used to predict cotton yield using the YOLO and SAM models. Section 3 delves into a comprehensive analysis and discussion of cotton boll image segmentation achieved with these models, highlighting the relationship between yield and segmentation results. Finally, Section 4 concludes the study, summarizing the key findings and their implications for agricultural research.

2. Materials and Methods

2.1. Study Area

The field experiments were conducted in Lubbock, Texas (33.59°N, 101.90°W) on a 2.2-acre cotton field (Figure 1). Cotton was harvested mechanically from each individual row, generating a dataset comprising 96 rows of yield data. Each row extended 200 feet, containing about 150 cotton plants, with a 40-inch distance between rows [8].

Figure 1. The cotton field on 9 November 2022.

2.2. UAV Image Acquisition and Processing

High-resolution aerial images of the cotton field were acquired on 9 November 2022 using a DJI Phantom 4 Pro (Shenzhen DJI Sciences and Technologies Ltd., Shenzhen, China) equipped with a 1 inch complementary metal oxide semiconductor sensor (CMOS). The flight was carried out at an altitude of 25 m above ground level (AGL) and lasted approximately 18 min. The UAV flight path was planned with an 80% side and front overlap, resulting in a total of 406 high-resolution aerial images.

Data processing was performed with Agisoft Metashape to generate an orthomosaic. The image processing workflow followed a structured pipeline, starting with the alignment of the photos by identifying key points. Ground Control Points (GCPs) were incorporated to improve spatial accuracy, followed by optimizing camera parameters to refine alignment. A dense point cloud was then created, which was further utilized to build a Digital Elevation Model (DEM). Finally, the orthomosaic was generated using the DEM and aligned images. A visual representation of this workflow is presented in Figure 2, which outlines the step-by-step process from image acquisition to the generation of final data products.

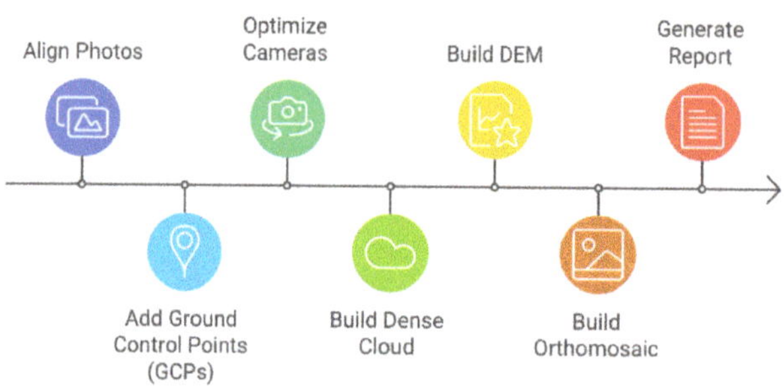

Figure 2. The step-by-step workflow from image acquisition to the generation of the orthomosaic image.

2.3. Image Annotation

The authors used the AnyLabeling software (Version: 0.4.15) to annotate images, leveraging SAM for automatic segmentation. The process involved running the encoder only once for each image, followed by running the decoder based on input prompts such as points or boxes to generate output masks. A post-processing step was automatically implemented to identify contours and created shapes like polygons or rectangles for labeling. AnyLabeling offers three versions of SAM. The original SAM, trained on 11 million images and 1 billion segmentation masks, excels at segmenting objects without prior training knowledge, making it ideal for autolabeling even with new objects. SAM 2, Meta's latest advancement, offers enhanced visual segmentation for both images and videos [31]. MobileSAM is a lightweight variant designed for mobile applications [32]. For our image annotation, we used the original SAM model, annotating around 1800 images. A demonstration of the original images and labeled masks is shown in Figure 3.

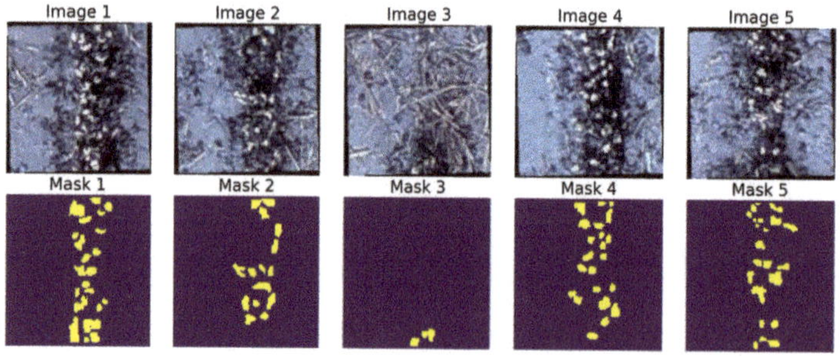

Figure 3. A demonstration of the original and mask images.

2.4. The Segment Anything and YOLO Models

SAM is a prompt-based model that, similar to language models, generates a valid segmentation mask from ambiguous prompts like points, text, or rough boxes. It comprises an image encoder, a pretrained masked autoencoder vision transformer that processes the image input, and a prompt encoder that handles sparse prompts by embedding them with learned embeddings and positional encodings. Text embeddings are generated using the contrastive language-image pretraining (CLIP) encoder, while dense prompts (masks) are processed through convolutional layers [33]. The decoder, inspired by Transformer models, uses self-attention within prompts and cross-attention between prompts and image features to update the embeddings. After two processing blocks, the image features are upscaled, and a neural network maps the output token to a classifier, which predicts the foreground mask, resulting in the final segmentation mask.

The YOLO v7 and YOLO v8 object detection algorithms were mainly explored for generating bounding boxes. The YOLO models have continuously evolved to address the limitations and disadvantages of their previous versions. Unlike traditional models such as RCNN and Fast RCNN [34], which use two separate outputs for object detection, one for classification and the other for bounding-box regression, the YOLO models perform both tasks in a single pass. This streamlined approach allows YOLO to predict class labels and bounding box coordinates simultaneously, enhancing both efficiency and performance [35]. The architecture of YOLO v7 is structured into three main components: the backbone, neck, and head [36]. The backbone is crucial for the extraction of features, playing a significant role in both the training of the model and its overall efficiency. It is responsible for capturing and representing detailed features from the input data. The neck functions as the feature aggregator, combining low-level spatial information with high-level semantic information to build rich feature maps at various levels. Finally, the head utilizes these aggregated features to perform final object detection, generating precise bounding boxes and class predictions. To enhance feature representation, YOLO v7 employs the efficient layer aggregation network (ELAN). Unlike traditional models that use stack convolutional layers, which limit detection capabilities for objects of various sizes, ELAN aggregates features from multiple layers. This approach allows YOLO v7 to combine information from different stages of processing, capturing features at multiple scales simultaneously. ELAN further extends this by integrating multiple paths for feature aggregation, improving the model's ability to handle a broad range of features without significantly increasing computational costs. YOLO v8 is the current superior object detection model built upon its previous versions to boost the performance and flexibility. Along with object detection, pose estimation, tracking and classification, this architecture also supports segmentation. Unlike earlier versions that relied on anchor boxes for detecting objects, YOLO v8 utilizes an anchor-free approach, which does not use predefined boxes [37]. It directly predicts the location and size of the objects. This reduces the computational complexity and makes the model simpler and faster, with fewer hyperparameters to tune. The head of the network is "decoupled"; it separates the processing for the objectness score, classification, and regression, which lets the model specialize in its specific task.

In this article, we propose an innovative two-stage approach for image segmentation that integrated object detection with advanced segmentation technique. In the first stage, we employed YOLO v7 and YOLO v8 as the object detector, which processed the input images to predict bounding boxes around the region of interest (Figure 4). These bounding boxes were then passed as input to the prompt encoder, which performed prompt encoding by embedding the box coordinates along with positional encodings. More specifically, from the AnyLabeling software, we extracted the bounding box coordinates for the bolls in the cotton images. These bounding boxes were then converted to the YOLO labeling format using Roboflow, which transformed the coordinates into the required format: (class index, normalized x center, normalized y center, normalized width, normalized height). Since all of the images contained only one class, cotton bolls, the class index was consistently set to 0. Out of the 1800 images, we used 1500 for training, 100 for validation, and the remaining 200 for testing the model. To further enhance the performance of the model, we resized the image resolution from 150 × 150 to 640 × 640 (on which originally YOLO models were trained) pixels to improve object detection for the small cotton bolls scattered across the field.

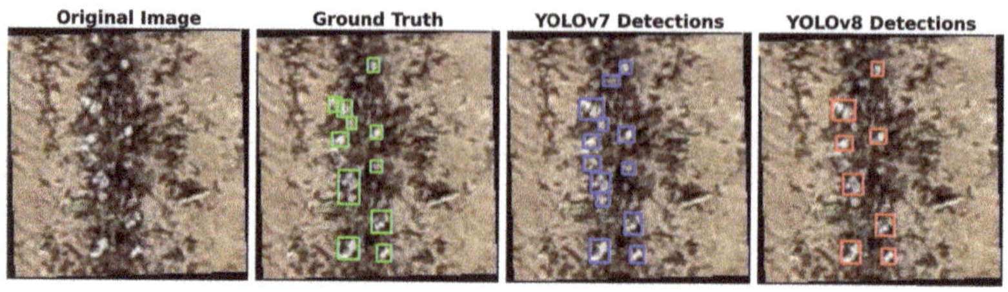

Figure 4. Comparison of cotton boll detection between YOLO v7 and YOLO v8 models.

In the second stage, the SAM encoded the entire image using a pretrained vision transformer. The bounding boxes, now encoded by the prompt encoder, were also fed into the SAM decoder (Figure 5). The SAM decoder performed a series of operations involving both self-attention within the prompts and cross-attention between the image features and the encoded prompts. This bidirectional attention mechanism enabled the model to refine and update the embeddings, ensuring that the segmentation was accurate and aligned with the detected objects. This information was processed to generate the final output, a segmented mask that precisely delineated the objects within the image. By combining YOLO v7 or v8 for object detection with SAM's powerful segmentation capabilities, our approach effectively handled images with complex scenarios, resulting in a robust and efficient solution for image segmentation tasks.

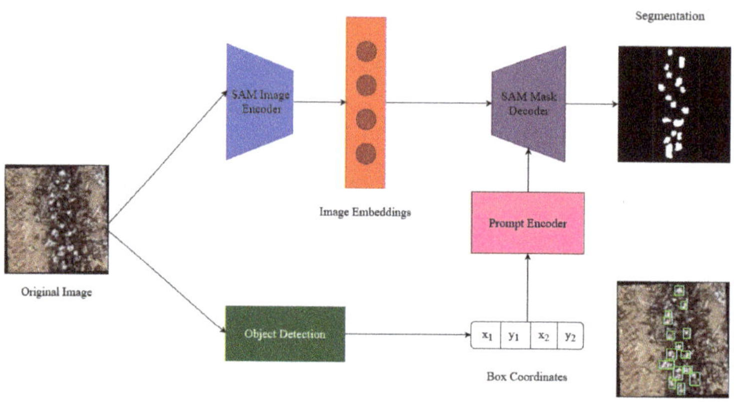

Figure 5. The proposed YOLO + SAM model architecture.

2.5. Model Evaluation Metrics

To evaluate the model's performance in detecting objects, we use the following metrics. Precision measures the proportion of objects detected by the model that are actually relevant (true positives). It is calculated as the ratio of true positives (TP) to the sum of true positives and false positives (FP). Precision reflects how accurate the detections are by indicating the percentage of correct detections (Equation (1)). Recall measures how well the model detects relevant objects in the dataset. It is calculated as the ratio of true positives to the sum of true positives and false negatives (FN). Recall indicates the model's ability to identify all relevant objects (Equation (2)). F1-Score finds the most optimal confidence score threshold where precision and recall give the highest F1 score. The F1 score calculates the balance between precision and recall. If the F1 score is high, precision and recall are high, and vice versa (Equation (3)):

$$\text{Precision} = \frac{TP}{TP + FP}, \tag{1}$$

$$\text{Recall} = \frac{TP}{TP + FN}, \tag{2}$$

$$\text{F1-Score} = 2 \times \frac{\text{Precision} \times \text{Recall}}{\text{Precision} + \text{Recall}}. \tag{3}$$

To evaluate the performance of our segmented images, we use the Intersection over Union (IoU) score, also known as the Jaccard coefficient (Equation (4)). The IoU score measures the overlap between the ground truth mask A and the predicted mask B, providing a quantitative measure of the precision of the segmentation of objects in the region of interest. The area of intersection is the area where the predicted mask and the ground truth mask overlap, and the area of union is the total area covered by both the predicted mask and the ground truth mask combined. For Mean Average Precision (mAP), the IoU threshold is set to 0.5, which means that a predicted object is considered a true positive if the overlap between the predicted bounding box and the ground truth bounding box is at least 50%. It indicates how competent the model is in localizing the objects:

$$\text{Jaccard}(A, B) = \frac{\text{Area of Intersection } (A \cap B)}{\text{Area of Union } (A \cup B)}. \tag{4}$$

3. Results and Discussions

3.1. The SAM and YOLO Model Performance

To achieve an optimal balance between precision and recall, we systematically evaluate the the F1 score on various decision thresholds. By plotting the F1 score as a function of the confidence threshold, we identify the threshold that maximizes the F1 score for each model on the validation dataset. For YOLO v7, the optimal confidence threshold is determined to be 0.3, while for YOLO v8, a threshold of 0.25 yields the highest F1 score. This analysis reveals that the maximum achievable F1 score in both models is 0.83. These thresholds ensure that the models achieve an optimal trade-off between precision and recall, enhancing their ability to accurately identify relevant objects while minimizing false positives and false negatives. The output of each object in the image includes the class number and the normalized coordinates of its bounding box, represented as $(x_{\text{center}}, y_{\text{center}}, w, h)$, where x_{center} and y_{center} are the normalized coordinates of the center of the bounding box, and w and h are the normalized width and height. To convert these normalized coordinates to absolute pixel coordinates, we use the following equations:

$$\begin{aligned} x_{\min} &= \left(x_{\text{center}} - \frac{w}{2}\right) \times W, \quad & y_{\min} &= \left(y_{\text{center}} - \frac{h}{2}\right) \times H; \\ x_{\max} &= \left(x_{\text{center}} + \frac{w}{2}\right) \times W, \quad & y_{\max} &= \left(y_{\text{center}} + \frac{h}{2}\right) \times H; \end{aligned} \tag{5}$$

where $(x_{\min}, y_{\min})$ and $(x_{\max}, y_{\max})$ represent the coordinates of the bottom-left and top-right corners of the bounding box in pixel coordinates, respectively. The width W and height H refer to the dimensions of the image in pixels. These absolute coordinates are then provided to the SAM as prompts for segmenting the regions around the bounding box. The SAM model uses these coordinates to accurately identify and segment objects of interest within the specified bounding box, ensuring precise and focused segmentation. From the comparative analysis presented in Table 1, it is evident that YOLO v7 outperforms YOLO v8 in several key metrics. YOLO v7 shows higher precision, recall, F1 score, and mAP@50 compared to YOLO v8. Furthermore, YOLO v7, when combined with SAM, achieves a higher IoU score of 0.683. This could be justified because there are more true positives detected in YOLO v7 compared to YOLO v8. These results indicate that YOLO v7 is more effective in accurately detecting and localizing objects, making it a more robust choice for object detection tasks. Also, from Figure 6, we observe that YOLO v7 not only detects all objects in the image but also identifies objects that are missed during labeling, further validating the generalization capabilities of our algorithm. Furthermore, YOLO v7 excels in locating smaller objects, which is a significant advantage over YOLO v8. YOLO v8 prioritizes certain optimizations such as faster inference or reduced model complexity, which leads to a slight compromise in detection accuracy, especially in specific tasks like detecting small objects. On the other hand, YOLO v7 incorporates architectural enhancements and reparameterization techniques that are particularly effective in improving precision, recall, and localization accuracy. This makes YOLO v7 better suited for cotton boll detection, where high detection accuracy and robust performance are more critical than other considerations like speed or model size. Therefore, while YOLO v8 brings advances, it might not always outperform YOLO v7 in terms of raw detection performance.

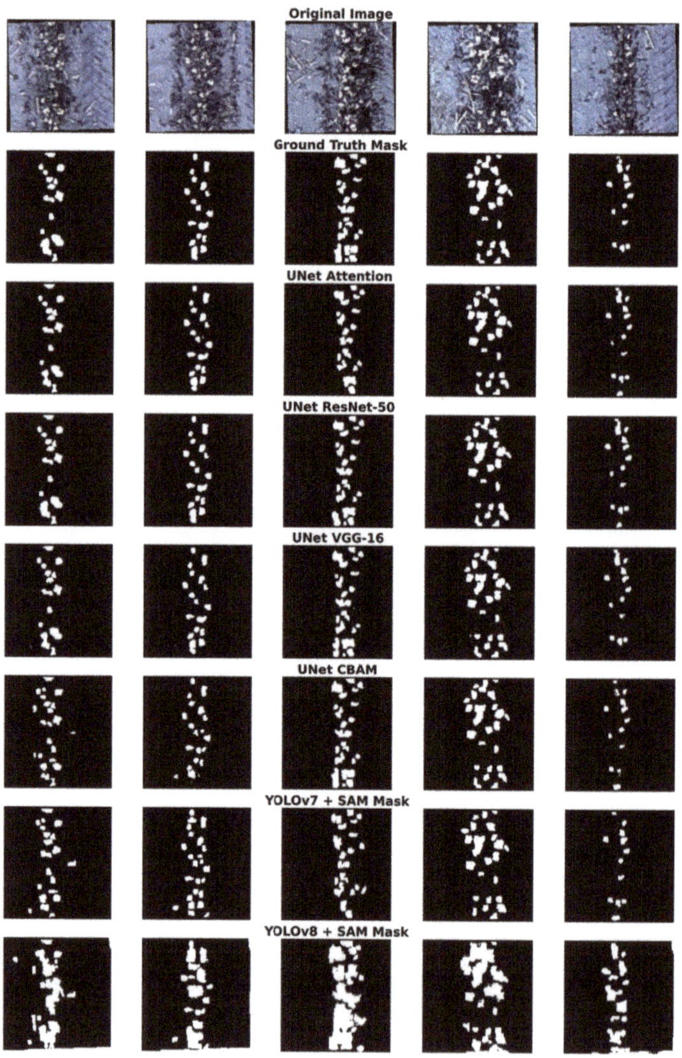

Figure 6. Comparison of cotton boll segmentation between different models.

Table 1. Comparison of YOLO v7 and YOLO v8 metrics.

Models	Precision	Recall	F1-Score	$mAP_{0.5}$	IoU
YOLO v7 + SAM	0.821	0.836	0.828	0.857	0.685
YOLO v8 + SAM	0.814	0.791	0.802	0.833	0.683

In addition to its superior object detection capabilities, the integration of YOLO v7 with SAM for segmentation tasks demonstrates significant improvements in both efficiency and accuracy compared to conventional segmentation models. Traditional segmentation models require a large amount of labeled data for training and struggle with small or overlapping objects. In contrast, our approach leverages the precise bounding boxes generated by YOLO v7 to provide focused and accurate segmentation prompts to SAM. This results in a more computationally efficient pipeline, as the model narrows down the regions of interest before applying segmentation. Furthermore, the high IoU score of 0.685 indicates that the YOLO v7 + SAM combination is not only faster but also more effective in identifying object boundaries, particularly for smaller objects that traditional models often miss. This combination of speed and accuracy makes our model a suitable choice for real-time applications requiring both object detection and segmentation, offering a substantial advantage over conventional segmentation models. We experimented with well-known U-Net architecture models for segmentation. U-Net with attention mechanisms leverages the encoder–decoder structure to focus on specific regions. Variants like U-Net with ResNet 50 and VGG 16 utilize models pretrained on the ImageNet dataset as their encoders, benefiting from feature extraction capabilities. Additionally, U-Net with a Convolutional Block Attention Module (CBAM) introduces channel and spatial attention in the encoder, further refining the focus on relevant features. Table 2 presents the efficiency metrics of each model configuration.

Table 2. Comparison of multiple segmentation models.

Models	IoU Score	Inference Time (s)
U-Net Attention	0.697	0.0087
U-Net ResNet 50	0.696	0.033
U-Net VGG 16	0.685	0.0128
U-Net CBAM	0.680	0.013
Yolov7 + SAM	0.685	0.526
Yolov8 + SAM	0.683	0.554

3.2. *Evaluation of Cotton Yield at Row Level*

We apply image segmentation using the proposed method to segment cotton bolls from field images. The segmentation process allows us to isolate cotton bolls as white pixels in the images. We then count the number of white pixels in each image, which represent the presence of cotton. To estimate the cotton yield for each row, we aggregate the white pixel counts across all images belonging to that specific row. This approach allows us to derive an estimated yield for the entire row on the basis of the segmented images. Next, we aim to quantify the relationship between the white pixel count and the actual cotton yield. A linear regression model is applied, where the white pixel count is used as the independent variable and the actual yield as the target variable. We use data from 58 rows to construct the linear regression model and test it on data from 38 rows. In Figure 7, we plot the pixel count for each row on the x axis and the actual yield on the y axis. We observe that the model successfully captures the correlation between these two values, achieving a high R^2 value of 0.913 with YOLOv8 + SAM, indicating that the model explains a large proportion of the variance in cotton yield. The model has a low mean absolute error (MAE) of 3.872 lbs/row, showing that the predictions are accurate.

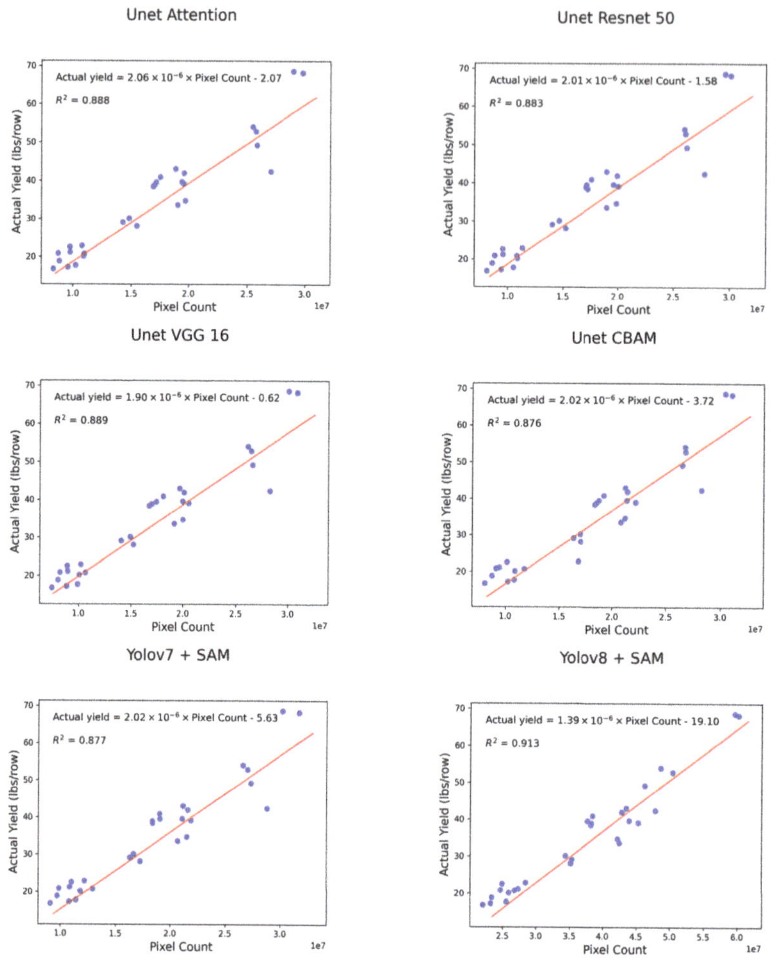

Figure 7. Linear relationship between pixel count and actual yield across different models. The YOLOv8 + SAM has the best yield prediction performance, with a R^2 of 0.913.

However, it is important to note that a higher IoU score, which typically indicates better segmentation quality, does not necessarily translate to higher accuracy in yield prediction (Figure 7). While a higher IoU reflects a better match between the predicted and ground truth segmentation, it does not always correlate with the overall cotton yield since other factors, such as boll distribution, growth stage, and environmental conditions, can influence yield independently of segmentation accuracy. Thus, a high IoU score might not fully capture the complexities of yield estimation. In contrast, our linear regression model focuses on the aggregated pixel counts, which are more directly related to the overall yield, demonstrating that effective yield prediction depends not only on accurate segmentation but also on how well the segmentation relates to actual cotton production. Future studies may explore additional features or hybrid models that consider both segmentation accuracy and environmental variables to improve yield predictions.

4. Conclusions

In this study, the authors proposed a two-stage approach for image segmentation that integrates object detection with an advanced segmentation technique to predict cotton yield. A comparison between two recent state-of-the-art object detection models, YOLO v7 and YOLO v8, was conducted. YOLO v7 was observed to perform better in all metrics, and this resulted in a high IoU score of 0.685. Using YOLO v8 and SAM in conjunction with UAV imagery, we were able to achieve a high correlation between the segmented output and actual yield data, reflected in an R^2 value of 0.913. This model provides a robust and reliable method for predicting cotton yields, significantly reducing the need for traditional labor-intensive methods. The proposed approach not only improves prediction accuracy but also offers a practical tool to optimize agricultural practices, resource allocation, and decision-making processes. It paves the way for exploring the potential of vision foundational models in agricultural applications. Additionally, the combination of object detection with segmentation in our approach aligns with the growing interest in zero-shot or few-shot segmentation, which enables models to adapt to new crop types with minimal labeled data. Future work could explore the application of this framework to other types of crops and further refine the model by incorporating additional variables, such as soil health and climatic conditions, to improve its predictive power and generalizability in diverse agricultural environments. In this study, we focused on a specific dataset collected in the Lubbock, Texas area, with parameters optimized based on the characteristics of this location, cotton variety, and growth stage. While this approach was effective within the defined scope, we agree that further research is needed to explore how segmentation parameters may vary across different planting sites, cotton species, growth stages, and imaging techniques. As part of future work, we intend to investigate how our segmentation model generalizes under varied conditions and to test parameter adjustments that may enhance its adaptability.

5. Research Reproducibility

We agree that generalizing to other regions and environmental conditions would be valuable, and we have noted this as a future research direction. We believe that the model's framework can be adapted to different regions by local researchers. To support this, we have provided a detailed description of our methodology and made the model code available, facilitating adaptation and validation in diverse contexts. All the research results in this article can be reproduced. The code is available in the author's Github: https://github.com/hniu-tamu/Cotton_yield_prediciton_with_YOLO_and_SAM, accessed on 10 November 2024.

Author Contributions: Conceptualization, H.N. and J.R.; methodology, H.N. and J.R.; software, H.N. and J.R.; validation, H.N. and J.R.; formal analysis, H.N. and J.R.; investigation, J.R.; resources, M.B., J.A.L., C.W.B. and N.D.; data curation, J.R. and J.L.L.S.; writing—original draft preparation, J.R., J.L.L.S. and H.N.; writing—review and editing, J.R., J.L.L.S. and H.N.; visualization, J.R., J.L.L.S. and H.N.; supervision, H.N. and N.D.; project administration, C.W.B., M.B., J.A.L. and N.D.; funding acquisition, C.W.B., M.B., J.A.L. and N.D. All authors have read and agreed to the published version of the manuscript.

Funding: This research was funded by Texas state allocated funds for the Water Exceptional Item through Texas A&M AgriLife Research facilitated by the Texas Water Resources Institute.

Institutional Review Board Statement: Not applicable.

Informed Consent Statement: Not applicable.

Data Availability Statement: The original contributions presented in the study are included in the article, further inquiries can be directed to the corresponding author.

Acknowledgments: The authors sincerely appreciate Murilo Maeda's invaluable assistance as the project lead responsible for collecting all the data used in this experiment during 2022.

Conflicts of Interest: The authors declare no conflicts of interest.

Abbreviations

The following abbreviations are used in this manuscript:

AGL	Above Ground Level
AVHRR	Advanced Very High Resolution Radiometer
CMOS	Complementary Metal-Oxide-Semiconductor
CNN	Convolutional Neural Network
CLIP	Contrastive Language-Image Pretraining
DEM	Digital Elevation Model
ELAN	Efficient Layer Aggregation Network
FN	False Negative
FP	False Positive
GCPs	Ground Control Points
GNDVI	Green Normalized Difference Vegetation Index
HSV	Hue, Saturation, and Value
IoU	Intersection over Union
LSTM	Long Short-Term Memory
MAE	Mean Absolute Error
mAP	Mean Average Precision
NDVI	Normalized Difference Vegetation Index
RGB	Red, Green and Blue
SAM	Segment Anything Model
SVR	Support Vector Regression
TP	True Positive
UAV	Unmanned Aerial Vehicle
US	United States
VCI	Vegetation Condition Index
ViT	Vision Transformer
YOLO	You Only Look Once

References

1. Muruganantham, P.; Wibowo, S.; Grandhi, S.; Samrat, N.H.; Islam, N. A systematic literature review on crop yield prediction with deep learning and remote sensing. *Remote Sens.* **2022**, *14*, 1990. [CrossRef]
2. Zhang, M.; Feng, A.; Zhou, J.; Lü, X. Cotton yield prediction using remote visual and spectral images captured by UAV system. *Trans. Chin. Soc. Agric. Eng.* **2019**, *35*, 91–98.
3. Khaki, S.; Pham, H.; Wang, L. Simultaneous corn and soybean yield prediction from remote sensing data using deep transfer learning. *Sci. Rep.* **2021**, *11*, 11132. [CrossRef] [PubMed]
4. Quarmby, N.; Milnes, M.; Hindle, T.; Silleos, N. The use of multi-temporal NDVI measurements from AVHRR data for crop yield estimation and prediction. *Int. J. Remote Sens.* **1993**, *14*, 199–210. [CrossRef]
5. Anastasiou, E.; Balafoutis, A.; Darra, N.; Psiroukis, V.; Biniari, A.; Xanthopoulos, G.; Fountas, S. Satellite and proximal sensing to estimate the yield and quality of table grapes. *Agriculture* **2018**, *8*, 94. [CrossRef]
6. Kogan, F.; Gitelson, A.; Zakarin, E.; Spivak, L.; Lebed, L. AVHRR-based spectral vegetation index for quantitative assessment of vegetation state and productivity. *Photogramm. Eng. Remote Sens.* **2003**, *69*, 899–906. [CrossRef]
7. Ali, A.M.; Abouelghar, M.; Belal, A.; Saleh, N.; Yones, M.; Selim, A.I.; Amin, M.E.; Elwesemy, A.; Kucher, D.E.; Maginan, S.; et al. Crop yield prediction using multi sensors remote sensing. *Egypt. J. Remote Sens. Space Sci.* **2022**, *25*, 711–716.
8. Niu, H.; Peddagudreddygari, J.R.; Bhandari, M.; Landivar, J.A.; Bednarz, C.W.; Duffield, N. In-season cotton yield prediction with scale-aware convolutional neural network models and unmanned aerial vehicle RGB imagery. *Sensors* **2024**, *24*, 2432. [CrossRef]
9. Niu, H.; Chen, Y. *Smart Big Data in Digital Agriculture Applications*; Springer: Berlin/Heidelberg, Germany, 2024.
10. Veenadhari, S.; Mishra, B.; Singh, C. Soybean productivity modelling using decision tree algorithms. *Int. J. Comput. Appl.* **2011**, *27*, 11–15. [CrossRef]
11. Ramesh, D.; Vardhan, B.V. Analysis of crop yield prediction using data mining techniques. *Int. J. Res. Eng. Technol.* **2015**, *4*, 47–473.
12. Khaki, S.; Wang, L. Crop yield prediction using deep neural networks. *Front. Plant Sci.* **2019**, *10*, 621. [CrossRef] [PubMed]
13. Aggarwal, A.K.; Jaidka, P. Segmentation of crop images for crop yield prediction. *Int. J. Biol. Biomed.* **2022**, *7*, 40–44.
14. You, J.; Li, X.; Low, M.; Lobell, D.; Ermon, S. Deep Gaussian process for crop yield prediction based on remote sensing data. In Proceedings of the AAAI Conference on Artificial Intelligence, San Francisco, CA, USA, 4–9 February 2017; Volume 31.

15. Wang, Q.; Nuske, S.; Bergerman, M.; Singh, S. Design of crop yield estimation system for apple orchards using computer vision. In Proceedings of the 2012 Dallas, Dallas TX, USA, 29 July–1 August 2012; American Society of Agricultural and Biological Engineers: St. Joseph, MI, USA, 2012; p. 1.
16. Sarkate, R.S.; Kalyankar, N.; Khanale, P. Application of computer vision and color image segmentation for yield prediction precision. In Proceedings of the 2013 International Conference on Information Systems and Computer Networks, Mathura, India, 9–10 March 2013; IEEE: Piscataway, NJ, USA, 2013; pp. 9–13.
17. Maji, A.K.; Marwaha, S.; Kumar, S.; Arora, A.; Chinnusamy, V.; Islam, S. SlypNet: Spikelet-based yield prediction of wheat using advanced plant phenotyping and computer vision techniques. *Front. Plant Sci.* **2022**, *13*, 889853. [CrossRef] [PubMed]
18. Peng, H.; Xue, C.; Shao, Y.; Chen, K.; Xiong, J.; Xie, Z.; Zhang, L. Semantic segmentation of litchi branches using DeepLabV3+ model. *IEEE Access* **2020**, *8*, 164546–164555. [CrossRef]
19. Palacios, F.; Diago, M.P.; Melo-Pinto, P.; Tardaguila, J. Early yield prediction in different grapevine varieties using computer vision and machine learning. *Precis. Agric.* **2023**, *24*, 407–435. [CrossRef]
20. Yu, C.; Lin, D.; He, C. ASE-UNet: An orange fruit segmentation model in an agricultural environment based on deep learning. *Opt. Mem. Neural Netw.* **2023**, *32*, 247–257.
21. Corò, F.; D'Angelo, G.; Velaj, Y.; et al. Recommending links to maximize the influence in social networks. In Proceedings of the Twenty-Eighth International Joint Conference on Artificial Intelligence (IJCAI 2019), Macao, China, 10–16 August 2019; AAAI Press: Washington, DC, USA, 2019; Volume 4, pp. 2195–2201.
22. Vaswani, A. Attention is all you need. *arXiv* **2017**, arXiv:1706.03762.
23. Chen, J.; Lu, Y.; Yu, Q.; Luo, X.; Adeli, E.; Wang, Y.; Lu, L.; Yuille, A.L.; Zhou, Y. Transunet: Transformers make strong encoders for medical image segmentation. *arXiv* **2021**, arXiv:2102.04306.
24. Silva, L.; Drews, P.; de Bem, R. Soybean weeds segmentation using VT-Net: A convolutional-transformer model. In Proceedings of the 2023 36th SIBGRAPI Conference on Graphics, Patterns and Images (SIBGRAPI), Rio Grande, Brazil, 6–9 November 2023; IEEE: Piscataway, NJ, USA, 2023; pp. 127–132.
25. Kirillov, A.; Mintun, E.; Ravi, N.; Mao, H.; Rolland, C.; Gustafson, L.; Xiao, T.; Whitehead, S.; Berg, A.C.; Lo, W.Y.; et al. Segment anything. In Proceedings of the IEEE/CVF International Conference on Computer Vision, Paris, France, 1–6 October 2023; pp. 4015–4026.
26. Zhang, L.; Liu, Z.; Zhang, L.; Wu, Z.; Yu, X.; Holmes, J.; Feng, H.; Dai, H.; Li, X.; Li, Q.; et al. Segment anything model (SAM) for radiation oncology. *arXiv* **2023**, arXiv:2306.11730.
27. Zhang, K.; Liu, D. Customized segment anything model for medical image segmentation. *arXiv* **2023**, arXiv:2304.13785.
28. Li, Y.; Wang, D.; Yuan, C.; Li, H.; Hu, J. Enhancing agricultural image segmentation with an agricultural segment anything model adapter. *Sensors* **2023**, *23*, 7884. [CrossRef] [PubMed]
29. Ridley, W.; Devadoss, S. Competition and trade policy in the world cotton market: Implications for US cotton exports. *Am. J. Agric. Econ.* **2023**, *105*, 1365–1387. [CrossRef]
30. Adhikari, P.; Ale, S.; Bordovsky, J.P.; Thorp, K.R.; Modala, N.R.; Rajan, N.; Barnes, E.M. Simulating future climate change impacts on seed cotton yield in the Texas High Plains using the CSM-CROPGRO-Cotton model. *Agric. Water Manag.* **2016**, *164*, 317–330. [CrossRef]
31. Ravi, N.; Gabeur, V.; Hu, Y.T.; Hu, R.; Ryali, C.; Ma, T.; Khedr, H.; Rädle, R.; Rolland, C.; Gustafson, L.; et al. SAM 2: Segment anything in images and videos. *arXiv* **2024**, arXiv:2408.00714.
32. Zhang, C.; Han, D.; Qiao, Y.; Kim, J.U.; Bae, S.H.; Lee, S.; Hong, C.S. Faster segment anything: Towards lightweight SAM for mobile applications. *arXiv* **2023**, arXiv:2306.14289.
33. Radford, A.; Kim, J.W.; Hallacy, C.; Ramesh, A.; Goh, G.; Agarwal, S.; Sastry, G.; Askell, A.; Mishkin, P.; Clark, J.; et al. Learning transferable visual models from natural language supervision. In Proceedings of the International Conference on Machine Learning, PMLR, Virtual, 18–24 July 2021; pp. 8748–8763.
34. Girshick, R. Fast r-cnn. In Proceedings of the IEEE International Conference on Computer Vision, Santiago, Chile, 7–13 December 2015; pp. 1440–1448.
35. Redmon, J.; Divvala, S.; Girshick, R.; Farhadi, A. You only look once: Unified, real-time object detection. In Proceedings of the IEEE Conference on Computer Vision and Pattern Recognition, Las Vegas, NV, USA, 27–30 June 2016; pp. 779–788.
36. Wang, C.Y.; Bochkovskiy, A.; Liao, H.Y.M. YOLOv7: Trainable bag-of-freebies sets new state-of-the-art for real-time object detectors. In Proceedings of the IEEE/CVF Conference on Computer Vision and Pattern Recognition, Vancouver, BC, Canada, 17–24 June 2023; pp. 7464–7475.
37. Varghese, R.; Sambath, M. YOLOv8: A novel object detection algorithm with enhanced performance and robustness. In Proceedings of the 2024 International Conference on Advances in Data Engineering and Intelligent Computing Systems (ADICS), Chennai, India, 18–19 April 2024; IEEE: Piscataway, NJ, USA, 2024; pp. 1–6.

remote sensing

MDPI

Article

Reconstruction of Fine-Spatial-Resolution FY-3D-Based Vegetation Indices to Achieve Farmland-Scale Winter Wheat Yield Estimation via Fusion with Sentinel-2 Data

Xijia Zhou [1,2], Tao Wang [1,2,*], Wei Zheng [1,2], Mingwei Zhang [1,2] and Yuanyuan Wang [1,2]

[1] Key Laboratory of Radiometric Calibration and Validation for Environmental Satellites, National Satellite Meteorological Center (National Center for Space Weather), China Meteorological Administration, Beijing 100081, China; zhouxijia@cma.gov.cn (X.Z.); zhengw@cma.gov.cn (W.Z.); zhangmw@cma.gov.cn (M.Z.); wangyuany@cma.gov.cn (Y.W.)

[2] Innovation Center for the FengYun Meteorological Satellite (FYSIC), Beijing 100081, China

* Correspondence: taowang@cma.gov.cn; Tel.: +86-18756963195

Abstract: The spatial resolution (250–1000 m) of the FY-3D MERSI is too coarse for agricultural monitoring at the farmland scale (20–30 m). To achieve the winter wheat yield (WWY) at the farmland scale, based on FY-3D, a method framework is developed in this work. The enhanced deep convolutional spatiotemporal fusion network (EDCSTFN) was used to perform a spatiotemporal fusion on the 10 day interval FY-3D and Sentinel-2 vegetation indices (VIs), which were compared with the enhanced spatial and temporal adaptive reflectance fusion model (ESTARFM). In addition, a BP neural network was built to calculate the farmland-scale WWY based on the fused VIs, and the Aqua MODIS gross primary productivity product was used as ancillary data for WWY estimation. The results reveal that both the EDCSTFN and ESTARFM achieve satisfactory precision in the fusion of the Sentinel-2 and FY-3D VIs; however, when the period of spatiotemporal data fusion is relatively long, the EDCSTFN can achieve greater precision than ESTARFM. Finally, the WWY estimation results based on the fused VIs show remarkable correlations with the WWY data at the county scale and provide abundant spatial distribution details about the WWY, displaying great potential for accurate farmland-scale WWY estimations based on reconstructed fine-spatial-temporal-resolution FY-3D data.

Keywords: FY-3D; vegetation index; spatiotemporal data fusion; farmland scale; winter wheat yield estimation; precision agriculture; deep learning; neural network

1. Introduction

Winter wheat (WW) is one of the main grain crops in northern China, and the WW growth indices at different growing stages exhibit varying correlations with the final yield (Y). Therefore, continuous time series growth monitoring during the WW growing period (from the regrowth stage to the milk stage) is important for Y estimation [1,2]. Compared with observation data from ground and meteorological stations, remote sensing (RS) technology can detect the continuous spatial distribution of crop growth. A vegetation index (VI), which is calculated based on multispectral reflectance data, is one of the most commonly used methods for estimating crop growth statuses and is characterized by its simplicity and ease of use. Currently, the most commonly employed multispectral VIs for estimating crop growth include the normalized difference vegetation index (NDVI) [3], soil-adjusted vegetation index (SAVI) [4], and enhanced vegetation index (EVI) [5]. The SAVI and EVI were developed based on the NDVI, and they consider soil backgrounds and atmospheric factors, respectively. Relevant studies have shown that the NDVI has a favourable linear or nonlinear relationship with vegetation coverage [6,7], while compared with the NDVI, the SAVI and EVI demonstrate better correlations with the leaf area index [8,9].

Over the past few decades, numerous VI products have been developed using high-temporal-resolution (HTR) satellite sensors, including the NOAA Advanced Very-High-Resolution Radiometer (AVHRR), Terra/Aqua Moderate Resolution Imaging Spectroradiometer (MODIS), and Fengyun-3 (FY-3) Medium Resolution Spectral Imager (MERSI). These products typically exhibit high precision in uniformly distributed farmland areas and can achieve a temporal resolution (TR) as high as 0.5 days. Among them, the FY-3 satellite is a second-generation polar-orbiting solar-synchronous meteorological satellite developed in China. Launched in November 2017, the FY-3 satellite is equipped with a MERSI containing 25 channels, including four visible and near-infrared channels, with a spatial resolution (SR) of 250 m. These channels are essential spectral bands for calculating VI products. The overpass time of FY-3D satellites is approximately 14:00 in the afternoon and the TR of FY-3D can reach 0.5 days. Currently, VIs such as the NDVI and EVI, which are calculated on the basis of satellite sensors such as MODIS, have been widely used in estimating the Y of various crops [10,11]. Johnson et al. [11] estimated the crop Y for barley, spring wheat, and rapeseed on Canadian prairies via the MODIS NDVI and EVI, and the results revealed that in July, the MODIS NDVI can be used to effectively predict the crop Y of the three crops mentioned above and that the combination with the MODIS EVI can further improve the precision of crop Y forecasts. Compared with crop Y estimation models based on VIs during a specific growing period, crop Y estimation models that integrate VIs across multiple growing stages may yield more accurate estimates [12–14]. Zhao et al. [12] estimated the winter wheat yield (WWY) in Henan Province based on the MODIS NDVI from March to May, and the results revealed that the multiple linear regression model constructed with the NDVI from multiple months achieved a high Y estimation precision for WW. Besides the application in crop Y estimation, the VI products based on HTR satellite sensors have also been widely used in crop growth monitoring, biomass estimation [15,16], drought monitoring [17,18] and phenology monitoring [19,20]. Compared with the more well-developed RS satellite products in the USA, VI products based on the FY-3 satellites developed relatively late but are now playing an increasingly significant role in agricultural monitoring and meteorological disaster prevention and mitigation in China and other countries in the world, especially the "Belt & Road" [21] countries in Asia, Africa, and South America, etc. [22–25]. To date, the application of FY-3 satellites in agricultural monitoring has formed a basic operational framework that provides regular services and data support to the relevant departments of countries around the world [26].

However, the SR (250–1000 m) of the above HTR satellite sensors is significantly coarser than the farmland scale (approximately 20 m) in most regions in China, such as the Weihe Plain in Shaanxi Province. Due to the existence of other vegetation and artificial land between farmlands in these regions, there are uncertainties in terms of the crop growth monitoring results based on the above satellite sensor data. In contrast, the SR (10–30 m) of the Landsat-5 Thematic Mapper (TM), the Landsat-8 Operational Land Imager (OLI), and the Sentinel-2 MultiSpectral Instrument (MSI) data are very close to the farmland scale in most regions of China, and croplands and artificial lands can be clearly distinguished in these satellite images. Among these satellites, Sentinel-2a and Sentinel-2b were launched in 2015 and 2017, respectively, and both are equipped with a multispectral instrument (MSI) containing 12 channels, including six visible and six near-infrared channels, with SRs of 10–60 m. The width of the Sentinel-2 imagery can reach 290 km, and the entire area of most provinces in China can be covered by a single orbit of Sentinel-2. Currently, VIs based on Landsat and Sentinel-2 data have been widely used in monitoring at the farmland scale [27–30]. However, due to the low TR (5–16 days) of the Landsat and Sentinel-2 images, it is difficult to obtain sufficient cloudless data during periods with few sunny days. Therefore, further development of methods for crop growth monitoring at the farmland scale based on high spatiotemporal resolution data are urgently needed.

The spatiotemporal data fusion (STDF) model can combine the advantages of fine-spatial-resolution (FSR) satellite imagery (such as Landsat) and HTR satellite images (such as MODIS), thereby realizing continuous time series monitoring of crop growth at the farmland scale. At present, the existing spatiotemporal fusion methods include weight function-based methods, unmixing-based methods, Bayesian-based methods, machine learning-based methods, deep learning-based methods, and hybrid methods [31,32]. The spatial and temporal adaptive reflectance fusion model (STARFM) [33] and its improved method, i.e., the enhanced STARFM (ESTARFM) [34], are widely employed STDF methods, and both are easy to use and have stable performance [35]. However, these methods exhibit high uncertainties in areas with varying land cover types and complex vegetation phenology variations. Several STDF models based on deep learning have emerged as the field has developed, including the deep convolutional spatiotemporal fusion network (DCSTFN) model, which is based on the convolutional neural network autoencoder [36]. The DCSTFN approach assumes that the variations in ground features in FSR imagery are the same as those in coarse-spatial-resolution (CSR) imagery. Therefore, the DCSTFN model still exhibits uncertainty in areas with complex and diverse land cover types. To improve the fusion precision of the DCSTFN model in areas with complex and diverse land cover types, the enhanced DCSTFN (EDCSTFN) model was proposed by Tan et al. [37]; this model uses a residual encoder to learn the variation in ground features in FSR imagery based on the input FSR and CSR imagery. The results showed that the fusion precision of the EDCSTFN is greater than that of other unmixing-based models, such as STARFM, in areas with complex land cover changes. At present, the STDF models have been widely used in different geographical regions of the world, such as the regions in China, South Asia, and the USA, and the application of the STDF involves many areas, such as agriculture, ecology, and land cover classification [31]. In agriculture, the STDF has been widely used in crop growth monitoring [38,39], crop Y estimation [40–42], biomass estimation [43,44], drought monitoring [45,46], and phenology monitoring [47,48], indicating the high application potential of the STDF in improving the precision agriculture monitoring. However, deep learning-based STDF models such as the EDCSTFN currently have relatively few applications in agricultural monitoring, which is important for achieving accurate farmland-scale agriculture monitoring during periods of complex variation in crop phenology.

At present, most existing STDF studies are based on satellite sensors in the USA, such as the Moderate Resolution Imaging Spectroradiometer (MODIS), and most studies involve fusions between satellites that overpass in the morning. Spatiotemporal fusion-based studies for FY-3 series meteorological satellites and FSR satellite data are rare, as are STDF studies for satellites that overpass in the morning and afternoon, such as spatiotemporal fusion between the VIs of the newly launched FY-3D satellite and Sentinel-2 satellite. To extend the application of the FY-3D satellite to farmland-scale precision agricultural monitoring, it is necessary to further construct the STDF model framework for both FY-3D and FSR satellite data, such as Sentinel-2. Among the various satellite data in STDF models, there are differences in the spectral response bands, satellite overpass times, precisions of atmospheric corrections and geometric corrections, all of which affect the precision of the STDF models to varying degrees. The Sentinel-2 satellite passes over at approximately 10:00 a.m. local time, while the FY-3D satellite passes over at approximately 2:00 p.m. In addition, the spectral responses of the corresponding bands between these two sensors are slightly different. Therefore, in terms of the spatiotemporal fusion of VIs (including the NDVI, SAVI and EVI) from Sentinel-2 and FY-3D, it is necessary to assess the consistency of the two indices, properly adjust the parameters of the spatiotemporal fusion models, and retrain the deep-learning-based fusion models to evaluate the performance of each STDF model under fusion periods of different lengths.

The main purpose of this study is to propose a method framework for reconstructing time-series farmland-scale FY-3D-based VI imagery (including the NDVI, SAVI and EVI) over the main WW growing period via fusion with Sentinel-2 data based on the newly deep-learning-based STDF model to achieve farmland-scale WWY estimation by using the reconstructed FY-3D VI imagery. The Weihe Plain in Shaanxi Province, which is a typical WWY planting area in China that has similar climate and farmland characteristics as other WWY planting areas in China, was selected as the study region. Specifically, the objectives of this paper include two steps: (1) constructing STDF models to fuse the 10 day interval VI imagery from Sentinel-2 and FY-3D based on the ESTARFM and EDCSTFN and comparing the precision levels of the two models under fusion periods of different lengths, thereby selecting the STDF model with the optimal precision to reconstruct farmland-scale VI imagery for each 10 day period during the WW main growing period; (2) establishing a linear regression model to estimate WWY imagery at a 500 m SR based on Aqua MODIS gross primary productivity (GPP) data and county-scale WWY data, thereby constructing a back propagation neural network (BPNN) model based on the reconstructed farmland-scale FY-3D-based VI imagery and the 500 m SR Y estimation result to realize farmland-scale WWY estimation.

2. Materials and Methods

2.1. Study Region

The study region is the Weihe Plain (106°22′–110°24′E, 33°57′–35°39′N) in the central part of Shaanxi Province (Figure 1), which has a length of approximately 360 km from east to west, a width of approximately 80 km from north to south, and an average altitude of approximately 500 m [49,50]. The topography is high in the west and low in the east. The water system of the Weihe Plain is well-developed, and the land is fertile. It is one of the main WW planting areas in China. The land-use types on the Weihe Plain include cultivated land, woodlands, water bodies, artificial lands and bare land, of which more than half of the land is farmland. The farmland area is divided into several small fields that are less than 1 km^2 in size, by artificial land and other vegetation areas. The cultivated land on the Weihe Plain can be classified into grainland, orchard and vegetable land areas. Before summer, the main crop grown in the grain land areas is WW, which is usually sown in October and harvested in early June of the following year. During this period, early March to mid-March constitutes the regrowth stage of WW on the Weihe Plain, late March to mid-April constitutes the jointing stage, late April to early May constitutes the heading-filling stage, and mid-May to late May constitutes the milk stage. The average annual temperature of the Weihe Plain varies between 6 and 13 °C, and the average annual precipitation varies between 500 and 700 mm [51,52]. The Weihe Plain has the same typical spring and early summer climatic and farmland-scale characteristics as the main WW planting areas in China, and the region's east–west length is very close to the width of the Sentinel-2 imagery; therefore, it is very suitable for use as a study region for the STDF and farmland-scale WWY estimation in China. The method framework for farmland-scale FY-3D-based VI imagery reconstruction and WWY estimation developed on the Weihe Plain in this study can be easily migrated to other WW planting areas in China in the future.

2.2. RS Data and Y Data

The RS data used in this study included 20 m SR Sentinel-2 MSI surface reflectance data, 250 m SR FY-3D MERSI surface reflectance data from early March to late May in 2020 to 2022, and 500 m SR Aqua MODIS GPP product (MYD17A2H) data at 8 day intervals from March to May 2014–2022. Among them, the Sentinel-2 data were obtained from the Sentinel Scientific Data Hub, the FY-3D data were obtained from the FY Satellite RS Data Service Network, and the MYD17A2H product was obtained from the Atmospheric Storage and Distribution System (LAADS) Distributed Active Storage Center (DAAC). According to the main WW growing period on the Weihe Plain, only the data from early March to late May were used for the STDF and WWY estimation. Table 1 lists the acquisition dates (AQDs) of the Sentinel-2 and FY-3D data, which were acquired under cloudless conditions. Since more Sentinel-2 data were available in 2020 and were evenly distributed between early March and late May, the EDCSTFN was trained on the 2020 data, and the trained model was subsequently applied to the 2021 and 2022 data. To obtain the geometrically corrected reflectance products, the Sentinel-2 data were processed using the Sentinel Application Platform (SNAP) 9.0, the FY-3D data were processed by using the SMART 3.2 software that can be downloaded from the FY Satellite RS Data Service Network and the MODIS data were processed by using the MODIS Reprojection Tool (MRT). After analyzing the Sentinel-2 and FY-3D imagery, the Sentinel-2 imagery exhibited high geometric correction precision, and no further mutual registration was needed. Moreover, the FY-3D imagery exhibited relatively low geometric correction precision, and geometric bias occurred between the images for different dates. Therefore, the obtained FY-3D images were further registered with the Sentinel-2 imagery. First, the cloudless Sentinel-2 surface reflectance imagery was selected as the reference imagery and resampled to the SR of the FY-3D imagery (approximately 250 m). Then, the ENVI 5.3 software was used to automatically register the FY-3D surface reflectance imagery with the Sentinel-2 imagery. To ensure stable registration, the first-order polynomial was selected for automatic registration. Based on the registered Sentinel-2 and FY-3D surface reflectance data, the NDVI, SAVI and EVI were calculated, and then, maximum value composites were generated to obtain the 10 day interval FY-3D VI. For any 10 day intervals lacking available FY-3D data (e.g., mid-May 2021), the average VI imagery of the first and next 10 day periods were used as the FY-3D VI imagery for these 10 day intervals. Since there was only one Sentinel-2 image for each 10 day interval during the study period, the Sentinel-2 VI imagery was directly used as the 10 day period VI imagery. In this study, the calculated 10 day period FY-3D and Sentinel-2 VI images were used for the STDF, and the obtained MYD17A2H products were used for WWY estimation. In the process of STDF, the FY-3D images were reprojected and resampled to the projection and SR (20 m) of the Sentinel-2 data.

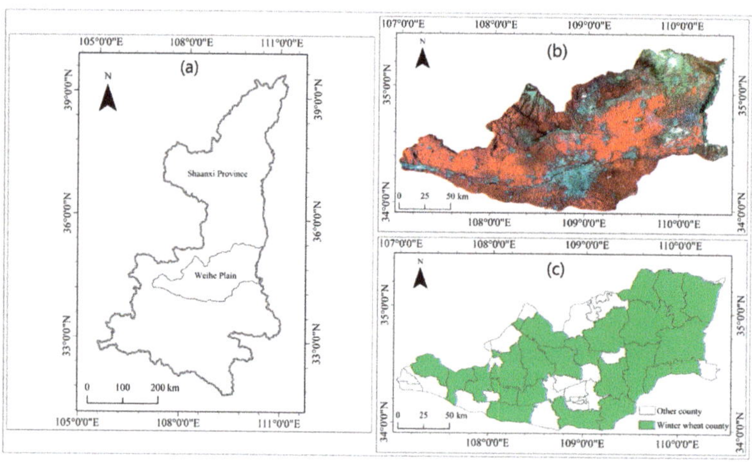

Figure 1. Overview of the study region: (**a**) location of the Weihe Plain; (**b**) FY-3D false colour composite image for 3 May 2020; and (**c**) locations of the county-scale WWY data points used in the WWY estimation.

Table 1. AQDs of the Sentinel-2 and FY-3D data under cloudless conditions.

Period	2020		2021		2022	
	Sentinel-2	FY-3D	Sentinel-2	FY-3D	Sentinel-2	FY-3D
Early March	4	4		3	4	2, 7, 10
Mid-March		14, 17–19		12, 13		13
Late March		23, 24	29	27, 28		28
Early April	8	8		8	8	6, 8, 10
Mid-April		12	18	12, 17, 18		13, 20
Late April	28	24–26, 28	28	28, 29, 30	23	21, 23, 25
Early May	3	1, 3	8	4, 5, 8, 9	3	2, 3, 6, 7
Mid-May	18	15, 16				16, 17, 19
Late May				24, 30, 31	23	22, 24

In addition, to construct and evaluate the precision of the WWY estimation model for the Weihe Plain, the WWY statistical yearbook data from 2020 to 2021 from 24 counties on the Weihe Plain (Figure 1c) covering numerous WW planting areas, most of which are located in the study region, were selected. The statistical yearbook data related to the counties on the Weihe Plain for 2022 were temporarily unavailable. Therefore, the WWY estimation results for 2022 were not verified in this study.

2.3. Methods

The method framework in this study involved two steps (Figure 2): spatiotemporal fusion of the 10 day interval VIs based on Sentinel-2 and FY-3D data and farmland-scale WWY estimation. (1) In the process of spatiotemporal fusion of the VIs, the ESTARFM and EDCSTFN were used to fuse the 10 day interval NDVI, SAVI and EVI images from Sentinel-2 and FY-3D, respectively, thereby reconstructing the 10 day interval VI imagery with a 20 m SR over the WW growing period. The fusion precisions of the two models were compared in this study. (2) In the process of farmland-scale WWY estimation, a BPNN WWY estimation model was constructed to estimate the WWY on the Weihe Plain with a 20 m SR based on spatiotemporally fused VI imagery, and the Aqua MODIS GPP product (MYD17A2H) and the county-scale WWY statistical data were used to construct the WWY estimation model. The key methods are described in this section.

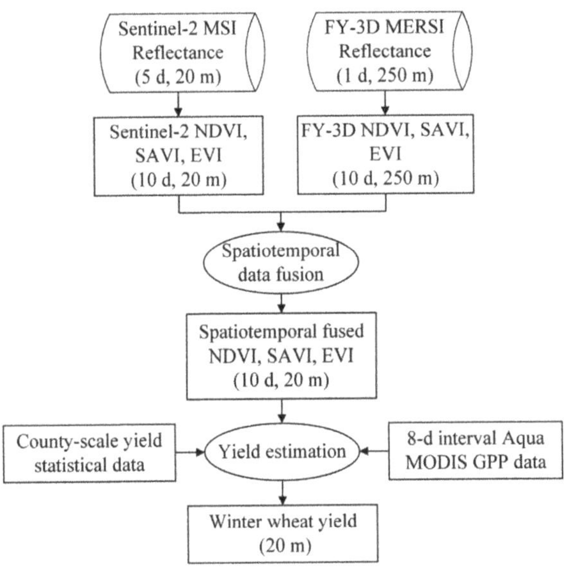

Figure 2. Flowchart of the 10 day interval VI imagery reconstruction and farmland-scale WWY estimation.

2.3.1. Spatiotemporal Fusion of VIs Based on Sentinel-2 and FY-3D Data

In accordance with the principles and uses of the ESTARFM and EDCSTFN described by Zhu et al. [34] and Tan et al. [37], in this section, the training scheme of the EDCSTFN and the usage and parameter setting scheme of the two models for spatiotemporal fusion of the 10 day interval Sentinel-2 and FY-3D VI imagery are described.

The EDCSTFN is a spatiotemporal fusion model based on convolutional neural network autoencoders with an encoder-feature fusion-decoder architecture. The model consists of two encoders: an encoder for FSR imagery and a residual encoder. The encoder for FSR imagery learns the features of the FSR reference imagery, and the residual encoder learns the feature changes in the FSR reference imagery between the reference and predicted imagery dates based on the input FSR and CSR imagery. The fusion process is performed in the feature space. The feature imagery of the FSR reference imagery is added to the feature imagery output by the residual encoder, and the resulting decoder is subsequently used to restore the fused feature imagery to the original pixel space. To evaluate the feature loss of the fusion results, an hourglass-shaped autoencoder is also built into the EDCSTFN to extract features from both the fusion and validation images. In the date-setting scheme of this study, Sentinel-2 NDVI, SAVI, and EVI images from 28 April 2020, were used to train the autoencoders to evaluate the feature loss of each VI. The main parameters of the EDCSTFN include the initial learning rate of the Adam optimization algorithm, the number of training epochs, the batch size, the size of the training images, and the size of the prediction images. In this study, the initial learning rate was set to 0.001. In addition, the number of training epochs was set to 100, and the training process was stopped when the learning rate decreased to 1×10^{-5}. The training batch size was set to 4 according to our hardware configuration. In addition, the structures of the encoders and decoders in the EDCSTFN and the settings of the hyperparameters and the regularization method of the Adam optimization algorithm are the same as those in the method discussed in Tan et al. [37].

During the EDCSTFN training stage, according to the method described in Zhou et al. [42], two areas of 1500 × 1500 pixels were selected from the cropland areas in the western and central parts of the Weihe Plain, two of which were randomly selected as training images and the other two as validation images. When the data AQD were selected for EDCSTFN training, the AQDs of the Sentinel-2 and FY-3D VI images for early March, early April, late April and mid-May 2020 were combined as reference and predicted imagery dates. In each training area, six sets of data were selected for EDCSTFN training during the main WW growing period on the Weihe Plain (Table 2). The EDCSTFN requires 1–2 pairs of FSR imagery and CSR imagery acquired on similar dates as the reference imagery. During the training process implemented in this study, only one pair of reference images was used, and the EDCSTFN trained on the 2020 data were used to predict the 10 day interval VI imagery during the main WW growing period at the farmland scale in 2021 and 2022.

Table 2. Data AQD for EDCSTFN training in 2020.

Data Group	Reference Imagery Date	Predicted Imagery Date
1	Early March	Early April
2	Early April	Early March
3	Early April	Late April
4	Late April	Early April
5	Late April	Mid-May
6	Mid-May	Late April

At the EDCSTFN prediction stage, the size of the prediction imagery was set to 7980, and the full imagery of the Weihe Plain was divided into two parts, each containing 7980 × 7980 pixels. To reconstruct the farmland-scale 10 day interval VI imagery during the main WW growing period, evaluate the impact of the STDF period length (i.e., the distance between the AQD of the two reference images) on the prediction precision, and considering the availability of Sentinel-2 data at each WW growing stage on the Weihe Plain from 2020 to 2022, the Sentinel-2 and FY-3D VI images obtained during different 10 day periods were selected as the reference imagery for the fusion model. The specific fusion scheme for each year is as follows: (1) In 2020, the VI imagery obtained in early March and late April and the VI imagery obtained in early April and mid-May were selected as reference imagery for the fusion model (with an interval of approximately 1.5 months) to predict the farmland-scale 10 day interval VI imagery between the AQDs of the two reference images. The prediction results were compared with those when the VI imagery acquired in early March and mid-May was selected as the reference imagery (with an interval of approximately 2.5 months). (2) In 2021, the VI imagery obtained in late March was selected as the reference imagery for predicting the farmland-scale VI imagery for early and mid-March. (3) In 2022, the VI imagery obtained in early and late April was selected as the reference imagery for predicting the farmland-scale VI imagery for early and mid-April, and the VI imagery obtained in early May was selected as reference imagery to predict farmland-scale VI imagery for mid- and late May. (4) In 2022, the VI imagery obtained in early and late April as well as the VI imagery obtained in late April and late May were selected as the reference imagery (with an interval of approximately 1.5 months) to predict the farmland-scale 10 day interval VI imagery between the AQDs of the two reference images. In addition, the prediction results were compared with those when the VI imagery acquired in early March and late May was selected as the reference imagery (with an interval of approximately 2.5 months).

The ESTARFM requires two FSR images as reference images. Thus, the fusion schemes of the ESTARFM from early March to mid-May 2020, from late March to late April 2021, and from early March to late May 2022 were the same as those of the EDCSTFN. In the ESTARFM parameter settings, according to the method from Zhou et al. [41], the number of categories was set to 10 according to the land cover in the study region. In addition, the moving window size was set to 25 Sentinel-2 VI pixels, corresponding to approximately two FY-3D VI pixels.

2.3.2. WWY Estimation Based on the Spatiotemporally Fused VIs

The WWY estimation framework comprises two parts: WWY estimation based on the MODIS cumulative GPP data (Figure 3a) and farmland-scale WWY estimation based on multiple parameters (Figure 3b). First, a linear regression WWY estimation model based on the MODIS cumulative GPP and county-scale WWY data were constructed to generate WWY data for the Weihe Plain with an SR of 500 m, and a BPNN WWY estimation model based on the spatiotemporally fused NDVI, SAVI and EVI data from across multiple growing stages was then constructed. The WWY data with an SR of 500 m were used to train the BPNN WWY estimation model, thereby estimating the farmland-scale WWY on the Weihe Plain. The two parts of the method are described in detail in this section.

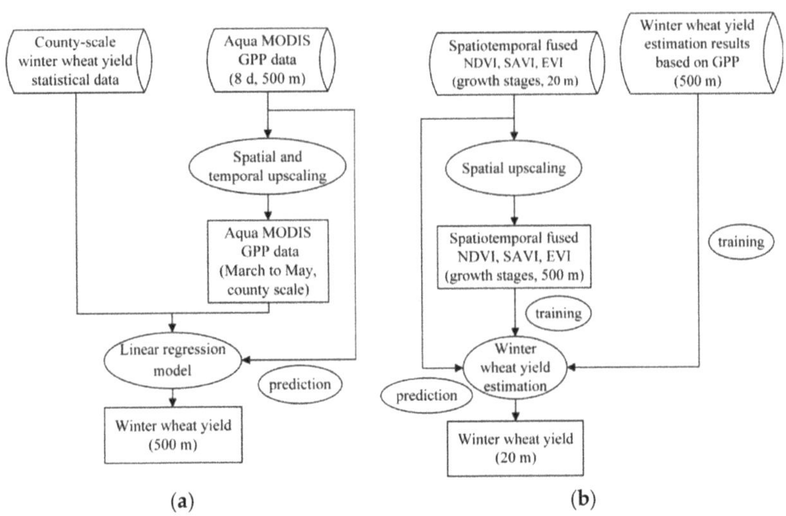

Figure 3. Flowchart of farmland-scale WWY estimation: (**a**) Y estimation model based on the cumulative GPP; and (**b**) farmland-scale Y estimation model based on multiple parameters.

In the process of constructing a WWY estimation model based on the cumulative GPP, the MODIS GPP data for the Weihe Plain from March to May were first collected to obtain the cumulative GPP data for the main WW growing period. According to the distribution of the WW planting areas on the Weihe Plain reported by Zhou et al. [45] and the county boundaries of the Weihe Plain, the 500 m SR cumulative GPP data for the overall WW planting area were averaged to the county scale, thereby obtaining the average cumulative GPP of WW for the main growing period of each county on the Weihe Plain. Many studies have shown that WWY data exhibit a favourable linear relationship with the cumulative GPP during the growing period [53,54]. In this study, a linear regression model was established between the average cumulative GPP and WWY statistics for each county on the Weihe Plain from 2014 to 2018. This model was used to estimate WWY data with a 500 m SR for 2020 to 2022. Due to the increase in cloud cover over the Weihe Plain from March to May 2019, the GPP data for 2019 were not used for WWY estimation. The R^2 value and the p-value of the hypothesis test were used to assess the correlation between the MODIS cumulative GPP data and WWY statistical data.

In the process of farmland-scale WWY estimation based on multiple parameters, the 10 day interval VI (including the NDVI, SAVI and EVI) images at the farmland scale reconstructed by the STDF model were averaged to obtain farmland-scale VI imagery for the 4 WW growing stages. Then, farmland-scale VI images for the 4 WW growing stages were used as input data, and the WWY data estimated from the MODIS cumulative GPP data were used as the target sample to construct a BPNN Y estimation model based on spatiotemporally fused VI imagery. In the BPNN Y estimation model, a single hidden layer neural network containing 5 neurons was used. The number of neurons in each layer of the model and the activation function are provided in Table 3. The specific steps used to construct a WWY estimation model based on the BPNN were as follows: (1) the farmland-scale VI images for the 4 WW growing stages were aggregated to a 500 m SR; (2) a BPNN Y estimation model based on the spatiotemporally fused VI imagery for the 4 growing stages and the WWY data estimated from the MODIS cumulative GPP data were constructed and trained at a 500 m SR, where the training and validation data accounted for 90% and 10% of the total data, respectively; (3) the farmland-scale VI imagery for each WW growing stage was entered into the BPNN Y estimation model, thereby estimating the WWY at the farmland scale; and (4) the WWY estimation results of the BPNN were evaluated by using the county-scale WWY statistical data for the Weihe Plain.

Table 3. Structure and parameters of the BPNN WWY estimation model.

Model Parameters	Network Layers		
	Input Layer	**Hidden Layer**	**Output Layer**
Number of neurons	12	5	1
Activation function		Hyperbolic tangent function	Linear function

2.3.3. Precision Evaluation of the STDF and WWY Estimation Model

The consistency of the FSR imagery and CSR imagery may affect the precision of the spatiotemporal fusion results [39]; therefore, it is necessary to verify the consistency of the 10 day interval VI imagery between Sentinel-2 and FY-3D before spatiotemporal fusion. To test the consistency of the 10 day interval Sentinel-2 and FY-3D VI (including NDVI, SAVI and EVI) imagery, the Sentinel-2 VI imagery during the study period was aggregated to 250 m and compared with the FY-3D VI imagery for the same 10 day period.

In the precision evaluation of the STDF model, the data from 2020 to 2022 were used to evaluate the impact of different reference date distances (at intervals of approximately 1.5 and 2.5 months) on the prediction precision of the ESTARFM and EDCSTFN. The fusion results were evaluated by using Sentinel-2 VI imagery that was not used in the STDF process. Due to the lack of data for the corresponding period in 2021, the data for 2021 were not used in the above test but were only used to reconstruct 10 day interval VI imagery at the farmland scale.

To evaluate the precision of the WWY estimation, the WWY statistical yearbook data from 2020 to 2021 for 24 counties on the Weihe Plain were used to evaluate the precision of the BPNN WWY model based on the spatiotemporally fused VI imagery. The WWY estimate results were averaged to the scales of each county, and the linear regression method was employed to correlate the WWY estimate results with the WWY statistical yearbook data at the county scale.

To quantify these precision evaluations, the coefficient of determination (R^2), root mean square error (RMSE), and average deviation (predicted results—validation data) were used to evaluate the precision of the STDF results and the WWY estimation results. These metrics are important indices for evaluating the differences between model estimation results and validation data. In this work, the random and systematic deviations between the fusion results and validation imagery, as well as the random and systematic deviations between the WWY estimation results and validation data, were evaluated by these indices.

3. Results

3.1. Consistency of the VIs Between Sentinel-2 and FY-3D

The consistency of the 10 day interval Sentinel-2 and FY-3D VI (including the NDVI, SAVI and EVI) imagery was tested by using the method described in Section 2.3.2. The consistency validation results for early March, early April, late April, and mid-May 2020 are displayed in Figure 4. The R^2 values between the aggregated Sentinel-2 VI imagery and the FY-3D VI imagery varied between 0.22 and 0.35, and the RMSE values of the fitting results varied between 0.09 and 0.14. The R^2 values of the NDVI, SAVI and EVI imagery were essentially consistent. These results also revealed large random deviations between the 10 day interval FY-3D VI imagery and the aggregated Sentinel-2 VI imagery, and the random deviations corresponding to the NDVI, SAVI and EVI imagery were basically the same. These random deviations may be related to differences in the satellite overpass time, satellite observation angle, spectral response characteristics of the bands, atmospheric correction and registration precision between Sentinel-2 and FY-3D [39], especially the registration precision of the FY-3D imagery. In this study, the FY-3D imagery still exhibited deviations of 1–2 pixels between the different dates after 1-degree polynomial registration. In addition, the average deviations between the FY-3D VI imagery and aggregated Sentinel-2 VI imagery varied mostly between −0.02 and 0.02, which indicates that the systematic deviations of the 10 day interval VI imagery between Sentinel-2 and FY-3D are small. Therefore, the STDF model could be used to fuse the 10 day interval VI imagery of Sentinel-2 and FY-3D. However, due to the large random deviations of the 10 day interval VI imagery between Sentinel-2 and FY-3D, it is necessary to select an STDF model that is insensitive to random deviations. The ESTARFM employs a sliding window to search for similar pixels and applies a weighted function to reduce the effect of random errors, and the EDCSTFN is an STDF model based on convolutional neural networks in which random errors can be reduced through the convolution operation. Therefore, both the ESTARFM and EDCSTFN can reduce the random deviations of the fusion data. The precision of the ESTARFM and EDCSTFN for the spatiotemporal fusion of the 10 day interval Sentinel-2 and FY-3D VI imagery is evaluated below.

3.2. VI Spatiotemporal Fusion Results

3.2.1. Training of the EDCSTFN

The established EDCSTFN based on the Sentinel-2 and FY-3D VIs (including the NDVI, SAVI and EVI) was trained according to the training data grouping scheme discussed in Section 2.3.1. After approximately 50 training iterations, the model gradually converged; the loss for the model converged at approximately 0.03, and the value of the training loss was very close to that of the validation loss. Table 4 displays the values of the final training and validation losses for training the EDCSTFN with different data grouping schemes. Among them, the values of the training and validation losses obtained by using the training datasets for early April and late April were the smallest among all of the groups, varying between 0.020 and 0.031, followed by the training datasets for early March and early April, with values varying between 0.024 and 0.030. The values of the training and validation losses corresponding to the training datasets for late April and mid-May were the largest, varying between 0.029 and 0.044, and the values of the training and validation losses obtained using all of the datasets for March to May were centred at all the groups, varying between 0.028 and 0.042. The above results are similar to the results based on the Sentinel-2 and Sentinel-3 images published by Zhou et al. [42]. These results indicate that the EDCSTFN provides greater precision for spatiotemporal fusion of the 10 day interval Sentinel-2 and FY-3D VI imagery for April, followed by March, with a lower spatiotemporal fusion precision for May. The model trained by using all of the datasets for March to May achieved high overall precision over the main WW growing period. Therefore, in practical applications, to increase the convenience of the EDCSTFN, the model trained by using all of the datasets for the period from March to May was uniformly employed to predict the 10 day interval VI imagery at the farmland scale during the main WW growing period.

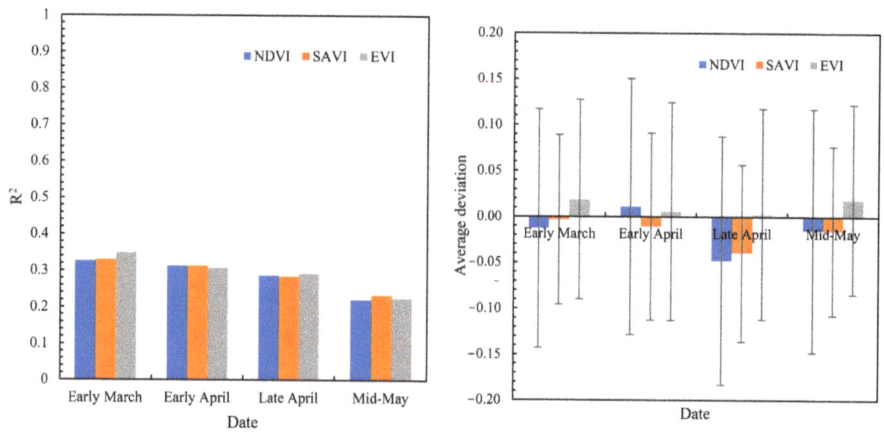

(**a**) R^2 (**b**) Average deviations

Figure 4. Results of the consistency analysis of the Sentinel-2 and FY-3D VIs: (**a**) R^2 values between the aggregated Sentinel-2 VI imagery and the FY-3D VI imagery at an SR of 250 m; and (**b**) average deviations and RMSE values of the fitting results between the aggregated Sentinel-2 VI imagery and FY-3D VI imagery. The error line in (**b**) denotes the RMSE of the fitting results.

Table 4. Training and validation losses for EDCSTFN training.

Period of Training Data	NDVI		SAVI		EVI	
	Training Loss	Validation Loss	Training Loss	Validation Loss	Training Loss	Validation Loss
Early March, early April	0.029	0.027	0.024	0.025	0.030	0.029
Early April, late April	0.021	0.021	0.021	0.020	0.031	0.026
Late April, mid-May	0.030	0.034	0.029	0.032	0.041	0.044
Early March, mid-May	0.030	0.031	0.028	0.028	0.042	0.041

3.2.2. Prediction by the EDCSTFN

The precisions of the EDCSTFN and ESTARFM were evaluated according to the date selection scheme employed during the prediction stage of the spatiotemporal fusion models described in Section 2.3.1, and farmland-scale 10 day interval VIs (including the NDVI, SAVI and EVI) were predicted during the main WW growing periods from 2020 to 2022. To assess the spatiotemporal fusion precision of the EDCSTFN and ESTARFM, at a distance of approximately 1.5 months between the AQDs of the two reference images, the Sentinel-2 and FY-3D VI images acquired in early March and late April 2020, early April and mid-May 2020, early March and late April 2022, and late April and late May 2022 were used as reference images to predict farmland-scale 10 day interval VI images between the AQDs of the reference images (including early April, late April and early May). The R^2 values and average deviations between the spatiotemporal fusion results and Sentinel-2 VI imagery used for validation are listed in Table 5. In 2020, the R^2 value between the predicted results of the EDCSTFN and ESTARFM and the Sentinel-2 VI data were approximately 0.9, and the absolute value of the average deviation in most 10 day periods did not exceed 0.03. These results are similar to the precision of the fusion results based on other satellite sensors (such as MODIS and Landsat) in previous studies [45,55–57], which indicates that both the EDCSTFN and ESTARFM can obtain high precisions when fusing FY-3D and Sentinel-2 VI imagery of the Weihe Plain when the distance between the acquisition and reference imagery dates is approximately 1.5 months. In addition, the R^2 values and average deviations corresponding to the predicted results of the EDCSTFN in 2022 are basically the same as those of the ESTARFM and are very close to the prediction precision of the EDCSTFN in 2020, which indicates that the EDCSTFN trained on single-year data still achieves satisfactory spatiotemporal fusion precision in the adjacent years.

Table 5. Fusion precision evaluation of the EDCSTFN and ESTARFM when the distance between the reference imagery AQD is approximately 1.5 months. $RMSE_{fit}$ represents the RMSE of the fitting results, and AD represents the average deviation.

Year	VI	Prediction Date	EDCSTFN			ESTARFM		
			R^2	$RMSE_{fit}$	AD	R^2	$RMSE_{fit}$	AD
2020	NDVI	Early April	0.91	0.06	−0.02	0.90	0.06	−0.03
		Late April	0.88	0.07	−0.05	0.89	0.07	0.00
	SAVI	Early April	0.91	0.04	−0.02	0.93	0.04	−0.01
		Late April	0.91	0.04	0.00	0.88	0.05	0.01
	EVI	Early April	0.92	0.05	−0.01	0.92	0.05	0.01
		Late April	0.90	0.05	−0.02	0.89	0.06	0.00
2022	NDVI	Early April	0.89	0.05	−0.02	0.89	0.05	0.00
		Early May	0.86	0.05	0.03	0.88	0.05	0.03
	SAVI	Early April	0.90	0.04	0.00	0.92	0.04	0.00
		Early May	0.92	0.04	0.01	0.92	0.04	0.00
	EVI	Early April	0.92	0.05	−0.02	0.92	0.05	0.00
		Early May	0.91	0.05	−0.02	0.93	0.05	0.00

To further analyze the spatiotemporal fusion precision of the EDCSTFN and ESTARFM when the distance between the AQDs of the two reference images is approximately 2.5 months, Sentinel-2 and FY-3D VI imagery acquired in early March and mid-May 2020 were used as reference imagery to predict the farmland-scale VI imagery for early and late April 2020, and Sentinel-2 and FY-3D VI imagery acquired in early March and late May 2022 were used as reference images to predict the farmland-scale VI imagery for early April and early May 2022. The R^2 values and average deviations of the spatiotemporal fusion results are listed in Table 6. Compared with the fusion precisions when the distance between the reference imagery AQD is approximately 1.5 months, the precisions of both the EDCSTFN and ESTARFM significantly decrease for a period of approximately 2.5 months between the reference imagery AQD. The R^2 values of the EDCSTFN are above 0.75, and the R^2 values of the ESTARFM are above 0.68. These results indicate that the performance of both the EDCSTFN and ESTARFM decreases when the spatiotemporal fusion period of the VIs increases. However, the deep learning-based EDCSTFN could obtain a higher prediction precision. Considering the influence of the distance between the reference imagery AQD and the precision of the spatiotemporal fusion models, the distance between the reference imagery AQD was ultimately maintained at approximately 1.5 months, and the EDCSTFN was employed to reconstruct the farmland-scale 10 day interval VI imagery for the main WW growing period on the Weihe Plain for the 2020–2022 period, thereby achieving WWY estimation at the farmland scale.

Table 6. Fusion precision evaluation of the EDCSTFN and ESTARFM when the distance between the reference imagery AQD is approximately 2.5 months. $RMSE_{fit}$ represents the RMSE of the fitting results, and AD represents the average deviation.

Year	VI	Prediction Date	EDCSTFN			ESTARFM		
			R^2	$RMSE_{fit}$	AD	R^2	$RMSE_{fit}$	AD
2020	NDVI	Early April	0.82	0.09	−0.03	0.86	0.08	−0.03
		Late April	0.92	0.06	−0.01	0.87	0.07	−0.01
	SAVI	Early April	0.84	0.06	−0.02	0.83	0.06	−0.01
		Late April	0.87	0.05	−0.01	0.83	0.06	0.01
	EVI	Early April	0.83	0.07	−0.01	0.78	0.08	0.00
		Late April	0.86	0.06	−0.03	0.81	0.07	0.00
2022	NDVI	Early April	0.77	0.07	−0.01	0.74	0.07	0.00
		Early May	0.84	0.05	−0.02	0.77	0.07	0.03
	SAVI	Early April	0.75	0.06	0.01	0.69	0.07	0.01
		Early May	0.84	0.05	0.00	0.81	0.06	0.02
	EVI	Early April	0.75	0.09	−0.01	0.68	0.10	0.01
		Early May	0.83	0.08	−0.04	0.79	0.09	0.00

3.2.3. Farmland-Scale EVI Images of WW During Each Growing Period

Notably, 20 m SR NDVI, SAVI and EVI imagery for the 4 WW growing stages were obtained by averaging the 10 day spatiotemporally fused NDVI, SAVI and EVI imagery for the main WW growing periods from 2020 to 2022. Under the influence of clouds, only one cloudless Sentinel-2 image of the Weihe Plain could be obtained for each WW growing period, and a single image cannot reflect the average crop growth during each growing stage. However, the average treatment of the reconstructed 10 day scale VI imagery for each WW growing stage could better reflect the average crop growth during each growing stage. Figure 5 shows the 20 m SR EVI imagery for the 4 growing stages on the Weihe Plain from 2020 to 2022, which was reconstructed by using the EDCSTFN. Between March and May of each year, the areas with notable changes in EVI values were mainly WW planting areas, and the changes in EVI values in forest areas were relatively small. Among the 4 WW growing stages, the EVI values in the WW planting area were usually the highest at the heading-filling stage, followed by those at the jointing and milk stages, and the EVI values at the regrowth stage were usually the lowest, as was the case for the EVI on the Weihe Plain in 2021 and 2022. However, the maximum EVI values in the WW planting areas in 2020 occurred earlier than those in 2021 and 2022, namely, from late March to mid-April, corresponding to the jointing stage in 2021 and 2022. In the WW planting area on the western Weihe Plain, the EVI values before late March were greater than those from late April to early May in 2020. These results indicate that the actual heading-filling stage of the WW on the western Weihe Plain in 2020 occurred earlier than that in 2021 and 2022, which could be related to the early WW sowing time in 2020. In addition, when comparing the EVI values in the WW planting areas in the different regions of the Weihe Plain from 2020 to 2022, the EVI values of the WW planting areas in the western Weihe Plain were greater than those in the central part of the Weihe Plain during the first three growing stages. Previous studies have indicated that the growth characteristics of WW on the Weihe Plain from late March to late April (including the jointing and heading-filling stages) are important for Y estimation [58]. The growth in the WW planting area in the western Weihe Plain was greater than that in the central part of the Weihe Plain, and the Y in the western Weihe Plain was also greater.

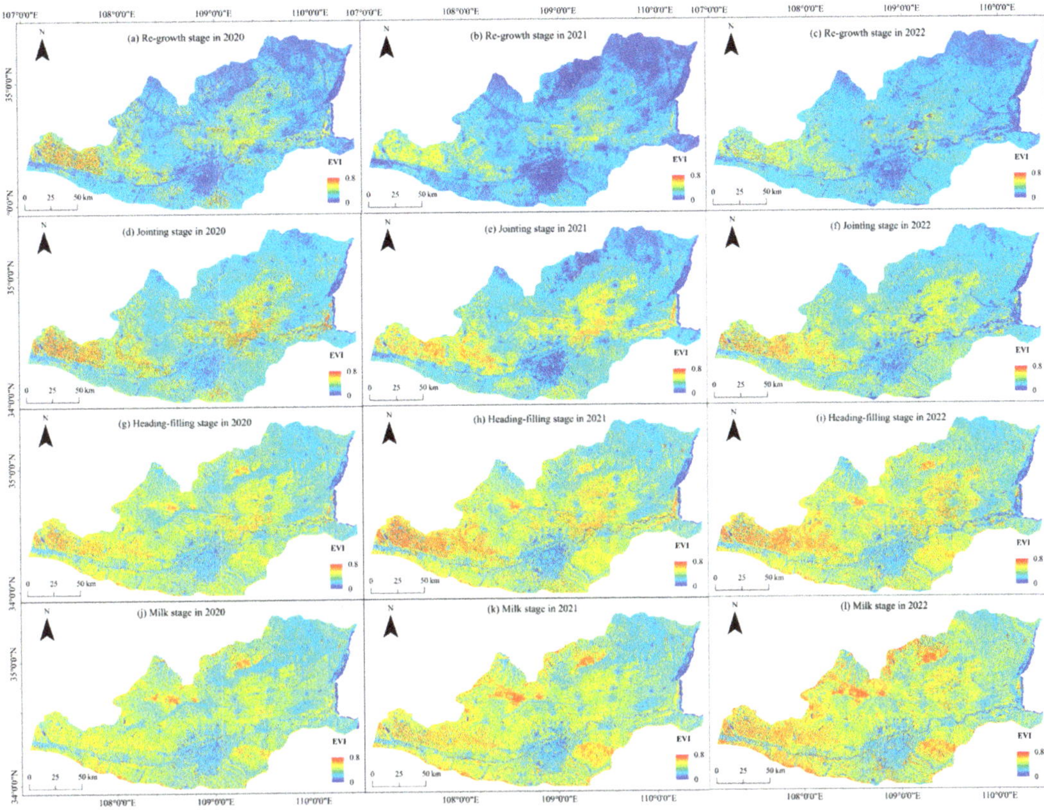

Figure 5. EVI at each WW growing stage from 2020 to 2022.

3.3. Farmland-Scale WWY Estimation Results and Analysis

3.3.1. WWY Estimation Results Based on the Cumulative GPP

According to the method description of WWY estimation based on the MODIS cumulative GPP data in Section 2.3.2, eight day-scale Aqua MODIS GPP data for the main WW growing period (from March to May of each year) from 2014 to 2018 were collected, and a linear regression model relating the cumulative GPP data for the main WW growing period over the five-year period and the county-scale WWY statistical data were constructed in this study (Figure 6a). The regression results reveal that the cumulative GPP data for the main WW growing period on the Weihe Plain from 2014 to 2018 were strongly correlated with the WWY statistical data at the county scale, with an R^2 value of 0.67 ($p < 0.01$). These results indicate that the WWY on the Weihe Plain with a 500 m SR could be accurately estimated based on the cumulative MODIS GPP data for multiple years. Therefore, the linear regression model for WWY estimation based on the MODIS cumulative GPP data were then used to estimate the WWY at a 500 m SR on the Weihe Plain from 2020 to 2022 (Figure 7). The high-Y areas of WW on the Weihe Plain were distributed mainly in the western and central parts, and the WWY in the western Weihe Plain was slightly higher than that in the central part of the Weihe Plain. These findings are similar to the analysis results of the spatial distribution of Y based on the EVI discussed in Section 3.2.3. Compared with that in 2020, the Y in the main WW planting areas of the Weihe Plain slightly increased in 2021, while the Y decreased in 2022 to values lower than those in 2020 and 2021. The estimation results of the WWY based on the MODIS cumulative GPP data were also verified using the WWY statistical data for the counties on the Weihe Plain from 2020 to

2021. The results reveal that the WWY estimation results were linearly correlated with the WWY statistical data at the county scale, with R^2 values reaching 0.65 and 0.63, respectively ($p < 0.01$) (Figure 6b and 6c, respectively). Therefore, the WWY estimation results at a 500 m SR could be used to train a farmland-scale WWY estimation model with multiple parameters.

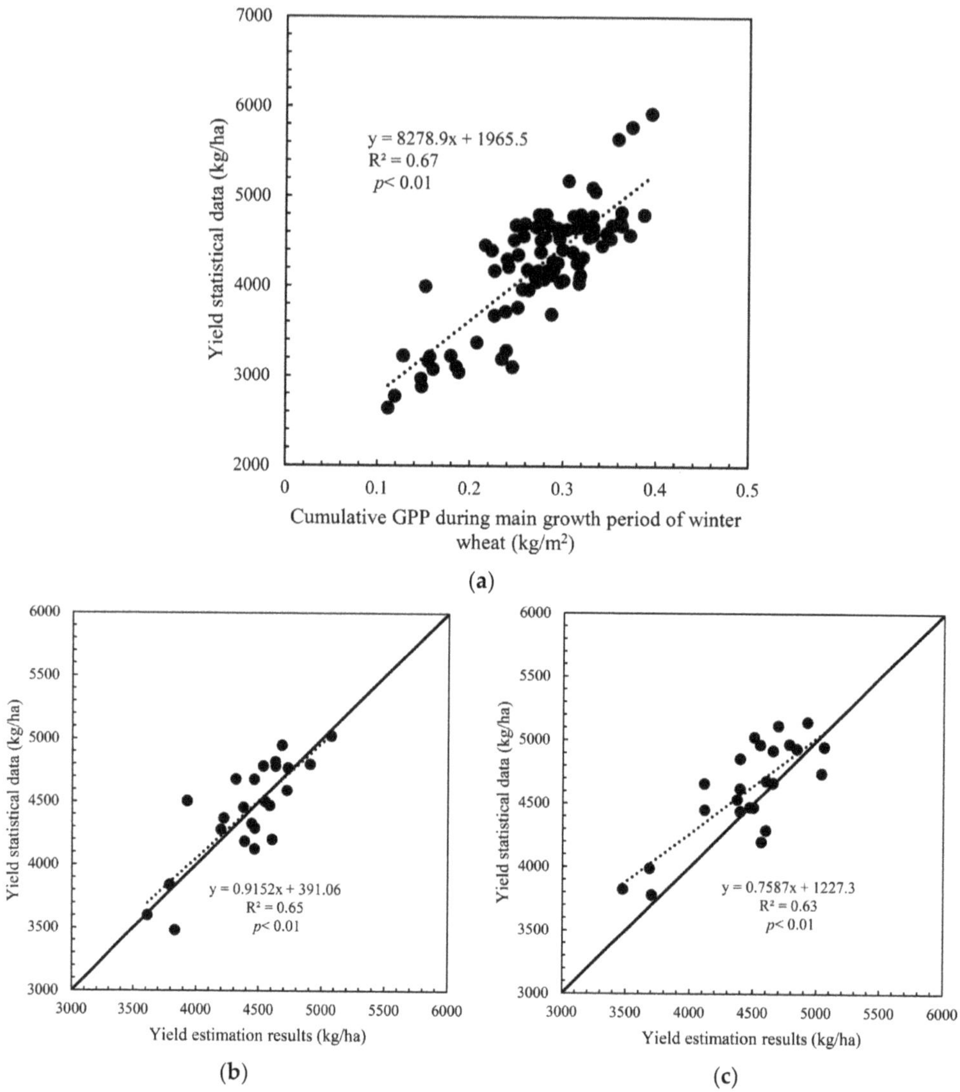

Figure 6. Y estimation model based on the cumulative GPP for the main WW growing period and the Y estimation precision evaluation results in 2020 and 2021. The dotted lines in the figures denote the fitted linear functions, which are close to the diagonal solid lines, indicating that the systematic deviation in the Y estimation results is small. (**a**) Linear regression model between the cumulative GPP data for the main WW growing period and the county-scale WWY from 2014 to 2018, (**b**) linear regression results between the WWY estimation results from 2020 based on the cumulative GPP and county-scale Y statistical data, and (**c**) linear regression results between the WWY estimation results in 2021 based on the cumulative GPP and county-scale Y statistical data.

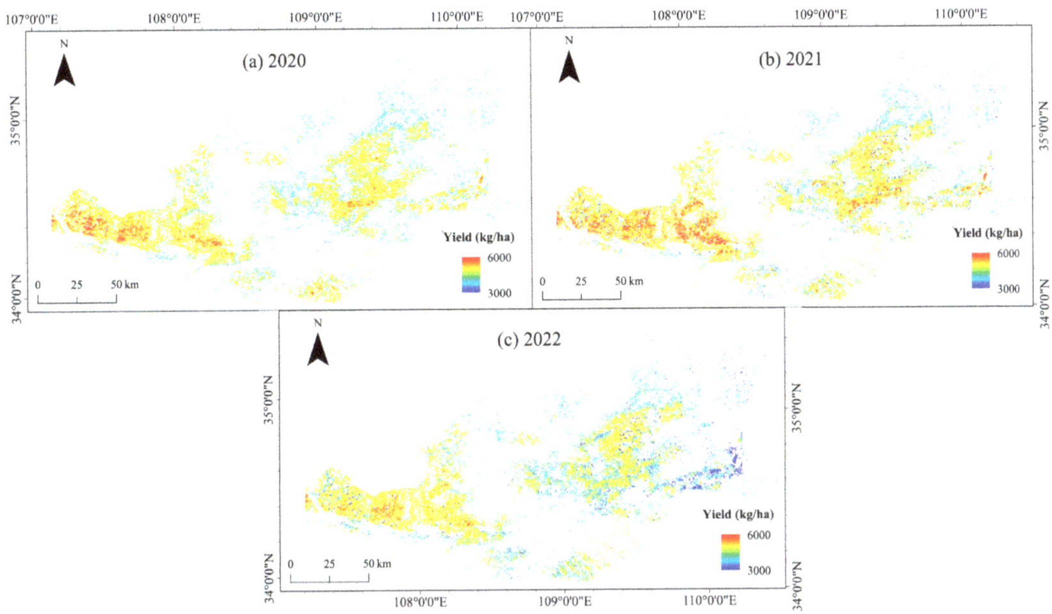

Figure 7. WWY estimation results for 2020 to 2022 based on the MODIS cumulative GPP data.

3.3.2. Training Results of the Farmland-Scale WWY Estimation Model Based on Multiple Parameters

The NDVI, SAVI and EVI imagery for the four WW growing stages from 2020 to 2022 were aggregated to the 500 m SR, and a BPNN model was constructed between the multiple VIs (including the NDVI, SAVI and EVI) at the four growing stages and the WWY at the 500 m SR. Table 7 lists the training and validation precisions of the BPNN WWY estimation model from 2020 to 2022. The R^2 values corresponding to the training and validation precisions of the BPNN model reached 0.67 ($p < 0.01$), indicating a satisfactory fitting ability. The BPNN Y estimation model constructed by integrating the VIs at the four WW growing stages could provide highly accurate WWY estimates. Compared with that in 2020, the training precision in 2021 and 2022 decreased, which could be related to the Y estimation errors based on the MODIS GPP data. The R^2 value between the Y estimates based on the cumulative GPP and county-scale WW statistical data in 2021 was slightly lower than that in 2020, and the systematic deviation was larger than that in 2020 (Figure 6). Over time, crop varieties and technologies will continue to improve [59], which will lead to increasing uncertainty in WWY estimation models based on GPP data from 2014 to 2018. Therefore, to obtain a higher Y estimation precision, a Y estimation model based on cumulative GPP data should be retrained using the latest yearly data in the future.

Table 7. Training and validation precisions of the multiparameter WWY estimation model based on the BPNN.

Year	Training Precision			Validation Precision		
	R^2	RMSE (kg/ha)	*p* Values	R^2	RMSE (kg/ha)	*p* Values
2020	0.67	279.85	<0.01	0.67	279.18	<0.01
2021	0.64	320.65	<0.01	0.64	319.99	<0.01
2022	0.56	333.83	<0.01	0.55	337.29	<0.01

3.3.3. Results and Analysis of WWY Estimation at the Farmland Scale

The farmland-scale spatiotemporally fused NDVI, SAVI and EVI imagery for the four WW growing stages from 2020 to 2022 were input into the BPNN WWY estimation model, which was used to estimate the farmland-scale WWY on the Weihe Plain from 2020 to 2022. Figure 8 shows the farmland-scale Y estimation results of the BPNN model based on the NDVI, SAVI and EVI from 2020 to 2022. The spatial distribution of the WWY estimation results and the variation pattern over 3 years were basically consistent with the 500 m SR results estimated based on the MODIS cumulative GPP data (Figure 7). The WWYs in the main planting areas varied mainly between 4000 and 6000 kg/ha, with the highest Y occurring in 2021, followed by 2020, and the lowest Y occurring in 2022. On the Weihe Plain, the WW planting area in Baoji City attained the highest Y, followed by the central part of the Weihe Plain, which is located at the junction of Xi'an City and Weinan City. The distribution of the WW planting areas in the other areas of the Weihe Plain was relatively sparse, and the Y was relatively low. These results are consistent with the corresponding annual statistical yearbook data. The R^2 values between the WWY statistical data at the county scale and the Y estimation results based on the BPNN model in 2020 and 2021 reached 0.78 and 0.68, respectively (Figure 9), revealing a favourable correlation between the Y estimation results and the Y statistical data at the county scale. These results indicate that the BPNN model, which is based on spatiotemporally fused VIs across multiple WW growing stages, provided suitable Y estimation precision for WW on the Weihe Plain, and the results were better than the Y estimation results based on the MODIS cumulative GPP data (with R^2 values of 0.65 and 0.63 in 2020 and 2021, respectively). In addition, the WWY estimation results at an SR of 20 m clearly reveal the difference in WWY between adjacent farmlands, indicating that the WWY estimation results based on the spatiotemporally fused VIs better reflect the detailed characteristics of the farmland-scale WWY distribution on the Weihe Plain.

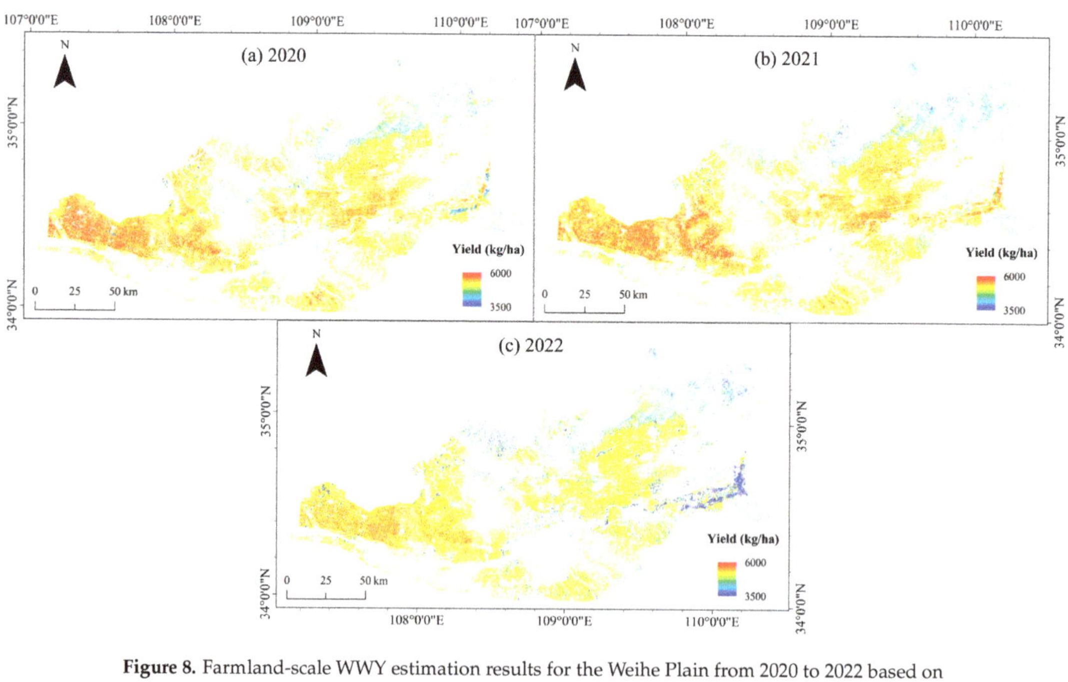

Figure 8. Farmland-scale WWY estimation results for the Weihe Plain from 2020 to 2022 based on multiple parameters.

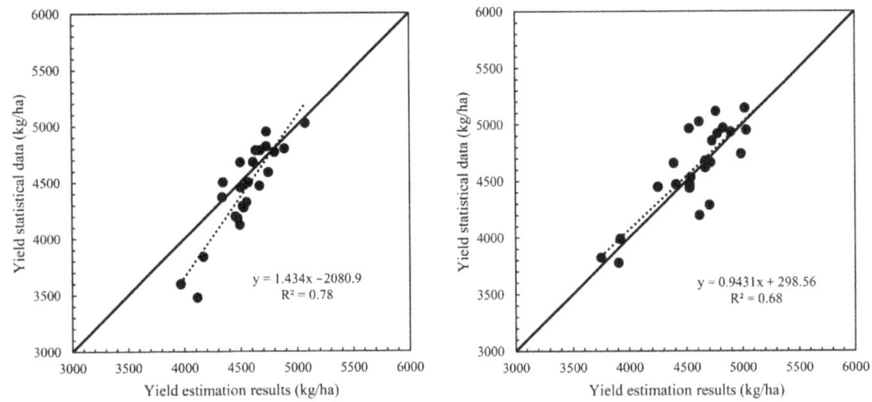

(**a**) 2020 (**b**) 2021

Figure 9. Linear regression results between the farmland-scale WWY estimation results and the Y statistical data in 2020 and 2021. The dotted lines in the figures denote the fitted linear functions, which are close to the diagonal solid lines, indicating that the systematic deviation of the Y estimation results is small.

4. Discussion

To address the problem of limited research based on the Chinese meteorological satellite FY-3 series (such as FY-3D) in the fields of STDF and farmland-scale WWY estimation, a method framework for 10 day interval VI (NDVI, SAVI and EVI) reconstruction based on the FY-3D MERSI and Sentinel-2 MSI data and farmland-scale WWY estimation based on reconstructed VI imagery was proposed in this paper. The ability of the deep-learning-based EDCSTFN to perform spatiotemporal fusion of 10 day interval VI imagery of the Weihe Plain based on FY-3D MERSI and Sentinel-2 MSI data were assessed in this study by compared with ESTARFM. The results in Section 3 showed that both the EDCSTFN and ESTARFM achieved satisfactory spatiotemporal fusion precision and could effectively provide farmland-scale WWY estimation. The precision evaluation results in this study are similar to the precision of the fusion results based on other satellite sensors (such as MODIS and Landsat) as well as the other study regions in previous studies [45,55–57]. In addition, when the date range of the STDF is long, the deep learning-based EDCSTFN achieves higher precision. Therefore, this study is meaningful for accurate farmland-scale growth monitoring and Y estimation of WW based on Chinese FY-3 meteorological satellite data. However, the STDF based on the FY-3D MERSI VIs and farmland-scale WWY estimations should still be investigated. The research prospects of these two components are described in this section.

4.1. Research Prospects of the STDF

4.1.1. Computational Efficiency

The computational efficiency of the STDF model is one of the important factors affecting its application prospects. The traditional weight function-based spatiotemporal fusion model spends much of its time searching for similar pixels by using sliding windows [31]. Approximately 3 h is needed to use the ESTARFM to fuse the Sentinel-2 and FY-3D imagery of the Weihe Plain once (an Intel Core i7-12700h CPU and 2.30 GHz were used in the test). Thus, it is difficult to apply this model in the WW planting areas of the whole country. The deep learning-based spatiotemporal fusion model does not use a sliding window to find similar pixels. After testing, the computational time required by the EDCSTFN at the prediction stage was much shorter than that of the ESTARFM. The EDCSTFN only needed approximately 10 min to fuse the Sentinel-2 and FY-3D images of the Weihe Plain with the same equipment configuration. In addition, the EDCSTFN trained on single-year data could still be used for near-future analysis, and the fusion precision remained basically unchanged. Therefore, the trained EDCSTFN exhibits greater application potential in the STDF of large areas and long time series. In the future, we will further research the method of transferring the EDCSTFN trained on the WW planting area on the Weihe Plain to other WW planting areas in China and assess its spatiotemporal fusion precision.

4.1.2. TR of FSR and CSR Imagery

To obtain high spatiotemporal fusion precision, it is usually necessary to acquire at least one Sentinel-2 image for each WW growing stage, especially the heading-filling and milk stages, when WW grows rapidly. However, due to the effects of clouds, the TR (5 days) of the Sentinel-2 satellite images can sometimes barely meet the above requirements. At present, the Chinese GF series satellites have undergone long-term development. The GF-6 satellites, which were launched in June 2018, can obtain multispectral imagery with an SR of 16 m, exhibiting high band consistency with the GF-1 satellite launched in 2013. With the networking of the GF-1 and GF-6 satellites, 16 m SR multispectral imagery with a TR of 2 days can be obtained [60]. In addition, since the swath widths of GF-1 and GF-6 reach 800 km, they are more suitable when employing the STDF in large areas. In the future, a spatiotemporal fusion method based on GF series data and FY-3 data will be developed.

On the other hand, the TRs of the CSR imagery are also important for time series' reconstructions of VI imagery at the farmland scale. Considering the effects of clouds, a 10 day interval VI reconstruction method at the farmland scale based on the FY-3D and Sentinel-2 satellite data were proposed in this paper. With the successful launch of the FY-3F satellite in August 2013 and the replacement of the FY-3C satellite, which has been operating for nearly 10 years, the monitoring capability of the FY-3 polar-orbiting meteorological satellites in the morning will continue to increase. In the future, a time series VI reconstruction method at the farmland scale based on the FY-3D/F and Sentinel-2 satellites will be developed, which will improve the TR of time series farmland-scale VI reconstructions (from 10 day intervals to 5 day intervals).

4.1.3. Registration Precision of the Imagery in the STDF

The registration precision of the RS imagery may notably influence the precision of the STDF model [31]. The results in Section 3 show that the STDF based on the Sentinel-2 and FY-3D data in this study yields high precision since the geometric correction precision of Sentinel-2 imagery is high, although there is a large random deviation between the Sentinel-2 and FY-3D VI imagery due to factors such as geometric correction errors in the FY-3D imagery. However, regarding the Chinese FSR satellites such as GF-1, due to the use of instruments and algorithms, the current mutual geometric correction precision cannot reach the level of Sentinel-2 [61]. Therefore, the STDF for the GF-1 satellites may involve high uncertainties. In the future, the fusion precision of Chinese FSR satellites and FY-3 satellites will be further analyzed, and a fusion model with low sensitivity to registration precision will be developed.

4.1.4. Difference in the SR Between the FSR and CSR Imagery

The difference in the SRs between the FSR and CSR imagery significantly affects the precision of the spatiotemporal data fusion models. Compared with FSR imagery, there are fewer spatial details of the ground features in CSR imagery, especially in regions with abrupt changes in land cover, such as regions with newly harvested WW and areas affected by floods [31]. To address the conversion relationship between the land surface features in FSR and CSR imagery at different scales, the ESTARFM and EDCSTFN build linear models and deep convolutional neural network models, respectively, to fit and learn the conversion relationship of the land surface features between the FSR and CSR imagery [34,37]. However, there are still uncertainties in areas with irregular abrupt changes in land cover type. Therefore, the reliability of the STDF has a negative correlation with the differences in the SRs between FSR and CSR imagery. Compared with the STDF, super-resolution technology can improve the SR of remote sensing imagery with fewer requirements, such as a sufficient amount of reference imagery [62]. Although the precision of super-resolution models may be lower than that of STDF models in terms of the reference imagery when the difference in the SR between the input image and target image is large, the super-resolution models can be used to appropriately improve the SR of CSR images, thereby providing a novel way to improve the precision of STDF models. In recent years, researchers have developed many super-resolution models, such as super-resolution generative adversarial networks (SRGANs) [63], enhanced SRGANs (ESRGANs) [64], the real-ESRGAN [65], and the improved SRGAN (ISRGAN) [62], which have been successfully applied to the spatial downscaling of satellite imagery, such as MODIS [62,66]. In the future, a super-resolution model will be introduced to the current FY-3D-based farmland-scale WWY estimation method framework, and improvements in the precision of the STDF model and WWY estimation model will be tested further.

4.2. *Research Prospects of the WWY Estimation Model*

There is a complex nonlinear relationship between the WWY and the various VIs, including the NDVI, SAVI, and EVI. Therefore, the construction of a WWY estimation model requires a large amount of training data. However, the amount of WWY data currently available cannot satisfy the training requirements of complex WWY estimation models. In addition, meteorological and soil factors such as temperature, precipitation and soil moisture also greatly impact the WWY [67]. Therefore, the impacts of meteorological and soil factors must be considered in multiyear WWY estimation. The WW time series GPP data consider the influence of meteorological factors on the growth of WW, and there was a high correlation between the cumulative GPP data for the main growing period and the WWY [53,54]. Therefore, the MODIS cumulative GPP and multiyear WWY data at the county scale were used to establish a linear regression model in this study. The model was subsequently used to estimate the multiyear WWY with an SR of 500 m and then employed to train the annual WWY estimation model based on the BPNN. The above method framework is often sub-optimal for WWY estimation compared to the WWY estimation model based directly on VI imagery and meteorological parameters. In the future, many ground-based measuring plots for the WWY will be constructed in the WW planting regions in China, which will provide support for the construction of complex machine learning models based on many feature parameters. In addition, sensitivity analyses will be performed to explore the impact of different parameters on the WWY, thereby selecting the effective parameters to construct the WWY estimation model.

In addition, continuous time series VI data provide a large amount of information related to the WWY on different dates during the growing period. Thus, traditional machine learning models may not fully utilize the effective information contained in continuous time series data and may even lead to overfitting. To reduce overfitting, the VI data from 10 day intervals were aggregated into 4 WW growing stages in this work, thereby greatly reducing the number of input parameters for the Y estimation model. However, much of the effective information related to the WWY was lost. Compared with traditional machine learning models, long short-term memory (LSTM) neural networks [68,69] provide greater learning abilities when using time series data; therefore, they are more suitable for crop Y estimation based on time series data. In the future, LSTM WWY estimation models based on continuous time series farmland-scale VI data will be developed to conduct more accurate WWY estimations at the farmland scale.

5. Conclusions

To reconstruct farmland-scale FY-3D-based 10 day interval VI (NDVI, SAVI and EVI) imagery over the main WW growing periods and achieve farmland-scale WWY estimation, a method framework for the STDF of VIs derived from FY-3D and Sentinel-2 based on the EDCSTFN and a farmland-scale WWY estimation method based on the BPNN model were proposed in this paper by using cumulative MODIS GPP data over the main WW growing periods as ancillary data.

(1) Similar to the ESTARFM, the EDCSTFN exhibited a satisfactory precision for the spatiotemporal fusion of Sentinel-2 and FY-3D VI imagery on the Weihe Plain, and the R^2 values of the fusion results reached over 0.9. (2) When the whole period of the STDF was close to 2.5 months, the EDCSTFN attained a higher fusion precision for VI imagery than the ESTARFM. (3) The cumulative MODIS GPP data over the main WW growing periods in multiple years were strongly correlated (R^2 = 0.67) with the county-scale WWY on the Weihe Plain. (4) Compared with the WWY estimation results with a 500 m SR based on the cumulative MODIS GPP data from 2020 to 2021 (R^2 = 0.65 and 0.63, respectively), the farmland-scale WWY estimation results based on the spatiotemporally fused VI imagery were better correlated with the county-scale WWY data (R^2 = 0.78 and 0.68, respectively) and provided rich and detailed farmland-scale Y distribution features for WW.

This study revealed that the proposed method can be employed to downscale FY-3D VI imagery to the farmland scale via fusion with Sentinel-2 data, thereby achieving farmland-scale WWY estimation on the Weihe Plain. Therefore, it has considerable significance in terms of accurate WWY estimation based on Chinese FY-3 meteorological satellites.

Author Contributions: X.Z.: Conceptualization, Methodology, Data curation, Writing—Original draft, Funding acquisition. T.W.: Data curation, Writing-Reviewing and Editing. W.Z.: Writing-Reviewing and Editing. M.Z.: Resources, Funding acquisition. Y.W.: Resources, Funding acquisition. All authors have read and agreed to the published version of the manuscript.

Funding: This work was funded by the Advanced Research on Civil Space Technology during China's 14th Five-Year Plan period (Grant No. D040405), the National Key R&D Program of China (Grant No. 2022YFC3002801), and the 2022 Youth Fund of the National Satellite Meteorological Center.

Data Availability Statement: The data that support the findings of this study are available from the corresponding author upon reasonable request.

Conflicts of Interest: The authors declare no conflicts of interest.

References

1. Huang, J.; Tian, L.; Liang, S.; Ma, H.; Becker-Reshef, I.; Huang, Y.; Su, W.; Zhang, X.; Zhu, D.; Wu, W. Improving winter wheat yield estimation by assimilation of the leaf area index from Landsat TM and MODIS data into the WOFOST model. *Agric. For. Meteorol.* **2015**, *204*, 106–121. [CrossRef]
2. Tian, H.; Wang, P.; Tansey, K.; Zhang, S.; Zhang, J.; Li, H. An IPSO-BP neural network for estimating wheat yield using two remotely sensed variables in the Guanzhong Plain, PR China. *Comput. Electron. Agric.* **2020**, *169*, 105180. [CrossRef]
3. Rouse, J.W.; Haas, R.H.; Schell, J.A.; Deering, D.W. Monitoring vegetation systems in the great plains with ERTS. *NASA Spec. Publ.* **1974**, *351*, 309.
4. Huete, A.R. A soil-adjusted vegetation index (SAVI). *Remote Sens. Environ.* **1988**, *25*, 295–309. [CrossRef]
5. Huete, A.; Didan, K.; Miura, T.; Rodriguez, E.P.; Gao, X.; Ferreira, L.G. Overview of the radiometric and biophysical performance of the MODIS vegetation indices. *Remote Sens. Environ.* **2002**, *83*, 195–213. [CrossRef]
6. Carlson, T.N.; Ripley, D.A. On the relation between NDVI, fractional vegetation cover, and leaf area index. *Remote Sens. Environ.* **1997**, *62*, 241–252. [CrossRef]
7. Xiao, J.; Moody, A. A comparison of methods for estimating fractional green vegetation cover within a desert-to-upland transition zone in central New Mexico, USA. *Remote Sens. Environ.* **2005**, *98*, 237–250. [CrossRef]
8. Baret, F.; Guyot, G. Potentials and limits of vegetation indices for LAI and APAR assessment. *Remote Sens. Environ.* **1991**, *35*, 161–173. [CrossRef]
9. Houborg, R.; Soegaard, H.; Boegh, E. Combining vegetation index and model inversion methods for the extraction of key vegetation biophysical parameters using terra and Aqua MODIS reflectance data. *Remote Sens. Environ.* **2007**, *106*, 39–58. [CrossRef]
10. Bolton, D.K.; Friedl, M.A. Forecasting crop yield using remotely sensed vegetation indices and crop phenology metrics. *Agric. For. Meteorol.* **2013**, *173*, 74–84. [CrossRef]
11. Johnson, M.D.; Hsieh, W.W.; Cannon, A.J.; Davidson, A.; Bédard, F. Crop yield forecasting on the Canadian prairies by remotely sensed vegetation indices and machine learning methods. *Agric. For. Meteorol.* **2016**, *218–219*, 74–84. [CrossRef]
12. Zhao, W.; He, Z.; He, J.; Zhu, L. Remote sensing estimation for winter wheat yield in Henan based on the MODIS-NDVI data. *Geogr. Res.* **2012**, *31*, 2310–2320. [CrossRef]

13. Son, N.T.; Chen, C.F.; Chen, C.R.; Minh, V.Q.; Trung, N.H. A comparative analysis of multitemporal MODIS EVI and NDVI data for large-scale rice yield estimation. *Agric. For. Meteorol.* **2014**, *197*, 52–64. [CrossRef]
14. Barriguinha, A.; Jardim, B.; Neto, M.D.; Gil, A. Using NDVI, climate data and machine learning to estimate yield in the Douro wine region. *Int. J. Appl. Earth Obs. Geoinf.* **2022**, *114*, 103069. [CrossRef]
15. Liying, G.; Tao, C.; Mingguo, M.; Junlei, T.; Haibo, W. Corn Biomass Estimation by Integrating Remote Sensing and Long-Term Observation Data Based on Machine Learning Techniques. *Remote Sens.* **2021**, *13*, 2352. [CrossRef]
16. Perry, E.M.; Morse-McNabb, E.M.; Nuttall, J.G.; O Leary, G.J.; Clark, R. Managing Wheat From Space: Linking MODIS NDVI and Crop Models for Predicting Australian Dryland Wheat Biomass. *IEEE J-STARS.* **2014**, *7*, 3724–3731. [CrossRef]
17. Zhu, X.; Li, Q.; Guo, C. Evaluation of the monitoring capability of various vegetation indices and mainstream satellite band settings for grassland drought. *Ecol. Inform.* **2024**, *82*, 102717. [CrossRef]
18. Sandeep, P.; Obi Reddy, G.P.; Jegankumar, R.; Arun Kumar, K.C. Monitoring of agricultural drought in semi-arid ecosystem of Peninsular India through indices derived from time-series CHIRPS and MODIS datasets. *Ecol. Indic.* **2021**, *121*, 107033. [CrossRef]
19. Gong, Z.; Ge, W.; Guo, J.; Liu, J. Satellite remote sensing of vegetation phenology: Progress, challenges, and opportunities. *Isprs. J. Photogramm.* **2024**, *217*, 149–164. [CrossRef]
20. Gan, L.; Cao, X.; Chen, X.; Dong, Q.; Cui, X.; Chen, J. Comparison of MODIS-based vegetation indices and methods for winter wheat green-up date detection in Huanghuai region of China. *Agric. For. Meteorol.* **2020**, *288–289*, 108019. [CrossRef]
21. Huang, Y. Understanding China's Belt & Road Initiative: Motivation, framework and assessment. *China Econ. Rev.* **2016**, *40*, 314–321. [CrossRef]
22. Hao, G.; Shihao, T.; Xiuzhen, H. China's Fengyun (FY) meteorological satellites, development and applications. *Sci. Technol. Rev.* **2021**, *39*, 9–22. [CrossRef]
23. Zhang, Z.; Lu, L.; Zhao, Y.; Wang, Y.; Wei, D.; Wu, X.; Ma, X. Recent advances in using Chinese Earth observation satellites for remote sensing of vegetation. *ISPRS J. Photogramm.* **2023**, *195*, 393–407. [CrossRef]
24. Zhao, J.; Li, J.; Liu, Q.; Xu, B.; Yu, W.; Lin, S.; Hu, Z. Estimating fractional vegetation cover from leaf area index and clumping index based on the gap probability theory. *Int. J. Appl. Earth Obs.* **2020**, *90*, 102112. [CrossRef]
25. Zhao, J.; Li, J.; Liu, Q.; Xu, B.; Chen, C.; Li, L. A Integrated Inversion Method for Estimating Global Leaf Area Index from Chinese FY-3A Mersi Data. In Proceedings of the IGARSS 2018, Valencia, Spain, 22–27 July 2018; pp. 1152–1155. [CrossRef]
26. Zhang, Y.; Zhang, M.; Sun, R.; Zhang, Y.; Zhang, L. Agricultural application using FY meteorological satellite. *Sci. Technol. Rev.* **2021**, *39*, 39–45. [CrossRef]
27. Boori, M.; Choudhary, K.; Kupriyanov, A. Crop growth monitoring through sentinel and landsat data based NDVI time-series. *Comput. Opt.* **2020**, *44*, 409–419. [CrossRef]
28. Son, N.T.; Chen, C.F.; Chen, C.R.; Guo, H.Y. Classification of multitemporal Sentinel-2 data for field-level monitoring of rice cropping practices in Taiwan. *Adv. Space Res.* **2020**, *65*, 1910–1921. [CrossRef]
29. Chen, Y.P.; Hu, J.; Cai, Z.W.; Yang, J.Y.; Zhou, W.; Hu, Q.; Wang, C.; You, L.Z.; Xu, B.D. A phenology-based vegetation index for improving ratoon rice mapping using harmonized landsat and sentinel-2 data. *J. Integr. Agric.* **2023**, *23*, 1164–1178. [CrossRef]
30. Gao, F.; Jennewein, J.; Hively, W.D.; Soroka, A.; Thieme, A.; Bradley, D.; Keppler, J.; Mirsky, S.; Akumaga, U. Near real-time detection of winter cover crop termination using harmonized landsat and sentinel-2 (HLS) to support ecosystem assessment. *Sci. Remote Sens.* **2023**, *7*, 100073. [CrossRef]
31. Zhu, X.; Cai, F.; Tian, J.; Williams, T.K.A. Spatiotemporal fusion of multisource remote sensing data: Literature survey, taxonomy, principles, applications, and future directions. *Remote Sens.* **2018**, *10*, 527. [CrossRef]
32. Chen, G.; Lu, H.; Zou, W.; Li, L.; Emam, M.; Chen, X.; Jing, W.; Wang, J.; Li, C. Spatiotemporal fusion for spectral remote sensing: A statistical analysis and review. *J. King Saud Univ. Comput. Inf. Sci.* **2023**, *35*, 259–273. [CrossRef]
33. Gao, F.; Masek, J.; Schwaller, M.; Hall, F. On the blending of the Landsat and MODIS surface reflectance: Predicting daily landsat surface reflectance. *IEEE Trans. Geosci. Remote Sens.* **2006**, *44*, 2207–2218. [CrossRef]
34. Zhu, X.; Chen, J.; Gao, F.; Chen, X.; Masek, J.G. An enhanced spatial and temporal adaptive reflectance fusion model for complex heterogeneous regions. *Remote Sens. Environ.* **2010**, *114*, 2610–2623. [CrossRef]
35. Chen, B.; Huang, B.; Xu, B. Comparison of spatiotemporal fusion models: A review. *Remote Sens.* **2015**, *7*, 1798–1835. [CrossRef]
36. Tan, Z.; Yue, P.; Di, L.; Tang, J. Deriving high spatiotemporal remote sensing images using deep convolutional network. *Remote Sens.* **2018**, *10*, 1066. [CrossRef]
37. Tan, Z.; Di, L.; Zhang, M.; Guo, L.; Gao, M. An enhanced deep convolutional model for spatiotemporal image fusion. *Remote Sens.* **2019**, *11*, 2898. [CrossRef]
38. Wu, M.; Wu, C.; Huang, W.; Niu, Z.; Wang, C. High-resolution leaf area index estimation from synthetic Landsat data generated by a spatial and temporal data fusion model. *Comput. Electron. Agric.* **2015**, *115*, 1–11. [CrossRef]
39. Gao, F.; Anderson, M.C.; Zhang, X.; Yang, Z.; Alfieri, J.G.; Kustas, W.P.; Mueller, R.; Johnson, D.M.; Prueger, J.H. Toward mapping crop progress at field scales through fusion of landsat and MODIS imagery. *Remote Sens. Environ.* **2017**, *188*, 9–25. [CrossRef]
40. Meng, L.; Liu, H.; Zhang, X.; Ren, C.; Ustin, S.; Qiu, Z.; Xu, M.; Guo, D. Assessment of the effectiveness of spatiotemporal fusion of multi-source satellite images for cotton yield estimation. *Comput. Electron. Agric.* **2019**, *162*, 44–52. [CrossRef]
41. Zhou, X.; Wang, P.; Tansey, K.; Zhang, S.; Li, H.; Tian, H. Reconstruction of time series leaf area index for improving wheat yield estimates at field scales by fusion of sentinel-2,-3 and MODIS imagery. *Comput. Electron. Agric.* **2020**, *177*, 105692. [CrossRef]

42. Zhou, X.; Zhang, Y.; Wang, P.; Zhang, S.; Li, H.; Tian, H. Reconstruction of time series LAI and winter wheat yield estimation at field scales based on sentinel satellites. *Trans. Chin. Soc. Agric. Mach.* **2022**, *53*, 173–185. [CrossRef]
43. Yajun, Z.; Tingxi, L.; Okke, B.; Limin, D.; Yixuan, W.; Xia, L.; Mingyang, L. Spatiotemporal fusion of multi-source remote sensing data for estimating aboveground biomass of grassland. *Ecol. Indic.* **2023**, *146*, 109892. [CrossRef]
44. Na, Z.; Honglin, H.; Xiaoli, R.; Li, Z.; Yuan, Z.; Jiangwen, F.; Yuzhe, L.; Zhongen, N.; Xiaobo, Z.; Qingqing, C. The utility of fusing multi-sensor data spatio-temporally in estimating grassland aboveground biomass in the three-river headwaters region of China. *Int. J. Remote Sens.* **2020**, *41*, 7068–7089. [CrossRef]
45. Zhou, X.; Wang, P.; Tansey, K.; Zhang, S.; Li, H.; Wang, L. Developing a fused vegetation temperature condition index for drought monitoring at field scales using sentinel-2 and MODIS imagery. *Comput. Electron. Agric.* **2020**, *168*, 105144. [CrossRef]
46. Xin, L.; Hongli, Z.; Yanyan, H.; Shuangmei, L.; Zelong, M.; Yunzhong, J.; Wei, Z.; Chuan, Z. Generating Daily Soil Moisture at 16 m Spatial Resolution Using a Spatiotemporal Fusion Model and Modified Perpendicular Drought Index. *Sensors* **2022**, *22*, 5366. [CrossRef] [PubMed]
47. Zheng, Y.; Wu, B.; Zhang, M.; Zeng, H. Crop Phenology Detection Using High Spatio-Temporal Resolution Data Fused from SPOT5 and MODIS Products. *Sensors* **2016**, *16*, 2099. [CrossRef] [PubMed]
48. Wu, M.; Huang, W.; Niu, Z.; Wang, C.; Li, W.; Yu, B. Validation of synthetic daily Landsat NDVI time series data generated by the improved spatial and temporal data fusion approach. *Inform. Fusion* **2018**, *40*, 34–44. [CrossRef]
49. Xue, L.; Ren, Z. Ecological security of Guanzhong region based on gridding GIS. *Geogr. Sin.* **2011**, *31*, 123–128. [CrossRef]
50. Chang, Y.; Liu, J.; Li, Y.; Sun, B.; Zhang, S.; Yang, X. Investigation and evaluation of fertilization under winter wheat and summer maize rotation system in Guanzhong Plain, Shaanxi Province. *J. Northwest Univ. Nat. Sci. Ed.* **2014**, *42*, 51–61. [CrossRef]
51. Sun, W.; Wang, P.X.; Zhang, S.Y.; Zhu, D.H.; Liu, J.M.; Chen, J.H.; Yang, H.S. Using the vegetation temperature condition index for time series drought occurrence monitoring in the Guanzhong Plain, PR China. *Int. J. Remote Sens.* **2008**, *29*, 5133–5144. [CrossRef]
52. Xie, Y.; Wang, P.; Bai, X.; Khan, J.; Zhang, S.; Li, L.; Wang, L. Assimilation of the leaf area index and vegetation temperature condition index for winter wheat yield estimation using Landsat imagery and the CERES-Wheat model. *Agric. For. Meteorol.* **2017**, *246*, 194–206. [CrossRef]
53. Prince, S.D.; Haskett, J.; Steininger, M.; Strand, H.; Wright, R. Net primary production of U.S. midwest croplands from agricultural harvest yield data. *Ecol. Appl.* **2001**, *11*, 1194–1205. [CrossRef]
54. Marshall, M.; Tu, K.; Brown, J. Optimizing a remote sensing production efficiency model for macro-scale GPP and yield estimation in agroecosystems. *Remote Sens. Environ.* **2018**, *217*, 258–271. [CrossRef]
55. Wang, S.; Cui, D.; Wang, L.; Peng, J. Applying deep-learning enhanced fusion methods for improved NDVI reconstruction and long-term vegetation cover study: A case of the Danjiang River Basin. *Ecol. Indic.* **2023**, *155*, 111088. [CrossRef]
56. Yuean, Q.; Junxiong, Z.; Jin, C.; Xuehong, C. Spatiotemporal fusion method to simultaneously generate full-length normalized difference vegetation index time series (SSFIT). *Int. J. Appl. Earth Obs.* **2021**, *100*, 102333. [CrossRef]
57. Zhang, K.; Zhu, C.; Li, J.; Shi, K.; Zhang, X. Reconstruction of dense time series high spatial resolution NDVI data using a spatiotemporal optimal weighted combination estimation model based on Sentinel-2 and MODIS. *Ecol. Inform.* **2024**, *82*, 102725. [CrossRef]
58. Wang, J.; Wang, P.; Tian, H.; Tansey, K.; Liu, J.; Quan, W. A deep learning framework combining CNN and GRU for improving wheat yield estimates using time series remotely sensed multi-variables. *Comput. Electron. Agric.* **2023**, *206*, 107705. [CrossRef]
59. Wu, B.; Zhang, M.; Zeng, H. Twenty years of cropwatch: Progress and prospect. *Natl. Remote Sens. Bull.* **2019**, *23*, 1053–1063. [CrossRef]
60. Guo, L.; Liu, Y.; He, H.; Lin, H.; Qiu, G.; Yang, W. Consistency analysis of GF-1 and GF-6 satellite wide field view multi-spectral band reflectance. *Optik* **2021**, *231*, 166414. [CrossRef]
61. Han, J.; Xie, Y. Study on the geo-positioning stability of GF-1 WFV Images. *Bull. Surv. Mapp.* **2018**, *2*, 50–54. [CrossRef]
62. Jin, H.; Qiao, Y.; Liu, T.; Xie, X.; Fang, H.; Guo, Q.; Zhao, W. A hierarchical downscaling scheme for generating fine-resolution leaf area index with multisource and multiscale observations via deep learning. *Int. J. Appl. Earth Obs.* **2024**, *133*, 104152. [CrossRef]
63. Ledig, C.; Theis, L.; Huszár, F.; Caballero, J.; Cunningham, A.; Acosta, A.; Aitken, A.; Tejani, A.; Totz, J.; Wang, Z.; et al. Photo-Realistic Single Image Super-Resolution Using a Generative Adversarial Network. In Proceedings of the IEEE Conference on Computer Vision and Pattern Recognition, Honolulu, HI, USA, 21–26 July 2017; pp. 105–114.
64. Wang, X.; Yu, K.; Wu, S.; Gu, J.; Liu, Y.; Dong, C.; Qiao, Y.; Loy, C.C. ESRGAN: Enhanced Super-Resolution Generative Adversarial Networks. In Proceedings of the European Conference on Computer Vision (ECCV) Workshops, Munich, Germany, 8–14 September 2019; Springer International Publishing: Cham, Switzerland, 2019; pp. 63–79. [CrossRef]
65. Wang, X.; Xie, L.; Dong, C.; Shan, Y. Real-ESRGAN: Training Real-World Blind Super-Resolution with Pure Synthetic Data. In Proceedings of the IEEE/CVF International Conference on Computer Vision, Montreal, BC, Canada, 11–17 October 2021; pp. 1905–1914. [CrossRef]
66. Hong, Y.; Que, X.; Wang, Z.; Ma, X.; Wang, H.; Salati, S.; Liu, J. Mangrove extraction from super-resolution images generated by deep learning models. *Ecol. Indic.* **2024**, *159*, 111714. [CrossRef]
67. Bocca, F.F.; Rodrigues, L.H.A. The effect of tuning, feature engineering, and feature selection in data mining applied to rainfed sugarcane yield modelling. *Comput. Electron. Agric.* **2016**, *128*, 67–76. [CrossRef]

68. Jiang, H.; Hu, H.; Zhong, R.; Xu, J.; Xu, J.; Huang, J.; Wang, S.; Ying, Y.; Lin, T. A deep learning approach to conflating heterogeneous geospatial data for corn yield estimation: A case study of the US Corn Belt at the county level. *Glob. Change Biol.* **2020**, *26*, 1754–1766. [CrossRef]
69. Tian, H.; Wang, P.; Tansey, K.; Zhang, J.; Zhang, S.; Li, H. An LSTM neural network for improving wheat yield estimates by integrating remote sensing data and meteorological data in the Guanzhong Plain, PR China. *Agric. For. Meteorol.* **2021**, *310*, 108629. [CrossRef]

remote sensing

MDPI

Article

Accurate Estimation of Gross Primary Production of Paddy Rice Cropland with UAV Imagery-Driven Leaf Biochemical Model

Xiaolong Hu [1], Liangsheng Shi [1,*], Lin Lin [1], Shenji Li [2], Xianzhi Deng [1], Jinmin Li [1], Jiang Bian [1], Chenye Su [1], Shuai Du [1], Tinghan Wang [1], Yujie Wang [1] and Zhitao Zhang [3]

1 State Key Laboratory of Water Resources Engineering and Management, Wuhan University, Wuhan 430072, China; xlhu@whu.edu.cn (X.H.); linlin.wh@whu.edu.cn (L.L.); 2022202060114@whu.edu.cn (X.D.); jinmlee@whu.edu.cn (J.L.); jiangbian@whu.edu.cn (J.B.); suchenye@whu.edu.cn (C.S.); 2018302060154@whu.edu.cn (S.D.); 2019302060215@whu.edu.cn (T.W.); yujie_wang@whu.edu.cn (Y.W.)

2 Urban Operation Management Center of Hengsha Township, Shanghai 201914, China; 15901916214@163.com

3 Key Laboratory of Agricultural Soil and Water Engineering in Arid and Semiarid Areas, Ministry of Education, Northwest A&F University, Yangling 712100, China; zhangzhitao@nwafu.edu.cn

* Correspondence: liangshs@whu.edu.cn

Abstract: Accurate estimation of gross primary production (GPP) of paddy rice fields is essential for understanding cropland carbon cycles, yet remains challenging due to spatial heterogeneity. In this study, we integrated high-resolution unmanned aerial vehicle (UAV) imagery into a leaf biochemical properties-based model for improving GPP estimation. The key parameter, maximum carboxylation rate at the top of the canopy ($V^{25}_{cmax,0}$), was quantified using various spatial information representation methods, including mean (μ_{ref}) and standard deviation (σ_{ref}) of reflectance, gray-level co-occurrence matrix (GLCM)-based features, local binary pattern histogram (LBPH), and convolutional neural networks (CNNs). Our models were evaluated using a two-year eddy covariance (EC) system and UAV measurements. The result shows that incorporating spatial information can vastly improve the accuracy of $V^{25}_{cmax,0}$ and GPP estimation. CNN methods achieved the best $V^{25}_{cmax,0}$ estimation, with an R of 0.94, an RMSE of 19.44 μmol m^{-2} s^{-1}, and an MdAPE of 11%, and further produced highly accurate GPP estimates, with an R of 0.92, an RMSE of 6.5 μmol m^{-2} s^{-1}, and an MdAPE of 23%. The μ_{ref}-GLCM texture features and μ_{ref}-LBPH joint-driven models also gave promising results. However, σ_{ref} contributed less to $V^{25}_{cmax,0}$ estimation. The Shapley value analysis revealed that the contribution of input features varied considerably across different models. The CNN model focused on nir and red-edge bands and paid much attention to the subregion with high spatial heterogeneity. The μ_{ref}-LBPH joint-driven model mainly prioritized reflectance information. The μ_{ref}-GLCM-based features joint-driven model emphasized the role of GLCM texture indices. As the first study to leverage the spatial information from high-resolution UAV imagery for GPP estimation, our work underscores the critical role of spatial information and provides new insight into monitoring the carbon cycle.

Keywords: gross primary production; maximum carboxylation rate; paddy rice cropland; UAV; texture features

1. Introduction

Gross primary production (GPP) is the total amount of carbon compounds produced through plant photosynthesis and indicates the carbon sequestration capacity of ecosystems [1]. In cropland ecosystems, GPP is integral to yield formation [2]. Paddy rice, as an important staple food crop, occupies approximately 165.58 million hectares globally and sustains over half of the world population [3]. Therefore, accurate estimation of GPP in paddy rice cropland not only benefits the quantification of the carbon sequestration but also contributes to crop yield prediction [4].

Some classical models for GPP estimation have been developed over the past few decades, which can be categorized into the statistical model [5,6], the leaf biochemical properties-based model [7–9], the light use efficiency model [10,11], and the solar-induced chlorophyll fluorescence model [12,13]. These models have significantly advanced our understanding of the global carbon cycle. Among them, the leaf biochemical properties-based model, on the foundation of the Calvin cycle theory, elucidates the three steps of carbon sequestration in photosynthetic carbon reactions and has received much attention for its excellent physical interpretability.

In the leaf biochemical properties-based model, a major source of error in GPP estimation arises from the uncertainty in a key parameter, the leaf maximum carboxylation rate (normalized to 25 °C, V_{cmax}^{25}) [14]. V_{cmax}^{25} indicates the enzyme activity of Rubisco. The substantial spatiotemporal variability of V_{cmax}^{25} has long been observed [15,16]. Consequently, extensive efforts have been invested in quantifying the dynamics of V_{cmax}^{25} to acquire an accurate GPP estimation. On the one hand, a number of biophysical and biochemical factors that mechanistically relate to V_{cmax}^{25}, such as leaf nitrogen, leaf phosphorus, specific leaf area, leaf chlorophyll content, and leaf area index, have been identified [17–20]. On the other hand, the spectral information, including vegetation index, multispectral reflectance, hyperspectral reflectance, and fluorescence spectral information, has also shown promising potential for retrieving V_{cmax}^{25} [21–25]. The abovementioned factors have been incorporated into the leaf biochemical properties-based model and significantly enhance the accuracy of GPP estimation [14].

At the field-scale, V_{cmax}^{25} is a macroscopic feature within the specific spatial range, theoretically upscaled from leaf-level V_{cmax}^{25} and thus determined by the spatial distribution of V_{cmax}^{25}-related properties on underlying surfaces. However, the spatial information of V_{cmax}^{25}-related properties on underlying surfaces has rarely been considered in previous studies. Traditional methods for measuring biophysical and biochemical factors, whether destructive or non-destructive, make it challenging to obtain precise spatial distributions. While satellite remote sensing offers widely accessible spectral images at a medium and low spatial resolution (10–30 m resolution) from platforms like Sentinel-2, Landsat 8 OLI and TIRS, Landsat ETM+, and Moderate-resolution Imaging Spectroradiometer (MODIS), this resolution is often insufficient for capturing the detailed spatial distribution of spectral properties. Moreover, satellite images are frequently disrupted by clouds, causing spatiotemporal discontinuities. These gaps in spatial information likely impede the accurate estimation of filed-scale V_{cmax}^{25}. In contrast, the unmanned aerial vehicle (UAV) can capture high-resolution images even in cloudy conditions, offering better capability to describe the spatial distribution of spectral information. Therefore, UAV imagery holds great potential for extracting valuable spatial information to enhance field-scale V_{cmax}^{25} estimation.

Due to the high-dimensional characteristics of UAV imagery, extracting meaningful information using traditional methods poses challenges. There are several approaches to extracting the spatial information from different perspectives. The standard deviation of the pixel values is the most straightforward indicator and is commonly used to quantify landscape heterogeneity [26]. However, it fails to provide other valuable information, such as the spatial distribution patterns. Texture analysis is a widely used approach in image processing. Texture is defined as the measure of coarseness, contrast, directionality, line-likeness, regularity, and roughness [27]. Various texture feature extraction methods have been proposed, including statistical approaches, structural approaches, transform-based approaches, model-based approaches, graph-based approaches, learning-based approaches, and entropy-based approaches [28]. Among these, the gray-level co-occurrence matrix (GLCM) is the most commonly used method [29] and has been widely applied in quantifying landscape patterns [30], as well as estimating various properties on underlying surfaces, such as leaf area index and biomass, from remote sensing images [31,32]. Other texture features, such as Fourier spectrum texture [33] and local binary pattern (LBP) texture [34,35], have also shown strong performance in quantifying specific landscape features, including leaf area index and orchard citrus fruits [36,37]. Despite the explicit and interpretable information extracted through texture analysis, this method may overlook large potential information in the raw images. The breakthrough in deep learning techniques, especially convolutional neural networks (CNNs), vastly enhances the capacity to extract image information. CNN utilizes the convolution and pooling operation to automatically detect the patterns across the entire image and thus can make full use of image information. CNN became a hot spot for its unparalleled performance regarding the image recognition issue [38]. Inspired by its success, the CNN technique is also employed to interpret remote sensing images. Notably, the combination of CNN and high-resolution UAV imagery shows a powerful capacity for retrieving valuable variables, such as crop yield [39] and leaf area index [40].

A few pioneering works have explored the retrieval of photosynthesis traits from images. Sun et al. [41] utilized several texture features, including the GLCM texture feature, LBP texture feature, and wavelet transform texture features, to monitor the leaf-scale net photosynthetic rate of potatoes. Deng et al. [42] proposed the optimal deep learning architecture to estimate the leaf-scale V_{cmax}^{25} from hyperspectral images. These works suggest the potential of images for accurate estimation of photosynthetic traits. However, to the best of our knowledge, few studies have examined the effect of spatial information on field-scale V_{cmax}^{25} estimation. As the first study that leveraged spatial information from high-resolution UAV imagery for V_{cmax}^{25} and GPP estimation, we extracted the spatial information of the underlying surface from high-resolution images using various representation methods, including texture features and deep learning, to quantify the macroscopic feature, field-scale V_{cmax}^{25}. The UAV-imagery-driven V_{cmax}^{25} model was integrated into a classical leaf biochemical properties-based model for better GPP estimation. Our findings demonstrate the critical role of spatial information and provide new insight into monitoring GPP. The specific objective of this study is to (1) evaluate whether spatial information can improve the accuracy of field-scale V_{cmax}^{25}, (2) build a UAV-imagery-driven hybrid framework for GPP estimation and validate its performance on the paddy rice field, and (3) see how spatial information contributes to field-scale V_{cmax}^{25} estimation.

2. Materials and Methods

The research workflow is depicted in Figure 1, delineated into three distinct stages: (1) data preprocessing, including generating high-resolution UAV RGB orthomosaics and reflectance maps of UAV images, and half-hourly high-quality carbon flux measurements; (2) texture feature extraction from UAV images using four methods, including standard deviation features, GLCM-based texture features, LBP-based texture features, and CNN-based features; (3) model assessment, including evaluating the performance of 8 UAV imagery-driven $V_{cmax,0}^{25}$ models and their influences on *GPP* estimation, and analyzing the feature contribution of each $V_{cmax,0}^{25}$ model using the Shaley value method.

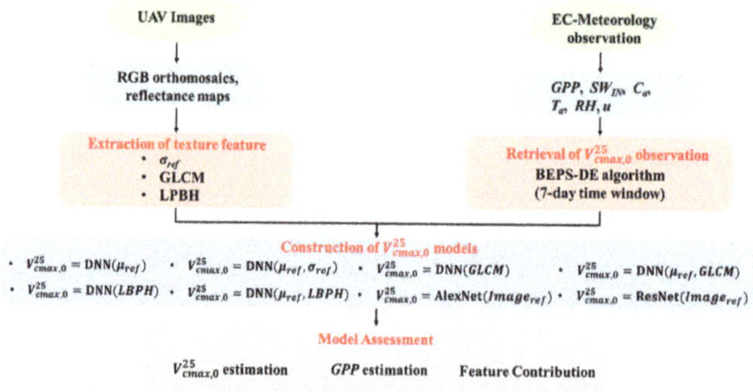

Figure 1. Flowchart of the study (μ_{ref}: the mean value of UAV reflectance; σ_{ref}: the standard deviation value of UAV reflectance; GLCM: gray-level co-occurrence matrix-based texture features; LBPH: local binary pattern histogram texture feature; *GPP*: gross primary production; SW_{IN}: incoming shortwave radiation; C_a: ambient CO_2 concentration; T_a: air temperature; *RH*: relative humidity; *u*: wind speed).

2.1. Experiment Site and Data Collection

A two-year experiment from 2022 to 2023 was conducted in an 18.93 ha paddy field, located in Hengsha Country, Shanghai City, China (Figure 2). Rice was transplanted on June 1st, 2022, and May 25th, 2023, and harvested on December 3rd, 2022, and October 20th, 2023. The soil texture was silt loam, with a mean dry bulk density of 1.37 g cm^{-3} and a mean saturated volumetric soil water content of 44.23%. The paddy field maintained a water layer with a 5–20 cm depth, except during the late tillering and full ripening stages. The plant density was 2.38×10^5 plants ha^{-1}.

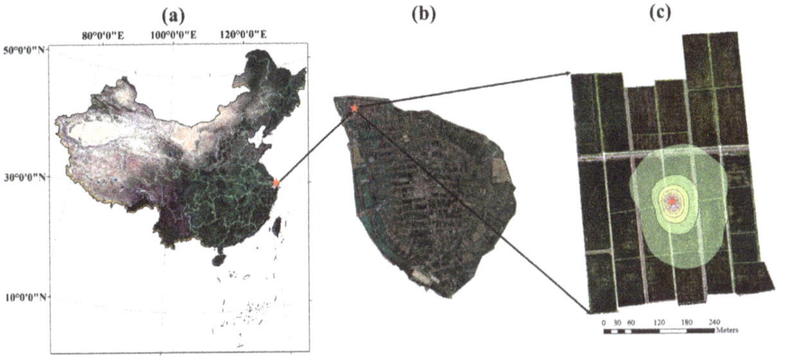

Figure 2. The location of the experiment site. The red five-pointed star represents the EC system. The contour lines of the footprint during the growing period in (3) are shown in steps of 10% from 10 to 90%. The background maps in (**a**,**b**) are RGB images from Google Earth. The background map in (**c**) is the RGB image from the UAV platform.

2.1.1. Eddy Flux and Micrometeorological Measurements

The eddy covariance system (IRGASON, Campbell Scientific, Logan, UT, USA) with an installation height of 2.85 m collected the 10 Hz three-dimensional wind speed and direction, ambient CO_2 and H_2O density, sonic temperature, and air pressure. The corresponding meteorological and soil elements, including air temperature and humidity (HMP155A, Vaisala Ojy, Vantaa, Finland), incoming outgoing shortwave radiations (CNR4-L, Kipp & Zonen, Delft, The Netherlands), four-layer soil moisture, and temperature at 5 cm, 10 cm, 20 cm, and 40 cm (TDR310N, Acclima, Meridian, ID, USA) were measured.

The 30 min eddy fluxes, including net ecosystem exchange (*NEE*), latent heat flux (*LE*), sensible heat flux (*H*), and friction velocity (u_*), were generated from 10 Hz raw data of the EC system using the EddyPro software v7.0.9 (www.licor.com/). The data quality of eddy flux was evaluated using the scheme proposed by Foken et al. [43], which classifies the flux data into 9 levels. The data labeled as levels 7–9 were regarded as low-quality and removed. In addition, the flux data were dropped when u_* was less than a specific threshold that indicated the period of low turbulent mixing. The u^* threshold was calculated for each season using the moving point method [43]. The seasonal u_* threshold is shown in Table S1 (see Supplementary Material Section S2). Then, the missing flux data were filled using the marginal distribution sampling method [44]. Finally, the 30 min *NEE* observation was partitioned into *GPP* and ecosystem respiration (R_{eco}) using the night-time flux partitioning method [44]. The processes of u_* filter, gap filling, and flux partitioning were conducted using the REddyProc package [45].

To determine the spatial extent of the EC system measurement, the flux footprint was calculated using the Kljun model [46], which requires the standard deviation of lateral velocity fluctuations (σ_y), friction velocity (u_*), Obukhov length (*L*), wind speed (*u*), and boundary layer height (*blh*) (see Supplementary Material, Section S2). σ_y, u_*, *L*, and *u* were measured from the EC system. There was no direct measurement of *blh*, so we adopted the *blh* data from ERA5 hourly data on singe-level products (https://cds.climate.copernicus.

eu/datasets/reanalysis-era5-single-levels, accessed on 17 October 2024) in this study. Then, an 80% footprint was regarded as the measurement range of the EC system.

2.1.2. UAV Data Collection

RGB and multispectral images were acquired using DJI Phantom 4 Multispectral (P4M) (SZ DJI Technology Co., Ltd., Shenzhen, China), equipped with one RGB sensor and five monochrome sensors to measure spectral response in the blue (450 nm ± 16 nm), green (560 nm ± 16 nm), red (650 nm ± 16 nm), red-edge (730 nm ± 16 nm) and nir (840 nm ± 16 nm) bands. Flight operations were conducted between 9:00 AM and 12:00 AM at an altitude of 120 m. The Pix4dMapper software v4.4.9 was employed to process the raw photos captured by UAV to generate the RGB orthomosaics and reflectance maps with a spatial resolution of 6.5 cm/pixel. The reflectance maps underwent correction using the images of the calibrated reflectance panel (MicaSense, RP06), which were captured during the UAV flight. Georeferencing was performed using ArcGIS software v10.8.2 to ensure pixel alignment in the UAV images. In total, 33 UAV observations were collected (Table 1). To build the daily $V_{cmax,0}^{25}$ model, the daily UAV images were generated using linear interpolation for each pixel. Then, daily UAV images were clipped using the external rectangle of the flux footprint climatology during the rice growth season and masked using daily flux footprint climatology (see Supplementary Material, Section S3).

Table 1. The date of the UAV flight.

2022	2023
June 6th, 2022	June 1st, 2023
July 20th, 2022	June 28th, 2023
July 27th, 2022	July 11th, 2023
August 3rd, 2022	July 19th, 2023
August 7th, 2022	July 24th, 2023
August 9th, 2022	July 27th, 2023
August 15th, 2022	August 11th, 2023
August 26th, 2022	August 16th, 2023
September 1st, 2022	August 24th, 2023
September 7th, 2022	August 31st, 2023
September 16th, 2022	September 6th, 2023
September 24th, 2022	September 9th, 2023
October 11th, 2022	September 17th, 2023
October 20th, 2022	September 26th, 2023
November 1st, 2022	October 1st, 2023
	October 6th, 2023
	October 11th, 2023
	October 16th, 2023

2.1.3. Canopy Parameters and Canopy Height Measurement

The plant area index (*PAI*) and clumping index (Ω) were measured using an LI-COR 2200 canopy analyzer. The leaf area index (*LAI*) was extracted from the *PAI* using an empirical method proposed by [47] (see Supplementary Material Section S4). Canopy height (h_c) was measured using a tape measure. *PAI*, Ω and h_c were all measured at fixed positions in the paddy field. There were 39 sampling positions in 2022 and 30 in 2023, respectively.

2.2. Leaf Biochemical Properties-Based Model

In this study, we employed the Boreal Ecosystem Productivity Simulator (BEPS) [48] for half-hourly GPP estimations. The BEPS model is based on leaf biochemical properties and follows a two-leaf scheme, which separated the canopy into sunlit leaves and shaded leaves using the following methods:

$$LAI_{sunlit} = 2\cos\theta\left(1 - e^{-0.5\Omega LAI/\cos\theta}\right) \quad (1)$$

$$LAI_{shaded} = LAI - LAI_{sunlit} \quad (2)$$

where θ is the solar zenith angle; Ω is the clumping index; LAI, LAI_{sunlit}, and LAI_{shaded} are the total leaf area index of the canopy, leaf area index of sunlit leaves, and leaf area index of shaded leaves, respectively.

The BEPS model calculates the instantaneous GPP as the sum of photosynthesis from the sunlit leaves and shaded leaves.

$$GPP = A_{sunlit}LAI_{sunlit} + A_{shaded}LAI_{shaded} \quad (3)$$

where A_{sunlit} and A_{shaded} are the instantaneous photosynthesis rates of the representative sunlit leaf and representative shaded leaf (μmol m^{-2} s^{-1}), respectively. The instantaneous photosynthesis rate is calculated using the Farquhar–von Caemmerer–Berry (FvCB) biochemical model [7].

$$A = \min(W_c, W_j) \quad (4)$$

where W_c is the RuBP carboxylation rate limited by Rubisco activity (μmol m^{-2} s^{-1}); W_j is the RuBP carboxylation rate limited by electron transport (μmol m^{-2} s^{-1}); R_d is the leaf dark respiration during the daytime (μmol m^{-2} s^{-1}).

$$W_c = \frac{V_{\text{cmax}}(C_i - \Gamma_*)}{C_i + K_c(1 + O/K_o)} \quad (5)$$

$$W_c = \frac{V_{\text{cmax}}(C_i - \Gamma_*)}{C_i + K_c(1 + O/K_o)} \quad (6)$$

$$R_d = 0.015V_{\text{cmax}} \quad (7)$$

where C_i is the CO_2 concentration in the intercellular air space (μmol mol^{-1}); O is the O_2 concentration in the intercellular air space and is set as 210 (mmol mol^{-1}); K_c is the Michaelis–Menten constant of Rubisco for CO_2 (μmol mol^{-1}); K_o is the Michaelis–Menten constant of Rubisco for O_2 (mmol mol^{-1}); Γ^* is the CO_2 compensation point in the absence of leaf dark respiration (μmol mol^{-1}); J is the electron transport rate (μmol e$^-$ m^{-2} s^{-1}), and is a function of the incoming photon flux density ($PPFD$, μmol photon m^{-2} s^{-1}) and J_{max}:

$$J = \frac{J_{\text{max}}PPFD}{PPFD + 2.1J_{\text{max}}} \quad (8)$$

The temperature dependence of K_o, K_c, Γ^* is described using the Arrhenius function [49]:

$$f(T_l) = k_{25}e^{\frac{(T_l-25)E_a}{298R(T_l+273)}} \quad (9)$$

The temperature dependence of V_{cmax} and J_{max} is described using the modified Arrhenius function [49]:

$$f(T_l) = k_{25}e^{\frac{(T_l-25)E_a}{298R(T_l+273)}}\frac{1 + e^{\frac{298\Delta S - H_d}{298R}}}{1 + e^{\frac{(T_l+273)\Delta S - H_d}{R(T_l+273)}}} \quad (10)$$

where k_{25} is the parameter value at 25 °C; T_l is the leaf temperature (°C); R is the universal gas constant (8.314 J K^{-1} mol^{-1}); E_a is the activation energy (J mol^{-1}); H_{d} is the deactivation energy (J mol^{-1}); ΔS is the entropy term (J K^{-1} mol^{-1}).

The classical BEPS model assumes that a linear relationship exists between the maximum carboxylation rate and leaf nitrogen content, and leaf nitrogen decreases exponentially from top to bottom in a canopy [48]. In this study, we assumed that the maximum carboxylation rate decreases exponentially from top to bottom in the canopy and eliminated the two parameters, namely nitrogen content at the top of the canopy and the relative change in V_{cmax}^{25} with leaf nitrogen content.

$$V_{\text{cmax}}^{25}(LAI) = V_{\text{cmax},0}^{25} e^{-k_n LAI} \tag{11}$$

Therefore, the V_{cmax}^{25} of representative sunlit and shaded leaves are calculated as follows (see the Supplementary Material Section S5):

$$V_{\text{cmax},sunlit}^{25} = V_{\text{cmax},0}^{25} \frac{k\left[1 - e^{-(k_n+k)LAI}\right]}{(k_n + k)\left(1 - e^{-kLAI}\right)} \tag{12}$$

$$V_{\text{cmax},shaded}^{25} = V_{\text{max},0}^{25} \frac{\frac{1-e^{-k_n LAI}}{k_n} - \frac{1}{k+k_n}\left[1 - e^{-(k_n+k)LAI}\right]}{LAI - \frac{1}{k}\left(1 - e^{-kLAI}\right)} \tag{13}$$

where $V_{cmax,0}^{25}$ is the V_{cmax}^{25} at the top of the canopy (μmol m^{-2} s^{-1}); k_n is the leaf nitrogen content decay rate with increasing depth into the canopy and is taken as 0.3; $k = G(\theta)\Omega/cos\theta$, and the projection coefficient $G(\theta)$ is taken as 0.5 for the assumption of spherical leaf angle distribution. The maximum electron transport rate for the sunlit and shaded leaves ($J_{\text{max, sunlit}}$, $J_{max,shaded}$, μmol m^{-2} s^{-1}) is obtained using the following equation [50]:

$$J_{\text{max},sunlit}^{25} = 2.39 V_{\text{cmax},sunlit}^{25} - 14.2 \tag{14}$$

$$J_{\text{max},shaded}^{25} = 2.39 V_{\text{cmax},shaded}^{25} - 14.2 \tag{15}$$

The leaf stomatal conductance for CO_2 (g_{sc}) is coupled with photosynthesis using the Ball–Woodrow–Berry equation [51]:

$$g_{sc} = m\frac{Ah_s}{C_s} + b \tag{16}$$

where h_s is the relative humidity at the leaf surface (%); C_s is the CO_2 concentration at the leaf surface (μmol mol^{-1}); m and b are the coefficients in the Ball–Woodrow–Berry equation, respectively. m and b are set as 6.08 and 0.019, respectively. These two values were obtained by fitting steady-state-condition A-C_i curves from Li-6800. More detailed information on the BEPS model can be found in the Supplementary Material, Section S6. The BEPS model also considers the influence of soil water stress on g_s [48]. Given that rice cultivation typically involves a flooded or saturated state, with minimal exposure to soil water stress, the impact of the soil water stress factor on g_s was not considered in this study.

2.3. Derivation of Field-Scale $V_{cmax,0}^{25}$ Measurement

This study employed a classical optimization algorithm, the differential evolution (DE) algorithm [52], to retrieve the daily $V_{cmax,0}^{25}$. The differential evolution algorithm contains three steps, including mutation, crossover, and selection. Detailed information about DE algorithm were displayed in the Supplementary Material Section S7. The differential evolution algorithm was implemented using the lmfit python package [53], with the mean square error (MSE) chosen as the objective function. Integrating the DE algorithm with BEPS, called BEPS-DE framework, was employed to derive the field-scale $V_{cmax,0}^{25}$ under the assumption of constant $V_{cmax,0}^{25}$ for 7 days. Only daytime GPP observation (when incoming shortwave radiation is larger than 0) were used to fit $V_{cmax,0}^{25}$ because the BEPS model sets nigh-time GPP as 0 while night-time GPP observation from the EC system are not 0.

2.4. Texture Extraction and $V^{25}_{cmax,0}$ Models

In this study, several types of texture features were evaluated for field-scale $V^{25}_{cmax,0}$ estimation.

2.4.1. Standard Deviation Texture

The first texture feature is the standard deviation of reflectance within the flux footprint climatology.

$$\sigma_{ref} = \sqrt{\frac{1}{N}\sum_{i=1}^{N}(x_i - \overline{x})^2} \tag{17}$$

where x_i is the reflectance value at each pixel within the flux footprint climatology; $\overline{x}$ is the mean reflectance value within the footprint climatology.

2.4.2. GLCM-Based Texture Features

GLCM is a square matrix that can examine the spatial distribution of the gray levels in the image. It describes how frequently a combination of pixels is present in an image in a given direction and distance. For a given image I with a size of $K \times K$ and maximum gray value of G, the element of a GLCM for a displacement vector $d(dx, dy)$ is defined as follows:

$$P(m,n) = \sum_{i=1}^{K}\sum_{j=1}^{K}\left\{ \begin{array}{l} 1, \text{ if } I(i,j) = m \text{ and } I(i+d_x, j+d_y) = n \\ 0, \text{ otherwise} \end{array} \right\} \tag{18}$$

$$P(m,n) = \frac{P(n,m)}{\sum_{i=0}^{G}\sum_{j=0}^{G} P(i,j)} \tag{19}$$

where $dx = dcos\theta$; $dy = dsin\theta$; d and θ are the predefined distance and direction, respectively. The GLCM is typically interpreted using texture indicators. Six GLCM-based texture features (Table 2) were extracted from UAV images.

Table 2. The list of GLCM-based texture features in this study.

Texture	Formula
Contrast (*con*)	$con = \sum_{i,j=0}^{M-1} P_{i,j}(i-j)^2$
Correlation (*cor*)	$cor = \sum_{i,j=0}^{G-1}\left[\frac{(i-\mu_i)(j-\mu_j)}{\sigma_i^2\sigma_j^2}\right]$
Dissimilarity (*dis*)	$dis = \sum_{i,j=0}^{G} P_{ij}\lvert i-j\rvert$
Homogeneity (*hom*)	$hom = \sum_{i,j=0}^{G}\frac{P_{i,j}}{1+(i-j)^2}$
Angular second moment (*asm*)	$asm = \sum_{i,j=0}^{G} P_{i,j}^2$
Entropy (*ent*)	$ent = \sum_{i,j=0}^{G} -\ln(P_{ij})P_{ij}$

The computation of GLCM is influenced by the pixel pair distance offsets and pixel pair angles. We pre-specified a series of distance offsets (1,2,4,8,16) and angles (0, $\pi/4$, $\pi/2$, $3\pi/4$). Then, 20 GLCMs, each representing a unique combination of distance offsets and angles, were employed to derive texture samples. Finally, the mean of 20 texture samples was set as the texture feature of the image.

2.4.3. LBP Texture

The LBP represents each image pixel with a local binary pattern operator that generates a binary code while comparing a neighboring pixel and its center patch gray unit. The LBP codes are computed as follows:

$$LBP_{P,R}(q_c) = \sum_{p=0}^{p-1} s(q_p - q_c)2^p \tag{20}$$

$$s(x) = \begin{cases} 1, \ x \geq 0 \\ 0, \ othersise \end{cases} \tag{21}$$

where q_c is the gray value of the center pixel; q_p is the gray value of P neighboring pixels with a specified radius R. The LBP is invariant against any monotonic transformation of image brightness. The local binary pattern histogram (LBPH) is a commonly used LBP-type texture feature. It employs the distribution of the LBP value to represent the image. A whole image is typically divided into several local regions, and the histograms from each subregion are extracted independently. Then, these subregional histograms are concatenated to build a global description. In this study, we divided the UAV images into 9 subregions (3 × 3), calculated the histogram with 10 bins, and then concatenated the 9 histograms together to obtain the LBPH texture with 90 features. Figure S3 shows the process of extracting the LBPH texture.

2.4.4. Convolutional Neural Network

In this study, we adopted two typically convolutional neural network architectures, including AlexNet [38] and ResNet [54], to extract spatial information from UAV images. ResNet is more complex and has a deeper structure than AlexNet (Figure S4). It is important to note that the input images should maintain the same size when fed into the CNN. Since images with a resolution of 6.5 cm/pixel are too large for the CNN to handle efficiently, UAV images were resampled to a 1m resolution and clipped by the external rectangle (see Section 2.1.2) to maintain the same size. Pixels outside the flux footprint climatology were set as 0. The five-band multispectral images were concatenated into an input tensor (430 × 420 × 5). The input images were preprocessed using the standardization and then fed into the CNN to estimate field-scale $V_{cmax,0}^{25}$.

The different combinations of texture features for $V_{cmax,0}^{25}$ were evaluated (Table 3). For the mean and standard deviation of the 5-band reflectance, GLCM-based texture features, and LBPH-driven models, the deep neural network (DNN) was employed to estimate field-scale $V_{cmax,0}^{25}$.

$$\begin{aligned} \boldsymbol{h}_1 &= f_0(\boldsymbol{w}_0\boldsymbol{x} + \boldsymbol{b}_0) \\ \boldsymbol{h}_l &= f_l(\boldsymbol{w}_{l-1}\boldsymbol{h}_{l-1} + \boldsymbol{b}_{l-1}) \\ \boldsymbol{y} &= \boldsymbol{w}_L h_L + \boldsymbol{b}_L \end{aligned} \tag{22}$$

where x is the input layer, namely the combination of the 5-band reflectance and corresponding texture features; h_l is the l-th hidden layer; w_l and b_l are the weights and bias of the l-th layer; f_l is the activation function of the l-th layer. The performance of the deep neural network highly relies on hyperparameters, including the number of hidden layers, the number of neurons, the selection of activation function, and the dropout value. We used the genetic programming algorithm to optimize the hyperparameters. The search space was set as follows: (1) the maximum number of hidden layer was set as 4; (2) the number of neurons for each hidden layer was selected from the set [16,32,64,128,256,512,1024]; (3) the activation function for each hidden layer was selected as either ReLU or sigmoid; (4) the dropout rate for each hidden layer was selected from an interval of 0 to 0.5. The daily $V_{cmax,0}^{25}$ and corresponding multispectral surface reflectance for each land cover type were randomly apportioned into training and validation sets, with an 80–20 split. The former was used to train the weights and biases, and the latter was employed to evaluate the generalization error.

Table 3. The list of $V_{cmax,0}^{25}$ models proposed in this study.

	$V_{cmax,0}^{25}$ Model	Description
M1	$V_{\text{cmax},0}^{25} = \text{DNN}\left(\mu_{ref}\right)$	μ_{ref}: mean value of reflectance within the flux footprint climatology. M1 is driven by the mean value of the 5-band reflectance.
M2	$V_{\text{cmax},0}^{25} = \text{DNN}\left(\mu_{ref}, \sigma_{ref}\right)$	σ_{ref}: standard deviation value of reflectance within the flux footprint climatology. M2 is jointly driven by the mean and standard deviation of the 5-band reflectance.
M3	$V_{c,\text{max}0}^{25} = \text{DNN}(GLCM)$	GLCM: GLCM-based texture features. M3 is driven by the GLCM-based texture features from the RGB images.
M4	$V_{c,\text{max}0}^{25} = \text{DNN}\left(\mu_{ref}, GLCM\right)$	M4 is jointly driven by the mean value of the 5-band reflectance and GLCM-based texture features from the RGB images.
M5	$V_{\text{cmax},0}^{25} = \text{DNN}(LBPH)$	LBPH: local binary pattern histogram. M5 is driven by the local binary pattern histogram feature from the RGB images.
M6	$V_{\text{cmax},0}^{25} = \text{DNN}\left(\mu_{ref}, LBPH\right)$	M6 is jointly driven by the mean value of the 5-band reflectance and the local binary pattern histogram feature from the RGB images.
M7	$V_{\text{cmax},0}^{25} = \text{AlexNet}\left(Image_{ref}\right)$	$Image_{ref}$: standardized UAV reflectance maps. M7 and M8 are driven by the 5-band reflectance maps.
M8	$V_{\text{cmax},0}^{25} = \text{ResNet}\left(Image_{ref}\right)$	

2.5. Feature Contribution Analysis Using the Shaley Value

The Shaley value (SHAP) was adopted to interpret $V_{cmax,0}^{25}$ models. In this study. SHAP [55] is a game-theoretic approach to explain the performance of the machine learning model. It can quantify the contribution of the input feature to the prediction for an individual sample. SHAP specifies the explanation using the additive feature attribution method:

$$f(x) = g(x') = \phi_0 + \sum_{i=1}^{M} \phi_i x' \tag{23}$$

where $f(x)$ is the original model, $g(x)$ is the explanation model with simplified input x'; $x' \in \{0,1\}^M$; x and x' are related via a mapping function, $x = h_x(x')$; M represents the number of input features; ϕ_i is the feature attribution for the feature i. The explanation model g has a unique solution:

$$\phi_i(f, x) = \sum_{z' \subseteq x'} \frac{|z'|!(M - |z'| - 1)!}{M!} \left[f_x(z') - f_x(z' \backslash i)\right] \tag{24}$$

where $|z'|$ represents the number of non-zero entries in z', and $z' \subseteq x'; f(x') = f(h_x(z')) = E[f(z) | z_S]$, and S is the set of non-zero indexes in z'.

3. Results

3.1. Seasonal Variation in Carbon Fluxes, Environmental Elements, and Biophysical Indicators on the Paddy Field

Figure 3a illustrates the seasonal variation in daily *NEE*, *GPP*, and R_{eco} during the rice growth period. The *NEE* displayed a one-valley pattern. The minimum value of daily *NEE* was −12.08 and −14.60 gC m^{-2} d^{-1} in 2022 and 2023, respectively. The two minimum values occurred on August 25th, 2022, and July 29th, 2023. The cumulative *NEE* in the growth period was 502 and 552 gC m^{-2}. The *GPP* exhibited a one-peak mode, with peak daily values of 21.6 and 20.16 gC m^{-2} d^{-1} in 2022 and 2023, respectively. The peak value occurred on August 21st, 2022, and July 29th, 2023. Total *GPP* during the growth period was 1429 gC m^{-2} in 2022 and 1337 gC m^{-2} in 2023. A strong fluctuation can be found in the dynamics of *NEE* and *GPP*, while daily R_{eco} changed smoothly.

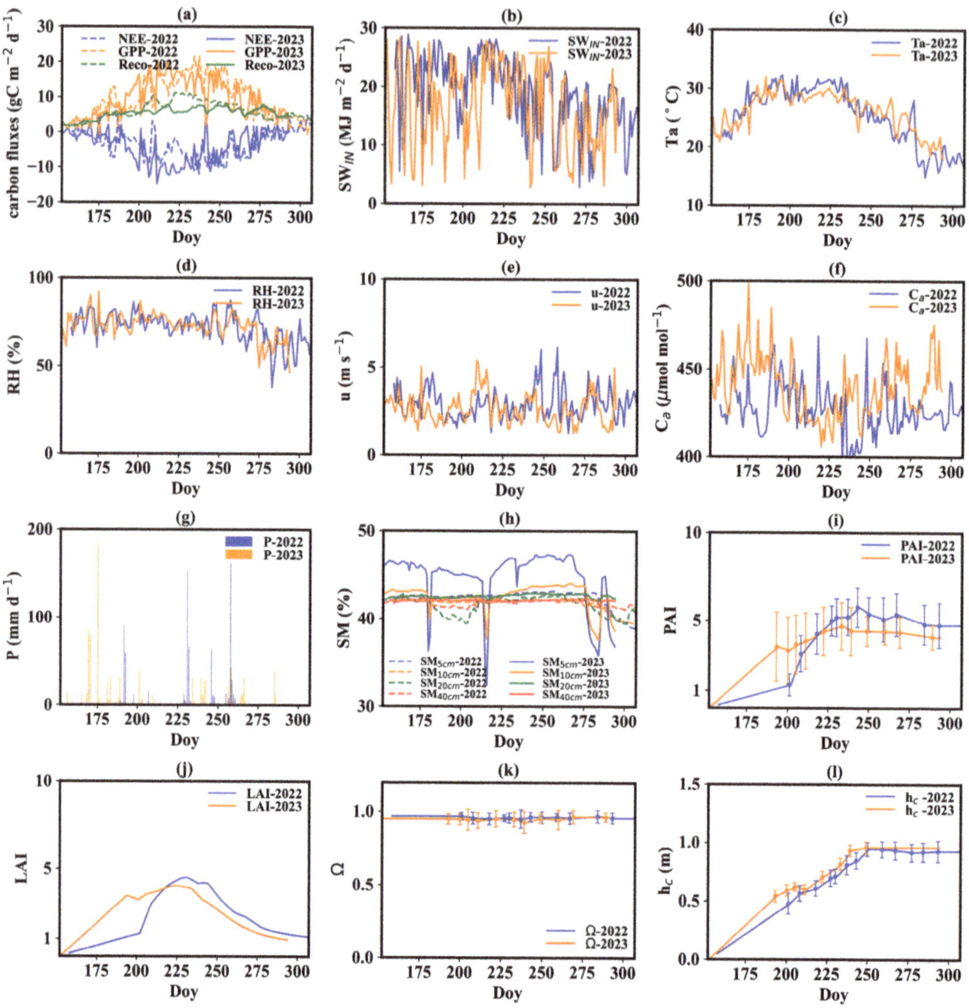

Figure 3. Seasonal variation in daily carbon fluxes, environmental elements and biophysical indicators. (**a**) Carbon fluxes, including net ecosystem exchange, gross primary production, and ecosystem respiration; (**b**) incoming shortwave radiation; (**c**) air temperature; (**d**) humidity; (**e**) wind speed; (**f**) ambient CO_2 concentration; (**g**) precipitation; (**h**) soil moisture; (**i**) plant area index; (**j**) leaf area index; (**k**) clumping index; (**l**) canopy height. The term 'Doy' means day of the year.

The summer of 2022 witnessed a severe meteorological drought along the Yangtze River. Accordingly, the incoming shortwave radiation (SW_{IN}) and air temperature (T_a) in 2022 were significantly higher than those in 2023 (Figure 3b,c). The daily mean SW_{IN} and T_a during the summer of 2022 were 21.35 MJ m^{-2} d^{-1} and 28.4 °C. The two values at the same time in 2023 were 18.00 MJ m^{-2} d^{-1} and 27.1 °C. The difference in T_a on the same days between 2022 and 2023 during the summer was as large as 6 °C. Moreover, 2022 experienced lower precipitation (P) and relative humidity (RH) compared to 2023, with mean RH values of 71% and 76, and accumulated P values of 759 and 922 mm during the growth period for 2022 and 2023, respectively (Figure 3d,g). Precipitation was more frequent in 2023, particularly during the summer. Differences in wind speed (u) and atmospheric CO_2 concentration (C_a) between the two years were minimal (Figure 3e,f), with wind speeds of 2.93 m s^{-1} in 2022 and 2.62 m s^{-1} in 2023, and CO_2 concentrations of 426 μmol mol^{-1} in 2022 and 439 μmol mol^{-1} in 2023. Due to the presence of a persistent water layer in the paddy field and a shallow groundwater table (about 0.5 m), only surface soil moisture (*SM*) went through several drying processes, while deep *SM* remained stable (Figure 3h). Therefore, rice rarely suffered soil water stress during the growth period.

When it comes to biophysical indicators, both *PAI* and h_c reached the plateau at the end of August (at the early heading and flowering stage) (Figure 3i,l). *LAI* also reached its peak at the end of August with a peak value of 4.52 and 4.03 in 2022 and 2023, respectively (Figure 3j). Then, *LAI* gradually declined due to leaf senescence. The *PAI* in 2022 was higher than that in 2023. The former reached 5.84 and the latter 4.44. The gap between 2022 and 2023 h_c values was small. The maximum h_c was about 0.95 m. The dynamics of Ω were stable during the growth period, with values varying from 0.93 to 0.96 (Figure 3k).

3.2. Seasonal Variation of $V_{cmax,0}^{25}$ Observation

Figure 4 presents the seasonal dynamics of field-scale $V_{cmax,0}^{25}$. The field-scale $V_{cmax,0}^{25}$ generally presented a single-peak pattern during the growth period, reaching its maximum in mid-July and declining after mid-August. The peak field-scale $V_{cmax,0}^{25}$ was approximately 170 μmol m^{-2} s^{-1}. There were two significant drops in field-scale $V_{cmax,0}^{25}$ in early July and in the end of July in 2022, likely due to weed removal activities. In 2022, weed prevention before rice sowing was ineffective, leading to the thriving of weeds. Consequently, a substantial portion of the carbon flux in the paddy field came from the contribution of weeds.

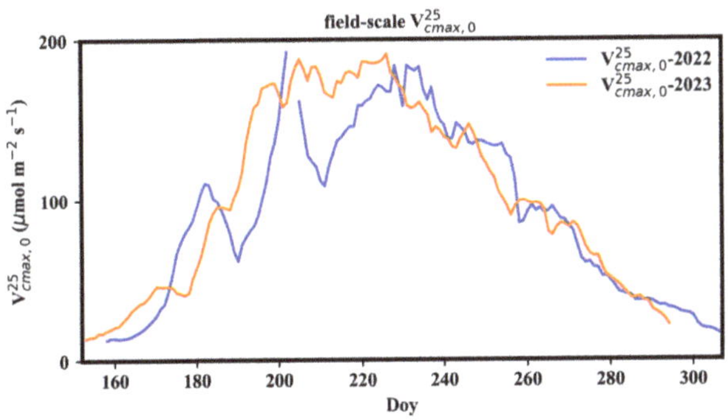

Figure 4. Seasonal variation in field-scale $V_{cmax,0}^{25}$.

3.3. Accuracy of $V_{cmax,0}^{25}$ and GPP Estimation

Figure 5 shows the performance of eight $V_{cmax,0}^{25}$ models. Three metrics, including correlation (R), root mean square error (RMSE), and medium absolute percentage error (MdAPE), were employed to assess the model performance. When employing traditional reflectance values as input (Figure 5a), the R, RMSE, and MdAPE of $V_{cmax,0}^{25}$ estimation on the test dataset were 0.87, 30.08 μmol m^{-2} s^{-1} and 23%, respectively. In contrast, the $V_{cmax,0}^{25}$ model driven solely by texture information displayed a noticeable performance degradation. The GLCM texture-driven model achieved $V_{cmax,0}^{25}$ estimation with an R of 0.80, RMSE of 36.50 μmol m^{-2} s^{-1} and MdAPE of 27% on the test dataset (Figure 5c). The LBPH-driven model exhibited the poorest accuracy, with R, RMSE, and MdAPE values of 0.81, 39.31, and 31% on the test dataset, respectively (Figure 5e). However, it can be seen that combining the reflectance value and texture information can significantly improve the accuracy of $V_{cmax,0}^{25}$ estimation, while the contribution of differing texture information varies significantly. Two joint-driven models, including μ_{ref} -GLCM texture feature and μ_{ref}-LBPH texture feature models, substantially improved the $V_{cmax,0}^{25}$ estimations, with R values reaching 0.93–0.94, and RMSE and MdAPE falling from 20.08 to 23.40 μmol m^{-2}

s^{-1} and 12% to 16%, respectively. However, the combination of μ_{ref} and σ_{ref} provided only a marginal improvement, with an R of 0.89, RMSE of 28.70 μmol m^{-2} s^{-1} and MdAPE of 22% (Figure 5b). Two CNN methods demonstrated excellent performance. ResNet is typically superior to AlexNet in the ImageNet classification challenge, whereas AlexNet slightly outperformed ResNet in this study. The former acquired the $V_{cmax,0}^{25}$ estimation with an R of 0.94, RMSE of 23.70 μmol m^{-2} s^{-1}, and MdAPE of 11%. The latter obtained the $V_{cmax,0}^{25}$ estimation with an R of 0.93, RMSE of 27.30 μmol m^{-2} s^{-1}, and MdAPE of 14%. A noteworthy observation is that the CNN methods showed an overwhelming advantage over other models on the training dataset, with R reaching 1, and RMSE and MdAPE being lower than 10 μmol m^{-2} s^{-1} and 10%. Although this result indicates that the CNN methods encountered a degree of overfitting, it still suggests a strong ability to discern and utilize spatial information embedded in UAV images. The ResNet, with a more complicated structure, displays significantly higher potential than the AlexNet. The overfitting issue of CNN methods can be alleviated by training with more data. As the dataset expands, there is a promising outlook for CNN methods to transcend these challenges and realize significant success.

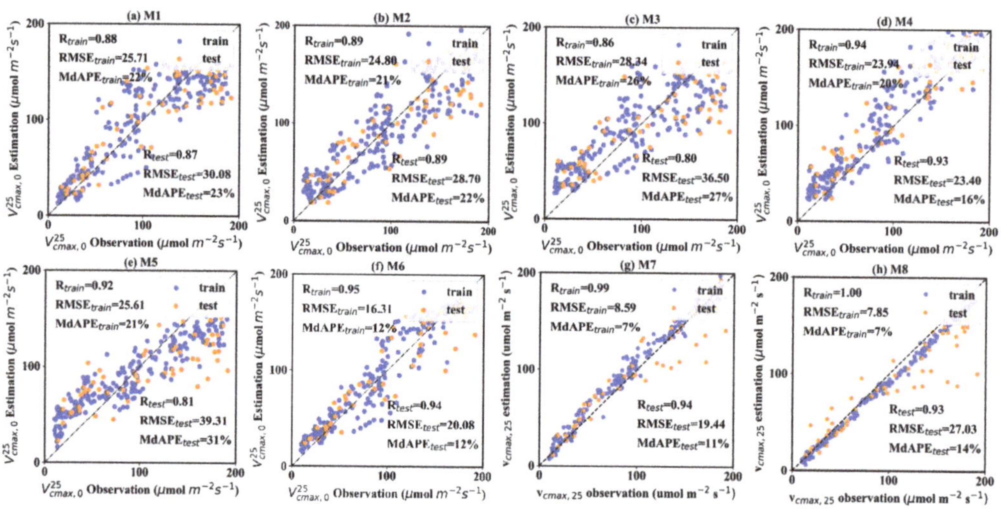

Figure 5. The accuracy of 8 field-scale $V_{cmax,0}^{25}$ models. The blue points represent the training samples, and the orange points represent the validation samples. The sample numbers of the training and validation dataset are 224 and 57, respectively.

Considering the texture information also contributes to the GPP estimation, it can be found that the superior $V_{cmax,0}^{25}$ models result in more accurate GPP estimation (Figure 6). Two CNN models acquired the best GPP estimation, with an R of 0.92, RMSE of 6.5–6.6 μmol m^{-2} s^{-1}, and MdAPE of 23–24%. The u_{ref} -GLCM texture feature and the u_{ref} -LBPH texture feature joint-driven models also achieved excellent GPP estimation. The former obtained the lowest RMSE, with a value of 6.43 μmol m^{-2} s^{-1}. The latter also obtained promising results, with an R of 0.92, RMSE of 6.74 μmol m^{-2} s^{-1}, and MdAPE of 26%. The μ_{ref}-σ_{ref} joint-driven model and LBPH solely driven model performed worst for GPP estimation. The RMSE of the two models was larger than 7 μmol m^{-2}, and the MdAPE was over 30%. The μ_{ref}-GLCM-based texture feature joint-driven model achieved better performance in $V_{cmax,0}^{25}$ estimation in the high-$V_{cmax,0}^{25}$ range (Figure 5d). In contrast, the two CNN models showed an obvious underestimation issue when $V_{cmax,0}^{25}$ was high (Figure 5g–h), and the underestimation of $V_{cmax,0}^{25}$ further resulted in a significant bias for high GPP estimation.

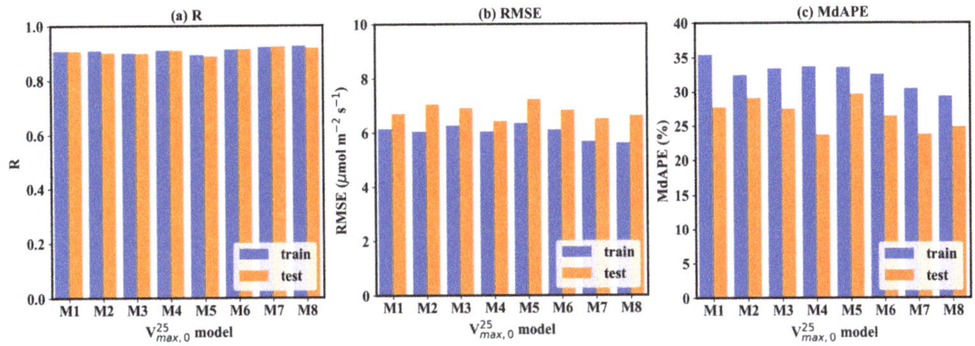

Figure 6. The accuracy of *GPP* estimation based on the 8 field-scale $V^{25}_{cmax,0}$ models. The blue bar represents the accuracy metrics on the training dataset, and the orange bar represents the accuracy metrics on the testing dataset.

4. Discussion

4.1. How Spatial Information Contributes to Field-Scale $V^{25}_{cmax,0}$ Estimation

This study assessed different forms of texture information for field-scale $V^{25}_{cmax,0}$ estimation and revealed significant differences in the ability of different texture features to describe field-scale $V^{25}_{cmax,0}$. We further elucidated the contribution of differing texture information using the SHAP method. Figure 7 shows the mean of the absolute SHAP (mean(|SHAP|)) for the six deep neural network-based models. Due to the high dimensionality of input data, only the top five most important features were presented. In the μ_{ref}-driven $V^{25}_{cmax,0}$ model (Figure 7a), the nir and red bands contributed most. This result is consistent with the current understanding of the relationship between $V^{25}_{cmax,0}$ and the spectrum. These two bands indicate the vegetation growth information and are the components of some classical vegetation indices, such as NDVI and EVI. In the GLCM-based texture-feature-driven-$V^{25}_{cmax,0}$ model (Figure 7c), the top 5 most important features were entropy in the green and red bands, dissimilarity in the blue band, contrast in the RGB band, and homogeneity in the green band. The five features exhibited nearly equal importance. In the LBPH-driven $V^{25}_{cmax,0}$ model (Figure 7e), the top five most important features were the histogram bars from subregions 1, 5, 1, 2, 3. In the μ_{ref}-σ_{ref} joint-driven model (Figure 7b), the top 5 most important features included mean value in red, green, and red-edge bands, and the standard deviation value in the green and nir bands. Here, the mean value information played a dominant role, and the standard deviation information played a subordinate one. Adding the standard deviation information diminished the importance of the nir band, which contributed most in the μ_{ref} -driven model. In the μ_{ref}-GLCM-based texture feature joint-driven model, the GLCM-based texture features contributed more than the mean reflectance. There were four GLCM textures in the top five most important features, including entropy in the blue, green and red-edge bands, and homogeneity. Conversely, the mean reflectance played a dominant role in the μ_{ref}-LBPH joint-driven model. The mean values of reflectance in the red, nir, green and blue bands were the top four most important features in this model. The most important LBPH feature was the histogram bar in subregion 8. The SHAP-based feature importance analysis suggests that entropy is a valuable feature for $V^{25}_{cmax,0}$ estimation, with dissimilarity and homogeneity also showing potential. The entropy, dissimilarity, and homogeneity describe the uncertainty of, local variation in, and smoothness of the image texture, respectively [56]. There perhaps exists some inherent link between $V^{25}_{cmax,0}$ and these GLCM-based texture features. While standard deviation is a simple and intuitive indicator of spatial heterogeneity, it offered limited information for $V^{25}_{cmax,0}$ estimation. The LBPH also failed to indicate the dynamics of $V^{25}_{cmax,0}$. Although this method is regarded as a powerful tool for describing local information and has achieved success in classification issues such as face recognition and weed detection, the histogram-format texture descriptor may be unsuitable for regression issues. LBPH failed to provide stable features for quantifying $V^{25}_{cmax,0}$. Despite this, the combination creating the μ_{ref}-LBPH joint-driven model outperforms the other two μ_{ref}-texture joint-driven model. A possible reason is that LBPH is a high-dimensional feature and contains more potential information. In summary, textures are important but may be better used as Supplementary Information to estimate field-scale $V^{25}_{cmax,0}$.

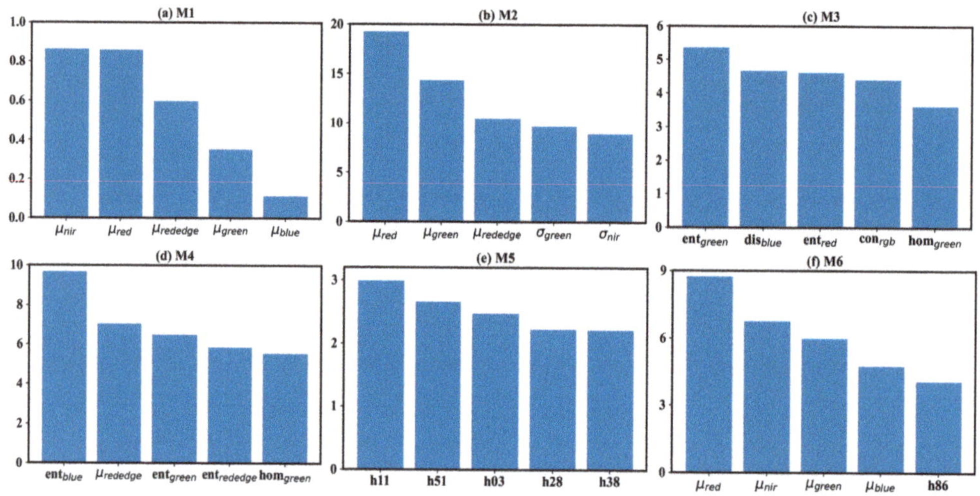

Figure 7. SHAP-based feature importance analysis for the DNN $V^{25}_{cmax,0}$ model. The feature importance is defined as the mean of the absolute value of the SHAP for each feature (mean(|SHAP|)). Only the top 5 most important features are displayed.

The SHAP method was also employed to interpret the AlexNet-based $V^{25}_{cmax,0}$. Figure 8 shows the spatial distribution of SHAP on 4 typical days, clearly revealing the spatial heterogeneity in the SHAP maps. The contributions from roadway and water bodies differed obviously from those from the paddy field. When the underlying surface was bare soil, the spatial variability in SHAPs across bands was relatively low. When the paddy field was covered by rice, the SHAP variability intensified, particularly in the nir and red-edge bands. Specifically, the regions with a high contribution in these bands were closely linked to areas with a mix of vegetation and bare soil or blends of well-growing and poor-growing vegetation. As the rice matured, the spatial variability in the SHAP in the nir band weakened, while its variability in the red band intensified. To sum up, the CNN model attached much importance to the subregion with high spatial heterogeneity in the nir and red-edge bands for $V^{25}_{cmax,0}$ estimation. This emphasis was particularly evident in regions characterized by a mix of vegetation and bare soil, and a mix of well-growing and poor-growing vegetation. The SHAP method was not used to analyze the ResNet-based model due to the complexity of ResNet, which made SHAP computation too resource-intensive. Therefore, the feature contribution of the ResNet-based model was not included in this study.

It should be noted that UAV images were resampled from 6.5 cm/pixel to 1 m/pixel before being fed into the CNN due to limitations to computation capacity. Resampling large images to a smaller, manageable resolution is the most common approach, and this was employed in this study. This approach retains the global spatial information but loses the fine-scale features. The loss of spatial detail will possibly affect the extraction of textural features. Alternative strategies, such as patch-based processing [57], and patch gradient descent [58], offer promising solutions to these challenges. Developing comprehensive methods to interpret ultra-high-resolution UAV images using deep learning is a key area for future exploration.

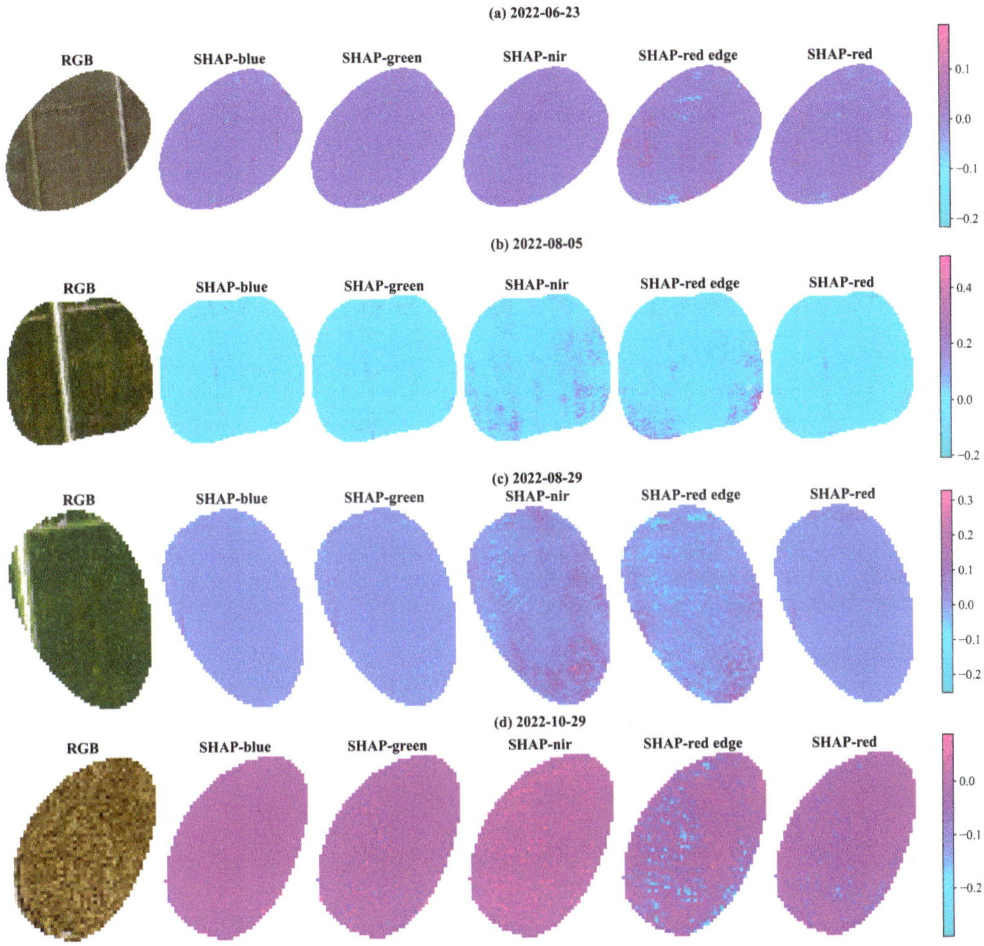

Figure 8. The spatial distribution of the AlexNet-model-based SHAP.

4.2. Spatial Heterogeneity: Implications for Parameters Conversion Across Different Scales

Our study underscores the significant influence of spatial heterogeneity field-scale $V_{cmax,0}^{25}$. The spatial heterogeneity originates from the diverse elements in paddy fields, including the bare soil, water body, and weeds, as well as rice in different growth states, all of which have distinct photosynthetic properties. Field-scale $V_{cmax,0}^{25}$, representing a macroscopic feature within the flux footprint climatology, is intricately linked to this spatial heterogeneity. In contrast, leaf-scale $V_{cmax,0}^{25}$ solely reflects the local feature. This inherent spatial heterogeneity contributes to the observed mismatch between field-scale and leaf-scale $V_{cmax,0}^{25}$, as depicted in Figure S6 (see Supplementary Material Section S11). Given the global prevalence of spatial heterogeneity, as evidenced in various studies [59], relying solely on the leaf-scale $V_{cmax,0}^{25}$ model for field-scale simulation introduces uncertainties into *GPP* estimation. Our study advocates for the direct use of field-scale parameters to model *GPP*, acknowledging the complexities introduced by spatial heterogeneity.

4.3. Limitations

In this study, we proposed high-resolution UAV-image-based models for field-scale $V_{cmax,0}^{25}$ and GPP estimation and demonstrated enhancement through texture information. However, some limitations remain. Firstly, the field-scale $V_{cmax,0}^{25}$ observation was retrieved from the BEPS-DE framework. The accuracy of $V_{cmax,0}^{25}$ observation was completely determined by the EC-based GPP observation and the structure of the BEPS model. We note that there are several different methods for conducting $V_{cmax,0}^{25}$ observation, including the data assimilation approach, sunlit GPP-based inverting method [60], and optimality photosynthesis hypothesis-based approach [15]. In addition, the BEPS model is highly parameterized, with many parameters predefined. The uncertainties in these predefined parameters will ultimately influence the retrieval of $V_{cmax,0}^{25}$. For example, Jiang et al. [61] revealed that the variation in the nitrogen distribution coefficient within the canopy (k_n) resulted in a large gap between retrieved $V_{cmax,0}^{25}$ values. Therefore, it is necessary to comprehensively evaluate the different $V_{cmax,0}^{25}$ retrieval methods. This study evaluates four typical spatial information extraction methods, including σ_{ref}, GLCM texture, LBPH texture, and CNN. Some newly proposed texture features, such as learning-based approaches and the entropy-based approaches, have garnered significant attention [28]. It is necessary to explore the capabilities of these new features for GPP estimation. Lastly, this study faced some challenges in sample size. The overwhelming advantage of CNN models in the training dataset demonstrates its strong ability to extract valuable spatial information. The performance of CNN algorithms primarily depends on big training data. The CNN-based $V_{cmax,0}^{25}$ models were trained and validated only using 281 daily RBG and five-band reflectance maps. Compared with the complex structure of the CNN, the sample size in this study is relatively insufficient, resulting in a degree of overfitting. It is urgent to expand our dataset to fully harness the potential of deep learning.

5. Conclusions

Accurate estimation of *GPP* estimation poses a considerable challenge due to the influence of spatial heterogeneity. High-resolution UAV images can capture spatial heterogeneity and offer new insights for *GPP* monitoring. In this study, we proposed high-resolution UAV-imagery-driven models to quantify $V_{cmax,0}^{25}$, a key parameter of the leaf biochemical properties-based *GPP* model. The reflectance and texture information were extracted from UAV images and employed to build $V_{cmax,0}^{25}$ models. A significant improvement in $V_{cmax,0}^{25}$ and *GPP* estimation was obtained. The main conclusions are drawn as follows:

(1) Combining reflectance and texture features can vastly improve the field-scale $V_{cmax,0}^{25}$ (R = 0.94, RMSE = 19.44 μmol m^{-2} s^{-1}, and MdAPE = 11%) and further result in high-accuracy GPP estimation (R = 0.92, RMSE = 6.5 μmol m^{-2} s^{-1}, and MdAPE = 23%). The performance of different texture features varies significantly. The CNN method shows the strongest ability to monitor $V_{cmax,0}^{25}$. The μ_{ref}-GLCM texture features and μ_{ref}-LBPH joint-driven models also acquire promising results. However, the standard deviation of reflectance contributes less to estimating $V_{cmax,0}^{25}$.

(2) The contribution of input features changes significantly among different models. The CNN model focuses on nir and red-edge bands and pays much attention to the subregion with high spatial heterogeneity. The μ_{ref}-LBPH joint-driven model mainly attaches importance to the reflectance information. On the other hand, the μ_{ref}-GLCM-based features joint-driven model emphasizes the role of GLCM texture indices, including contrast, dissimilarity and homogeneity.

(3) The strong spatial heterogeneity results in the large gap difference between field-scale and leaf-scale $V_{cmax,0}^{25}$ values and hinders the conversion of parameters across different scales

Our study confirms the necessity of considering spatial heterogeneity information for monitoring $V_{cmax,0}^{25}$ and GPP, which is ignored in the current understanding of the GPP model. The following perspectives should be the focus of future work. First, the fusion of satellite and UAV data offers great potential. Medium-resolution satellites, such as Landsat and Sentinel-2, can capture remote sensing images with spatial resolutions of 10–60 m, providing valuable spatial information about the underlying surface. Additionally, integrating multi-scale remote sensing images can achieve information complementarity, which is likely to enhance the accuracy of GPP monitoring. Second, further exploration of new interpretation methods is needed to better elucidate the mechanistic link between spatial heterogeneity and field-scale parameters.

Supplementary Materials: The following supporting information can be downloaded at https://www.mdpi.com/article/10.3390/rs16203906/s1, Table S1: The seasonal u^* threshold for filtering NEE; Table S2: The fixed parameter values in the BESP model; Table S3: The date of leaf gas-exchange experiment; Figure S1: Dynamics of daily RGB and Reflectance maps within the flux footprint climatology; Figure S2: The coefficient of stomatal conductance model; Figure S3: The process of generating LBPH; Figure S4: The architectures of AlexNet and ResNet. The blue block indicates the convolution layer; Figure S5: Dynamics of spatial distribution of SHAP value based on the AlexNet model; Figure S6. Seasonal variation of filed-scale and leaf-scale $V_{cmax,0}^{25}$. References [32,51,62] are cited in the supplementary materials.

Author Contributions: Conceptualization, X.H. and L.S.; methodology, X.H., X.D., J.B. and Z.Z.; software, X.H.; validation, L.L., L.S. and J.B.; formal analysis, X.H.; investigation, X.H.; resources, S.L.; data curation, X.H., X.D., J.L., C.S., S.D., T.W. and Y.W.; writing—original draft preparation, X.H.; writing—review and editing, S.L., L.L. and S.L.; visualization, X.H.; supervision, L.S.; project administration, L.S.; funding acquisition, X.H. and L.L. All authors have read and agreed to the published version of the manuscript.

Funding: This work was funded by the National Natural Science Foundation of China (Grants No. 52309058 and No. U2243235), and the China Postdoctoral Science Foundation 2023M732702.

Data Availability Statement: The data are available from the corresponding author on reasonable request. The data are not publicly available because there are still graduate students using it for research.

Acknowledgments: The authors gratefully acknowledge Shanghai Hengsha Agricultural and Forestry Industrial Development Co., Ltd. for their support in this experiment.

Conflicts of Interest: The authors declare no conflicts of interest.

References

1. He, B.; Chen, C.; Lin, S.; Yuan, W.; Chen, H.W.; Chen, D.; Zhang, Y.; Guo, L.; Zhao, X.; Liu, X.; et al. Worldwide impacts of atmospheric vapor pressure deficit on the interannual variability of terrestrial carbon sink. *Natl. Sci. Rev.* **2022**, *9*, nwab150. [CrossRef] [PubMed]
2. Marshall, M.; Tu, K.; Brown, J. Optimizing a remote sensing production efficiency model for macro-scale GPP and yield estimation in agroecosystems. *Remote Sens. Environ.* **2018**, *217*, 258–271. [CrossRef]
3. FAO. *World Food and Agriculture-Statistical Yearbook 2022*; FAO: Rome, Italy, 2022.
4. Hwang, Y.; Ryu, Y.; Huang, Y.; Kim, J.; Iwata, H.; Kang, M. Comprehensive assessments of carbon dynamics in an intermittently-irrigated rice paddy. *Agric. For. Meteorol.* **2020**, *285–286*, 107933. [CrossRef]
5. Joiner, J.; Yoshida, Y.; Zhang, Y.; Duveiller, G.; Jung, M.; Lyapustin, A.; Wang, Y.; Tucker, C.J. Estimation of terrestrial global gross primary production (GPP) with satellite data-driven models and eddy covariance flux data. *Remote Sens.* **2018**, *10*, 1346. [CrossRef]
6. Jung, M.; Schwalm, C.; Migliavacca, M.; Walther, S.; Camps-Valls, G.; Koirala, S.; Anthoni, P.; Besnard, S.; Bodesheim, P.; Carvalhais, N.; et al. Scaling carbon fluxes from eddy covariance sites to globe: Synthesis and evaluation of the FLUXCOM approach. *Biogeosciences* **2020**, *17*, 1343–1365. [CrossRef]
7. Farquhar, G.D.; von Caemmerer, S.; Berry, J.A. A biochemical model of photosynthetic CO_2 assimilation in leaves of C3 species. *Planta* **1980**, *149*, 78–90. [CrossRef]
8. Collatz, G.; Ribas-Carbo, M.; Berry, J. Coupled Photosynthesis-Stomatal Conductance Model for Leaves of C4Plants. *Aust. J. Plant Physiol.* **1992**, *19*, 519.
9. Chen, J.; Liu, J.; Cihlar, J.; Goulden, M. Daily canopy photosynthesis model through temporal and spatial scaling for remote sensing applications. *Ecol. Modell.* **1999**, *124*, 99–119. [CrossRef]
10. Monteith, J.L. Solar Radiation and Productivity in Tropical Ecosystems. *J. Appl. Ecol.* **1972**, *9*, 747. [CrossRef]

11. Running, S.W.; Nemani, R.R.; Heinsch, F.A.; Zhao, M.; Reeves, M.; Hashimoto, H. A continuous satellite-derived measure of global terrestrial primary production. *BioScience* **2004**, *54*, 547. [CrossRef]
12. Sun, Y.; Frankenberg, C.; Jung, M.; Joiner, J.; Guanter, L.; Köhler, P.; Magney, T. Overview of Solar-Induced chlorophyll Fluorescence (SIF) from the Orbiting Carbon Observatory-2: Retrieval, cross-mission comparison, and global monitoring for GPP. *Remote Sens. Environ.* **2018**, *209*, 808–823. [CrossRef]
13. Campbell, P.K.E.; Huemmrich, K.F.; Middleton, E.M.; Ward, L.A.; Julitta, T.; Daughtry, C.S.T.; Burkart, A.; Russ, A.L.; Kustas, W.P. Diurnal and seasonal variations in chlorophyll fluorescence associated with photosynthesis at leaf and canopy scales. *Remote Sens.* **2019**, *11*, 488. [CrossRef]
14. Luo, X.; Croft, H.; Chen, J.M.; He, L.; Keenan, T.F. Improving estimates of global terrestrial photosynthesis using information on leaf chlorophyll content. *Glob. Change Biol.* **2019**, *25*, 2499–2514. [CrossRef] [PubMed]
15. Smith, N.G.; Keenan, T.F.; Colin Prentice, I.; Wang, H.; Wright, I.J.; Niinemets, Ü.; Crous, K.Y.; Domingues, T.F.; Guerrieri, R.; Zhou, S.X.; et al. Global photosynthetic capacity is optimized to the environment. *Ecol. Lett.* **2019**, *22*, 506–517. [CrossRef] [PubMed]
16. Chen, J.M.; Wang, R.; Liu, Y.; He, L.; Croft, H.; Luo, X.; Wang, H.; Smith, N.G.; Keenan, T.F.; Prentice, I.C.; et al. Global datasets of leaf photosynthetic capacity for ecological and earth system research. *Earth Syst. Sci. Data* **2022**, *14*, 4077–4093. [CrossRef]
17. Evans, J.R. Nitrogen and photosynthesis in the flag leaf of wheat (*Triticum aestivum* L.). *Plant Physiol.* **1983**, *72*, 297–302. [CrossRef]
18. Ryu, Y.; Baldocchi, D.D.; Kobayashi, H.; Van Ingen, C.; Li, J.; Black, T.A.; Beringer, J.; Van Gorsel, E.; Knohl, A.; Law, B.E.; et al. Integration of MODIS land and atmosphere products with a coupled-process model to estimate gross primary productivity and evapotranspiration from 1 km to global scales. *Glob. Biogeochem. Cycles* **2011**, *25*, GB4017. [CrossRef]
19. Walker, A.P.; Beckerman, A.P.; Gu, L.; Kattge, J.; Cernusak, L.A.; Domingues, T.F.; Woodward, F.I. The relationship of leaf photosynthetic traits—Vcmax and Jmax—to leaf nitrogen, leaf phosphorus, and specific leaf area: A meta-analysis and modeling study. *Ecol. Evol.* **2014**, *4*, 3218–3235. [CrossRef]
20. Croft, H.; Chen, J.; Luo, X.; Bartlett, P.; Chen, B.; Staebler, R. Leaf chlorophyll content as a proxy for leaf photosynthetic capacity. *Glob. Change Biol.* **2017**, *23*, 3513–3524. [CrossRef]
21. Doughty, C.E.; Asner, G.P.; Martin, R.E. Predicting tropical plant physiology from leaf and canopy spectroscopy. *Oecologia* **2011**, *165*, 289–299. [CrossRef]
22. Zhou, Y.; Ju, W.; Sun, X.; Hu, Z.; Han, S.; Black, T.A.; Jassal, R.S.; Wu, X. Close relationship between spectral vegetation indices and V_{cmax} in deciduous and mixed forests. *Tellus B* **2014**, *66*, 23279. [CrossRef]
23. Zhang, Y.; Guanter, L.; Berry, J.A. Estimation of vegetation photosynthetic capacity from space-based measurements of chlorophyll fluorescence for terrestrial biosphere models. *Glob. Change Biol.* **2014**, *20*, 3727–3742. [CrossRef] [PubMed]
24. Zhang, Y.; Guanter, L.; Joiner, J.; Song, L.; Guan, K. Spatially-explicit monitoring of crop photosynthetic capacity through the use of space-based chlorophyll fluorescence data. *Remote Sens. Environ.* **2018**, *210*, 362–374. [CrossRef]
25. Song, G.; Wang, Q.; Jin, J. Estimation of leaf photosynthetic capacity parameters using spectral indices developed from fractional-order derivatives. *Comput. Electron. Agric.* **2023**, *212*, 108068. [CrossRef]
26. Stoy, P.C.; Mauder, M.; Foken, T.; Marcolla, B.; Boegh, E.; Ibrom, A.; Arain, M.A.; Arneth, A.; Aurela, M.; Bernhofer, C.; et al. A data-driven analysis of energy balance closure across FLUXNET research sites: The role of landscape scale heterogeneity. *Agric. For. Meteorol.* **2013**, *171–172*, 137–152. [CrossRef]
27. Tamura, H.; Mori, S.; Yamawaki, T. Texture features corresponding to visual perception. *IEEE Trans. Syst. Man Cybern.* **1978**, *8*, 460–473. [CrossRef]
28. Humeau-Heurtier, A. Texture feature extraction methods: A survey. *IEEE Access* **2019**, *7*, 8975–9000. [CrossRef]
29. Haralick, R.M.; Shanmugan, K.; Dinstein, I. Texture features for image classification. *IEEE Trans. Syst. Man. Cybern.* **1973**, *3*, 610–621. [CrossRef]
30. Park, Y.; Guldmann, J. Measuring continuous landscape patterns with Gray-Level Co-Occurrence Matrix (GLCM) indices: An alternative to patch metrics? *Ecol. Indic.* **2020**, *109*, 105802. [CrossRef]
31. Fu, Y.; Yang, G.; Song, X.; Li, Z.; Xu, X.; Feng, H.; Zhao, C. Improved estimation of winter wheat aboveground biomass using multiscale textures extracted from UAV-based digital images and hyperspectral feature analysis. *Remote Sens.* **2021**, *13*, 581. [CrossRef]
32. Yang, N.; Zhang, Z.; Zhang, J.; Guo, Y.; Yang, X.; Yu, G.; Bai, X.; Chen, J.; Chen, Y.; Shi, L.; et al. Improving estimation of maize leaf area index by combining of UAV-based multispectral and thermal infrared data: The potential of new texture index. *Comput. Electron. Agric.* **2023**, *214*, 108294. [CrossRef]
33. Reddy, B.S.; Chatterji, B.N. An FFT-based technique for translation, rotation, and scale-invariant image registration. *IEEE Trans. Image Process.* **1996**, *5*, 1266–1271. [CrossRef] [PubMed]
34. Ojala, T.; Pietikäinen, M.; Harwood, D. A comparative study of texture measures with classification based on featured distributions. *Pattern Recognit.* **1996**, *29*, 51–59. [CrossRef]
35. Ahonen, T.; Hadid, A.; Pietikainen, M. Face description with local binary patterns: Application to face recognition. *IEEE Trans. Pattern Anal. Mach. Intell.* **2006**, *28*, 2037–2041. [CrossRef]
36. Duan, B.; Liu, Y.; Gong, Y.; Peng, Y.; Wu, X.; Zhu, R.; Fang, S. Remote estimation of rice LAI based on Fourier spectrum texture from UAV image. *Plant Methods* **2019**, *15*, 124. [CrossRef]

37. Zhuang, J.J.; Luo, S.M.; Hou, C.J.; Tang, Y.; He, Y.; Xue, X.Y. Detection of orchard citrus fruits using a monocular machine vision-based method for automatic fruit picking applications. *Comput. Electron. Agric.* **2018**, *152*, 64–73. [CrossRef]
38. Krizhevsky, A.; Sutskever, I.; Hinton, G.E. ImageNet classification with deep convolutional neural networks. *Adv. Neural Inf. Process. Syst.* **2012**, *25*, 1106–1114. [CrossRef]
39. Yang, Q.; Shi, L.; Han, J.; Zha, Y.; Zhu, P. Deep convolutional neural networks for rice grain yield estimation at the ripening stage using UAV-based remotely sensed images. *Field Crops Res.* **2019**, *235*, 142–153. [CrossRef]
40. Ilniyaz, O.; Du, Q.; Shen, H.; He, W.; Feng, L.; Azadi, H.; Kurban, A.; Chen, X. Leaf area index estimation of pergola-trained vineyards in arid regions using classical and deep learning methods based on UAV-based RGB images. *Comput. Electron. Agric.* **2023**, *207*, 107723. [CrossRef]
41. Sun, S.; Zhu, L.; Liang, N.; He, Y.; Wang, Z.; Chen, S.; Liu, J.; Chen, H.; He, Y.; Zhou, Z. Monitoring drought induced photosynthetic and fluorescent variations of potatoes by visible and thermal imaging analysis. *Comput. Electron. Agric.* **2023**, *215*, 108433. [CrossRef]
42. Deng, X.; Zhang, Z.; Hu, X.; Li, J.; Li, S.; Su, C.; Du, S.; Shi, L. Estimation of photosynthetic parameters from hyperspectral images using optimal deep learning architecture. *Comput. Electron. Agric.* **2023**, *216*, 108540. [CrossRef]
43. Foken, T.; Göckede, M.; Mauder, M.; Mahrt, L.; Amiro, B.; Munger, W. Post-field data quality control. In *Handbook of Micrometeorology. A Guide for Surface Flux Measurement and Analysis*; Lee, X., Massman, M., Law, B., Eds.; Kluwer Academic: Boston, MA, USA, 2004; pp. 181–208.
44. Reichstein, M.; Falge, E.; Baldocchi, D.; Papale, D.; Aubinet, M.; Berbigier, P.; Bernhofer, C.; Buchmann, N.; Gilmanov, T.; Granier, A.; et al. On the separation of net ecosystem exchange into assimilation and ecosystem respiration: Review and improved algorithm. *Glob. Change Biol.* **2005**, *11*, 1424–1439. [CrossRef]
45. Wutzler, T.; Lucas-Moffat, A.; Migliavacca, M.; Knauer, J.; Sickel, K.; Šigut, L.; Menzer, O.; Reichstein, M. Basic and extensible post-processing of eddy covariance flux data with REddyProc. *Biogeosciences* **2018**, *15*, 5015–5030. [CrossRef]
46. Kljun, N.; Calanca, P.; Rotach, M.W.; Schmid, H.P. The simple two-dimensional parameterisation for Flux Footprint Predictions FFP. *Geosci. Model Dev.* **2015**, *8*, 3695–3713. [CrossRef]
47. Yang, Q.; Shi, L.; Han, J.; Zha, Y.; Yu, J.; Wu, W.; Huang, K. Regulating the time of the crop model clock: A data assimilation framework for regions with high phenological heterogeneity. *Field Crop. Res.* **2023**, *93*, 108847. [CrossRef]
48. Chen, J.; Mo, G.; Pisek, J.; Liu, J.; Deng, F.; Ishizawa, M.; Cha, D. Effects of foliage clumping on the estimation of global terrestrial gross primary productivity. *Glob. Biogeochem. Cycles* **2012**, *6*, GB1019. [CrossRef]
49. Medlyn, B.E.; Dreyer, E.; Ellsworth, D.; Forstreuter, M.; Harley, P.C.; Kirschbaum, M.U.F.; Roux, X.L.; Montpied, P.; Strassemeyer, J.; Loustau, D.; et al. Temperature response of parameters of a biochemically based model of photosynthesis, I.I. A review of experimental data. *Plant Cell Environ.* **2002**, *25*, 1167–1179. [CrossRef]
50. Medlyn, B.E.; Badeck, F.-W.; De Pury, D.G.G.; Barton, C.V.M.; Broadmeadow, M.; Ceulemans, R.; De Angelis, P.; Forstreuter, M.; Jach, M.E.; Kellomäki, S.; et al. Effects of elevated [CO2] on photosynthesis in European forest species: A meta-analysis of model parameters. *Plant Cell Environ.* **1999**, *22*, 1475–1495. [CrossRef]
51. Ball, M.C.; Woodrow, I.E.; Berry, J.A. A model predicting stomatal conductance and its contribution to the control of photosynthesis under different environmental conditions. In *Progress in Photosynthesis Research*; Biggins, J., Ed.; Martinus Nijhoff Publishers: Dordrecht, The Netherlands, 1987; pp. 221–224.
52. Storn, R.; Price, K. Differential evolution—A simple and efficient heuristic for global optimization over continuous spaces. *J. Glob. Optim.* **1997**, *11*, 341–359. [CrossRef]
53. Newville, M.; Stensitzki, T.; Allen, D.B.; Rawlik, M.; Ingargiola, A.; Nelson, A. LMFIT: Non-Linear Least-Square Minimization and Curve-Fitting for Python. Astrophysics Source Code Library: 2016; p. ascl-1606. Available online: https://ui.adsabs.harvard.edu/abs/2016ascl.soft06014N/abstract (accessed on 17 October 2024).
54. He, K.; Zhang, X.; Ren, S.; Sun, J. Deep residual learning for image recognition. In Proceedings of the IEEE Conference on Computer Vision and Pattern Recognition, Las Vegas, NV, USA, 27–30 June 2016; pp. 770–778.
55. Lundberg, S.M.; Lee, S.I. A unified approach to interpreting model predictions. *Adv. Neural Inf. Process. Syst.* **2017**, *30*, 4765–4774.
56. Hlatshwayo, S.T.; Mutanga, O.; Lottering, R.T.; Kiala, Z.; Imsail, R. Mapping forest aboveground biomass in the reforested Buffelsdraai landfill site using texture combinations computed from SPOT-6 pan-sharpened imagery. *Int. J. Appl. Earth Obs. Geoinf.* **2019**, *74*, 65–77. [CrossRef]
57. Zhang, P.; Dai, X.; Yang, J.; Xiao, B.; Yuan, L.; Zhang, L.; Gao, J. Multi-scale vision longformer: A new vision transformer for high-resolution image encoding. In Proceedings of the IEEE/CVF International Conference on Computer Vision, Montreal, QC, Canada, 10–17 October 2021; pp. 2998–3008.
58. Gupta, D.K.; Mago, G.; Chavan, A.; Prasad, D.K. Patch gradient descent: Training neural networks on very large images. *arXiv* **2023**, arXiv:2301.13817.
59. Chu, H.; Luo, X.; Ouyang, Z.; Chan, W.S.; Dengel, S.; Biraud, S.C.; Torn, M.S.; Metzger, S.; Kumar, J.; Arain, M.A.; et al. Representativeness of eddy-covariance flux footprints for area surrounding AmeriFlux sites. *Agric. For. Meteorol.* **2021**, *301–302*, 108350. [CrossRef]
60. Zheng, T.; Chen, J.; He, L.; Arain, M.A.; Thomas, S.C.; Murphy, J.G.; Geddes, J.A.; Black, T.A. Inverting the maximum carboxylation rate (V_{cmax}) from the sunlit leaf photosynthesis rate derived from measured light response curves at tower flux sites. *Agric. For. Meteorol.* **2017**, *236*, 48–66. [CrossRef]

61. Jiang, C.; Ryu, Y.; Wang, H.; Keenan, T.F. An optimality-based model explains seasonal variation in C3 plant photosynthetic capacity. *Glob. Change Biol.* **2020**, *26*, 6493–6510. [CrossRef]
62. Saathoff, A.J.; Welles, J. Gas exchange measurements in the unsteady state. *Plant Cell Environ.* **2021**, *44*, 3509–3523. [CrossRef]

remote sensing MDPI

Article

At Which Overpass Time Do ECOSTRESS Observations Best Align with Crop Health and Water Rights?

Benjamin D. Goffin [1,*], Carlos Calvo Cortés-Monroy [2], Fernando Neira-Román [2], Diya D. Gupta [3] and Venkataraman Lakshmi [1]

1 Department of Civil and Environmental Engineering, University of Virginia, Charlottesville, VA 22904, USA; vlakshmi@virginia.edu
2 Centro de Información de Recursos Naturales, Providencia, Santiago 7501556, Chile
3 College of Arts and Sciences, University of Virginia, Charlottesville, VA 22904, USA
* Correspondence: benjamingoffin@virginia.edu

Abstract: Agroecosystems are facing the adverse effects of climate change. This study explored how the ECOsystem Spaceborne Thermal Radiometer Experiment on Space Station (ECOSTRESS) can give new insight into irrigation allocation and plant health. Leveraging the global coverage and 70-m spatial resolution of the Evaporative Stress Index (ESI) from ECOSTRESS, we processed over 200 overpasses and examined patterns over 3 growing seasons across the Maipo River Basin of Central Chile, which faces exacerbated water stress. We found that ECOSTRESS ESI varies substantially based on the overpass time, with ESI values being systematically higher in the morning and lower in the afternoon. We also compared variations in ESI against spatial patterns in the environment. To that end, we analyzed the vegetation greenness sensed from Landsat 8 and compiled the referential irrigation allocation from Chilean water regulators. Consistently, we found stronger correlations between these variables and ESI in the morning time (than in the afternoon). Based on our findings, we discussed new insights and potential applications of ECOSTRESS ESI in support of improved agricultural monitoring and sustainable water management.

Keywords: ECOSTRESS; Thermal Infrared (TIR); Evaporative Stress Index (ESI); crop-water stress; irrigation water allocation; Normalized Difference Vegetation Index (NDVI); plant greenness; Maipo River Basin

1. Introduction

Agriculture is the primary consumer of water withdrawals globally [1]. Nevertheless, the agricultural sector remains highly sensitive to climate variations [2]. If water is available, crops can use water (i.e., actual evapotranspiration, AET) at the maximum rate of atmospheric demand (i.e., potential evapotranspiration, PET) [3,4]. Once the demand for water from the leaves exceeds the supply from the roots, plants experience water stress and close their stomates to conserve water [5,6]. Because ET fluxes are restricted, the temperature of the canopy increases, causing potential heat-related impacts [7]. When this water deficit extends over time, the crop's growth and yield are adversely affected [8,9]. In the face of climate change, crops are at a greater risk of facing droughts and limited water availability [10,11].

Therefore, it is becoming increasingly important for food and water security to monitor crop stress from local to global scales with high accuracy [12]. Reduction in AET relative to PET can indicate plant-water stress [13]. The ratio of AET to PET is referred to as the Evaporative Stress Index (ESI) and serves as a key indicator of rapidly evolving drought conditions that pose a threat to agricultural areas [14–20]. Unlike other drought indices relying on ground observations, ESI can incorporate remotely sensed data from satellites and monitor drought conditions at the global scale. Previously, ESI was limited by the spatial resolution of available geostationary satellites, on the order of 5–10 km [21].

Since its launch in 2018, the ECOsystem Spaceborne Thermal Radiometer Experiment on Space Station (ECOSTRESS) models plant-water dynamics at a 70-m spatial resolution [22]. Aboard the International Space Station (ISS), ECOSTRESS follows an irregular orbit (i.e., non-sun-synchronous, non-geostationary), overpassing a site every 1–5 days at different revisiting times. This unique vantage point allows ECOSTRESS to sample the diurnal cycle in critical regions [23]. Specifically, ECOSTRESS measures the Thermal Infrared (TIR) radiation emitted from Earth's surface to calculate the brightness temperature of plants and derive ET. The algorithm for global ET retrieval is based on the Priestley-Taylor Jet Propulsion Lab (PT-JPL) method [24–26]. Past studies confirmed the ability of the PT-JPL model to characterize variation in ET across regions and seasonal scales [27–31]. On ECOSTRESS, ET estimates based on PT-JPL also showed high agreement with observations from 82 flux towers around the world [22].

More recent assessments highlighted potential inaccuracies and biases in the PT-JPL *ET* product, particularly during the summer season or morning time [31–34]. But to this day, there has been very limited research on the PT-JPL *ESI* product (one of the core products of the ECOSTRESS mission). A prior study by Pascolini-Campbell et al. (2022) applied ESI from ECOSTRESS (among other variables) to predict fire burn severity in California [35]. ESI was found to be among the leading predictors (in certain vegetation types), but those results were sensitive to the time of day of the ECOSTRESS acquisition. No paper has reviewed how ECOSTRESS overpasses can affect ESI measurements and to what extent they can capture patterns in the environment. To the best of our knowledge, only one preliminary study has indicated that ESI from ECOSTRESS could exhibit significant variability or "noise" between overpasses [36]. It is critical for agricultural monitoring to better understand where these remotely sensed data have discrepancies and why [37]. Therefore, our work aims to:

- Examine how ESI observations vary based on their acquisition time;
- Evaluate how closely ESI aligns with crop health and water rights;
- Assess the potential for ESI to capture sub-optimal crop conditions.

2. Data and Methods

2.1. Study Area

Our study focused on the Maipo River Basin located in Central Chile (Figure 1a). Per its regulatory delineation, the basin collects water from over 15,200 sq km, from the Andes Mountains to the Pacific Ocean [38]. It encompasses the Región Metropolitana de Santiago, Chile's most densely populated region. It also extends over parts of the Región de Valparaíso (to the west) and Región del Libertador General Bernardo O'Higgins (to the south). We combined the most recent land registry information available from the three administrative regions studied to build a comprehensive inventory of all the 180,000 hectares of agricultural fields in the basin [39–41].

The basin is characterized by a Mediterranean climate, with most of the annual precipitation concentrated during the winter months [42]. Note that this precipitation regime results in low cloud cover for the rest of the year and optimal conditions for remote sensing monitoring. During the growing season, the basin supplies water to around 135,000 ha of irrigated agricultural land [43]. Irrigation is achieved by means of consumptive water-use rights, allowing some farmers to meet the water demand of both annual and perennial crops [44].

However, the basin has experienced long downward trends in precipitation and rising temperatures [11,45]. In addition, a multi-year dry spell (referred to as megadrought) began in 2010 and further reduced runoff and the quantity (and quality) of surface water and groundwater levels across the region [42,46,47]. The decade-long megadrought and changing climate have already impacted socio-economic systems and exacerbated conflicts across water users [48–54].

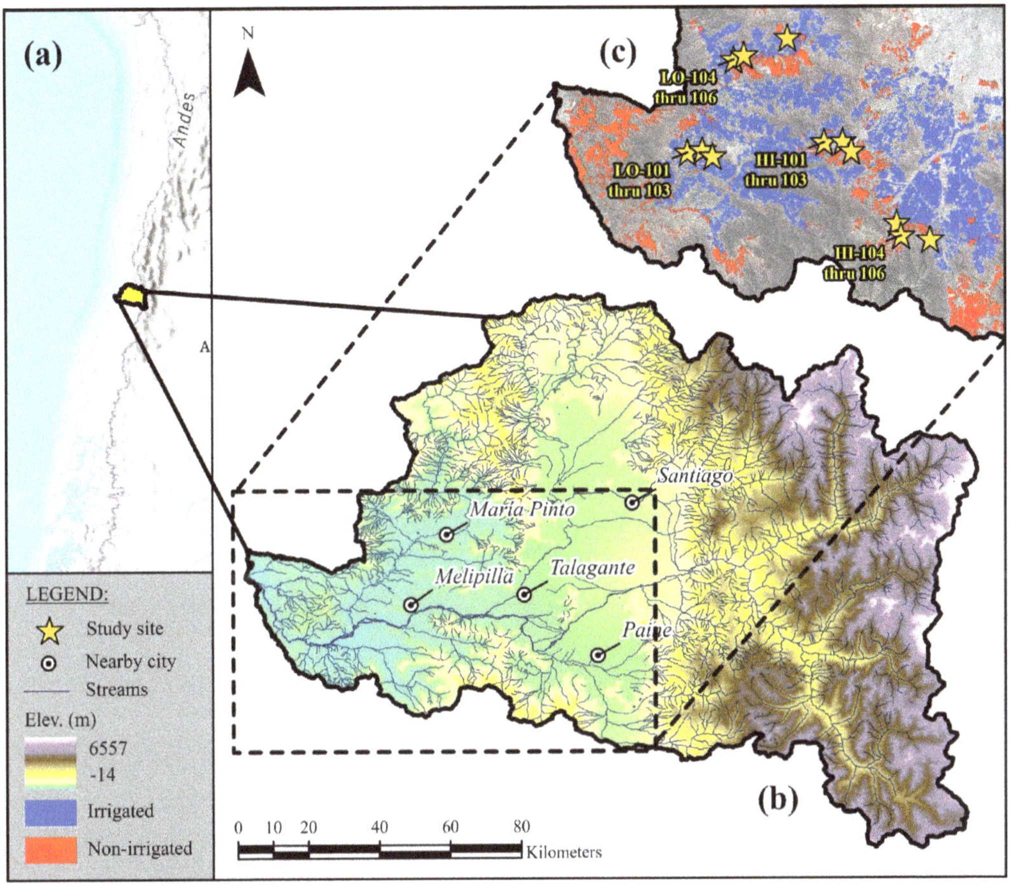

Figure 1. Our study focuses on the Maipo River Basin located in (**a**) Central Chile and characterized by (**b**) a complex terrain and river network, and (**c**) extensive areas with surface- and ground-water allocations.

2.2. Earth Observations

In this context, we leveraged different Earth observations from satellites to capture crop conditions on the ground. Using NASA's open-access Application for Extracting and Exploring Analysis Ready Samples (AppEEARS) portal, we downloaded all ECOSTRESS PT-JPL ESI Level 4 data available over our study area [55]. ESI is one of the core products of ECOSTRESS, which evaluates the evaporative stress based on the ratio in Equation (1) below [13]:

$$\text{ESI} = \frac{\text{AET}}{\text{PET}} \tag{1}$$

We aligned each ECOSTRESS snapshot on a 0.0006° grid (i.e., ~70 m spacing), encompassing approximately 4.1 M pixels over the study area (EPSG 32719). We compiled a total of 316 observations taken at various acquisition time over 242 days, spanning from 21 July 2018 to 19 January 2023 (Figure S1). Due to its irregular orbit aboard the ISS, ECOSTRESS returns to a given location at various times of day with inconsistent intervals. Each ESI value represents an instantaneous measure at the time of ECOSTRESS overpass [56]. We converted the Coordinated Universal Time (UTC) reported for each observation to the local solar time. To account for the potential variations between the different revisit times, we differentiated ECOSTRESS overpasses based on the following time periods:

- the morning hours from 5:00 a.m. to 9:59 a.m.; and
- the afternoon hours from 12:00 p.m. to 4:59 p.m.

We also utilized surface reflectance data (Level 2, Collection 2, Tier 1) from the ongoing Landsat 8 mission [57,58]. Image processing and acquisition was completed through Google Earth Engine using the JavaScript Application Programming Interface (API). Pixels affected by cloud cover or cloud shadow were masked to retain only the highest quality information (Figure S2). We used the Red, Green, and Blue (RGB) bands of the Operational Land Imager (OLI) with the prescribed scaling factors to create natural color composites of the study area. We also computed the Normalized Difference Vegetation Index (NDVI), a measure of vegetation greenness already in use by the Chilean government for drought monitoring [59]. NDVI is calculated based on the Red and Near-Infrared (NIR) bands using Equation (2) [60]:

$$\mathrm{NDVI} = \frac{\mathrm{NIR} - \mathrm{Red}}{\mathrm{NIR} + \mathrm{Red}} \tag{2}$$

We primarily examined ESI (and NDVI) during the growing season, which is defined as the five months from November to March, when crops in the basin are in their most active growing stages and water demand is the highest [43]. To get representative measure of ESI (and NDVI) for a growing season, we took the median value of the ESI (and NDVI) observed during those hours over the 5 months studied. Given the launch of ECOSTRESS in 2018, we considered a total of three growing seasons for which ECOSTRESS data had been released over:

- 2019 to 2020;
- 2020 to 2021;
- 2021 to 2022.

2.3. Selection of Maize Site at the Parcel Level

First, we examined these remotely sensed data at discrete locations with consistent crop cover. Maize (*Zea mays*) is a particularly relevant crop to study since it is the most produced cereal grain in the basin (with an estimated 14,100 ha in 2020) and is expected to require more irrigation due to climate change [61]. Maize is also well suited for remote sensing monitoring since it develops over the course of approximately 170 days, allowing for many satellite overpasses [62]. Based on that, we selected twelve maize fields across the Maipo River Valley (Table 1). Six of the parcels (labeled LO-101 through LO-106) were located near the cities of Melipilla and María Pinto, while the other six (labeled HI-101 through HI-106) were selected at a relatively higher elevation in the vicinity of Talagante and Paine, thus encompassing a range of topographic and climatic conditions (Figure 1b). The elevation at each location was obtained from NASA's Shuttle Radar Topography Mission [63]. To minimize the effects of outside-pixel contamination, each parcel under study was about 10 hectares or more [39–41]. The planting dates of maize at each field were determined based on the least-squares fit of NDVI against FAO crop development curves [62]. We calculated the total rainfall and average daytime land surface temperature for each site using the Climate Hazards Group InfraRed Precipitation with Stations (CHIRPS) and Terra Moderate Resolution Imaging Spectroradiometer (MODIS), respectively [64,65]. To get a representative value of ESI at one location, we computed the median of a 3 × 3 grid around the middle of the field. With that, we examined ESI as a function of the overpass time of ECOSTRESS.

Table 1. Characteristics of a dozen maize fields selected for study.

Site ID	Lat.	Lon.	Elev. (m.)	Area (ha.)	Planting Date			Precipitation *1 (mm)			Temperature *1 (°C)		
					2019	2020	2021	2019–2020	2020–2021	2021–2022	2019–2020	2020–2021	2021–2022
LO-101	−33.6885	−71.3262	159	82.0	2 Oct.	6 Oct.	20 Oct.	19	12	13	38.8	35.9	36.3
LO-102	−33.6818	−71.2950	157	61.7	9 Oct.	28 Sep.	10 Sep.	21	12	31	36.6	33.9	34.2
LO-103	−33.6850	−71.2780	170	130.9	13 Oct.	10 Oct.	-- *2	16	12	-- *2	35.9	33.4	-- *2
LO-104	−33.5180	−71.2085	158	24.2	4 Nov.	20 Oct.	21 Oct.	7	7	13	33.9	32.4	34.0
LO-105	−33.5090	−71.2060	157	40.7	5 Nov.	20 Oct.	20 Oct.	12	7	16	35.3	35.0	36.5
LO-106	−33.4785	−71.0937	173	55.6	15 Oct.	15 Sep.	24 Aug.	13	12	24	35.2	34.0	33.6
HI-101	−33.6735	−70.0155	278	9.7	31 Oct.	21 Nov.	7 Nov.	9	39	37	33.6	29.7	31.2
HI-102	−33.6733	−70.9728	308	11.4	4 Oct.	25 Oct.	21 Oct.	18	7	10	36.1	33.6	34.1
HI-103	−33.6883	−70.9559	309	11.0	26 Sep.	29 Sep.	25 Sep.	18	5	12	35.1	32.5	33.1
HI-104	−33.8263	−70.8487	357	14.4	25 Sep.	27 Sep.	5 Dec.	17	11	59	35.2	32.5	29.8
HI-105	−33.8518	−70.8402	355	26.2	3 Oct.	15 Oct.	1 Oct.	18	13	11	35.5	32.2	33.7
HI-106	−33.8591	−70.7745	379	59.4	20 Sep.	30 Sep.	29 Sep.	18	12	12	34.4	32.0	32.6

*1 Total precipitation amount and mean land surface temperature calculated for the growing cycle specific to each field. *2 Maize not planted at this site during this season.

2.4. Compilation of Water Allocations at the Regional Scale

Beyond a dozen maize fields, we also examined variations in ECOSTRESS observations over the irrigated land of the Maipo River Basin (Figure 1c). We analyzed how ESI varies across areas that are allocated various amounts of irrigation water. Previous studies compared remotely sensed data with information on water use available in various parts of Chile [66–68]. We conducted an extensive compilation of selected surface- and groundwater allocations for 41 specific agricultural areas in the Maipo River Basin (Table 2). In total, these irrigated areas constituted around 126,000 hectares (i.e., over 70% of the 180,000 hectares of all agricultural land in the Maipo River Basin). The selected areas also made up for a combined 190 m^3/s of surface- and ground-water withdrawal allocations of the roughly 237 m^3/s used for agriculture in the whole basin [69]. Additional irrigated land exists in the Maipo River Basin, among which several areas hold "eventual" (i.e., intermittent) water-use rights. Due to the extended drought conditions, these surface water allocations were negligible during the period studied (before 2023), and therefore not included in this study.

Agricultural systems are known to experience various degrees of temperature-related damage based on the irrigation and transpiration allowed [7]. Given that each of the 41 entities has different stakeholders and water practices, it made sense to investigate how ESI varied across these geographic entities as a function of agricultural areas, total allocated waters, and crop greenness. To get a representative measure of ESI (and NDVI) in every spatial entity, we took the median value of the ESI (and NDVI) pixels contained within each of these irrigated areas. We also considered the spatial variability in ESI within each area by calculating the Relative Standard Deviation (RSD) across ESI pixels.

Table 2. Major irrigated areas of the Maipo River Basin with documented surface- and ground-water allocations.

ID	Primary Water Source	Irrigation Channel	Agri-Cultural Area (ha.)	Surface Water Right (m^3/s)	Ground-Water Right *1 (m^3/s)	Total Allocated Water	
						Rate (L/s/ha)	Depth (mm/yr)
1	Estero Cholqui o Chocalan	Wodehouse [70]	1473	2.30	0.04	1.59	5010
2	Estero Puangue	Cancha de Piedra [70]	1084	0.27	0.24	0.47	1472
3	Estero Puangue	San Diego [71]	1603	1.98	0.00	1.24	3896
4	Estero Puangue	Santa Emilia o Rulano [71]	990	0.88	0.07	0.97	3045
5	Estero Puangue	Toma del Toro [71]	910	0.06	0.03	0.10	315
6	Rio Angostura	Hospital [72]	995	0.58	0.05	0.64	2005
7	Rio Angostura	Unificado Aguila Norte Aguila Sur [73]	900	0.58	0.05	0.70	2212
8	Rio Clarillo	Clarillo [71]	1036	0.50	0.10	0.57	1813
9	Rio Colina	Colina [74]	5691	0.55	1.01	0.27	864
10	Rio Maipo 1st Section	Comun Asociacion Canales del Maipo [74]	2235	35.00	0.62	3.44	10,839
11	Rio Maipo 1st Section	Derivado El Carmen Uno [74]	10,912	5.71	1.07	0.51	1596
12	Rio Maipo 1st Section	Eyzaguirre [75]	1771	11.47	0.07	6.52 *2	20,554
13	Rio Maipo 1st Section	Fernandino [76]	5331	2.90	0.33	0.61	1909
14	Rio Maipo 1st Section	Huidobro [77]	8483	16.07	0.58	1.96	6191
15	Rio Maipo 1st Section	La Isla [78]	2577	1.50	0.05	0.60	1900
16	Rio Maipo 1st Section	Lo Herrera [75]	2288	2.00	0.12	0.93	2918
17	Rio Maipo 1st Section	Lonquen Isla [75]	2443	0.61	0.88	0.61	1919
18	Rio Maipo 1st Section	Paine [76]	2660	1.90	0.34	0.84	2655
19	Rio Maipo 1st Section	Quinta [76]	4913	3.38	0.64	0.82	2582
20	Rio Maipo 1st Section	Santa Rita [76]	1387	0.56	0.04	0.43	1350
21	Rio Maipo 1st Section	Santa Rita Uno [76]	2332	5.34	0.06	2.32	7304
22	Rio Maipo 1st Section	Viluco [76]	3992	2.75	0.49	0.81	2561
23	Rio Maipo 2nd Section	San Antonio de Naltahua [75]	2404	2.46	0.22	1.11	3515
24	Rio Maipo 3rd Section	Carmen Alto [70]	3204	8.00	0.16	2.55	8027
25	Rio Maipo 3rd Section	Chocalan [79]	2223	5.00	0.06	2.28	7177
26	Rio Maipo 3rd Section	Cholqui [79]	2407	2.00	0.06	0.86	2699
27	Rio Maipo 3rd Section	Codigua [79]	1871	4.80	0.00	2.57	8091
28	Rio Maipo 3rd Section	Culipran [79]	3119	5.00	0.03	1.61	5085
29	Rio Maipo 3rd Section	Hualemu [70]	1234	2.00	0.04	0.51	1608
30	Rio Maipo 3rd Section	Huechun [70]	2803	4.20	0.04	1.51	4770
31	Rio Maipo 3rd Section	Isla De Huechun [75]	1165	2.29	0.01	1.98	6229
32	Rio Maipo 3rd Section	Puangue [70]	2946	3.60	0.00	1.22	3858
33	Rio Maipo 3rd Section	San Jose [75]	6340	3.26 *3	0.04	0.52	1640
34	Rio Mapocho	Castillo [80]	1823	3.43	0.16	1.97	6203
35	Rio Mapocho	Chinihue [81]	1783	2.70	0.02	1.52	4804
36	Rio Mapocho	El Paico [75]	1068	1.60	0.00	1.50	4725
37	Rio Mapocho	Esperanza Bajo [75]	1333	1.80	0.01	1.36	4275
38	Rio Mapocho	Las Mercedes [82]	11,777	10.20	7.28	1.48	4680
39	Rio Mapocho	Mallarauco [83]	8552	8.69	0.29	1.05	3311
40	Rio Mapocho	San Miguel [75]	1767	4.00	0.10	2.32	7317
41	Rio Peuco	Chada Tronco [84]	2142	1.93	0.21	1.00	3143

*1 Subsurface water rights were obtained from several publications by Dirección General de Aguas [85–87]. *2 This estimated rate appeared to be an outlier and was not included in further analysis. *3 This reference number varied based on the source and was not included in further analysis.

We evaluated the level of correlation between ESI (at various times of day) and crop greenness, total allocated waters, and agricultural areas. To characterize the monotonic association between two variables, we calculated Kendall's τ and Spearman's ρ coefficients [88–90]. Unlike Pearson's correlation, these two tests are non-parametric and robust to potential outliers, non-normally distributed data, and non-linear relationships [91]. A coefficient of 1 indicates a perfect association of ranks. Lastly, we explored variations in the distribution of ESI between crop productions. Specifically, we considered the 4 major crop types in the basin (by irrigated surface area), namely grapevine (~16,200 ha), walnut (~16,100 ha), maize (~12,200 ha), and avocado (~5700 ha).

3. Results

3.1. Temporal Patterns in ESI at Selected Maize Fields

For different sites across the Maipo River Basin, we found that valid observations of ECOSTRESS were distributed across local times between 5:00 a.m. and 7:00 p.m. (Figure 2). The number of observations taken each hour fluctuated throughout the day and from year to year. We found that ESI at these selected fields ranged generally from around 0.1 to 1.0 based on the time of day and growing season. The higher ESI values (i.e., less stress) were produced around 5:00 a.m. and ESI reached a consistent low (i.e., high stress) around 2:00 p.m., regardless of the site or season. Transient conditions can be observed in the later parts of the morning and the afternoon. Fluctuations in ESI were also noted between seasons with some higher ESI values during the 2020–2021 growing cycle, particularly in the afternoon. The curve of ESI as a function of acquisition time is noticeably flatter at higher elevations (e.g., HI-104 through 106) in part due to relatively higher ESI in the afternoon hours.

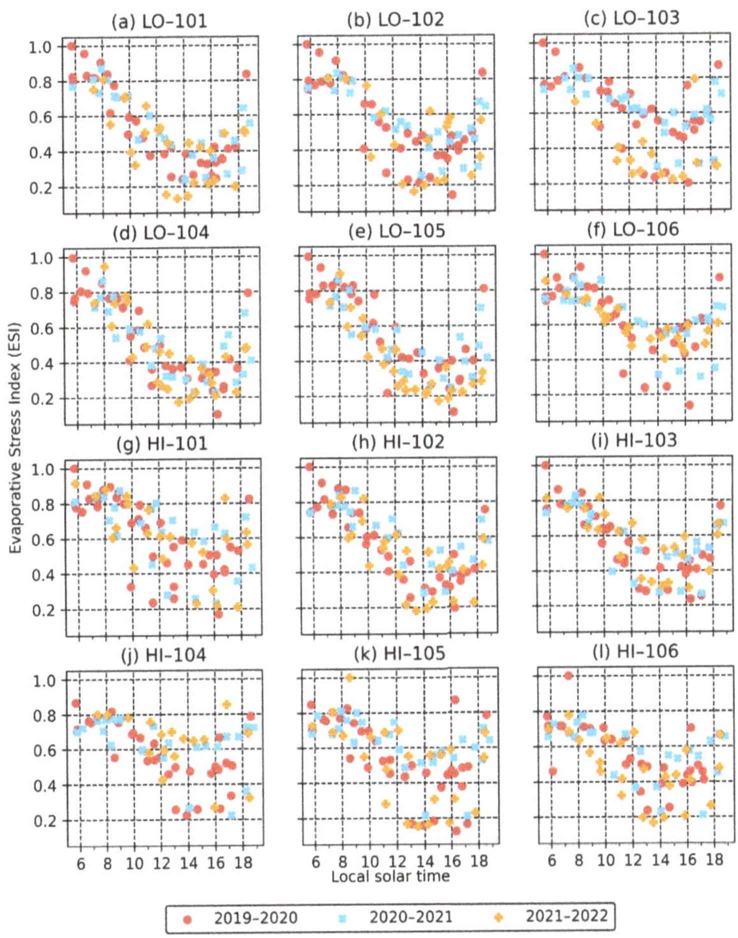

Figure 2. Variations in Evaporative Stress Index (ESI) from ECOSTRESS over the hours of the day for a dozen maize fields across the Maipo River Basin. Values of ESI approaching 1 (e.g., early morning hours) indicate a close match between available water and atmospheric demand, resulting in minimal plant stress. Note that no maize was grown at LO-103 in 2021–2022.

3.2. Spatial Patterns in ESI throughout the Agricultural Land

ESI varied substantially across the irrigated land of the Maipo River Basin, regardless of the time of day or growing season (Figure 3). However, a stark contrast in ESI can be seen across the acquisition times, with ESI up to four times higher in the morning hours than in the afternoon. The all-hours ESI showed attenuated details compared to the specific hourly groupings. In addition, we found contrasting spatial patterns in ESI across the different acquisition times. For instance, morning ESI appeared to increase in a westward direction (i.e., light blue to purple) while ESI in the later hours of the day increased in the eastward direction (i.e., red to green). Comparing the growing seasons, we observed important shifts from year to year with afternoon ESI increasing on average by 21% from the 2019–2020 cycle to the next one (Figure 3d,e). However, the inter-seasonal shifts in morning ESI were not as pronounced, varying on average by only 1% over the same timeframe (Figure 3a,b).

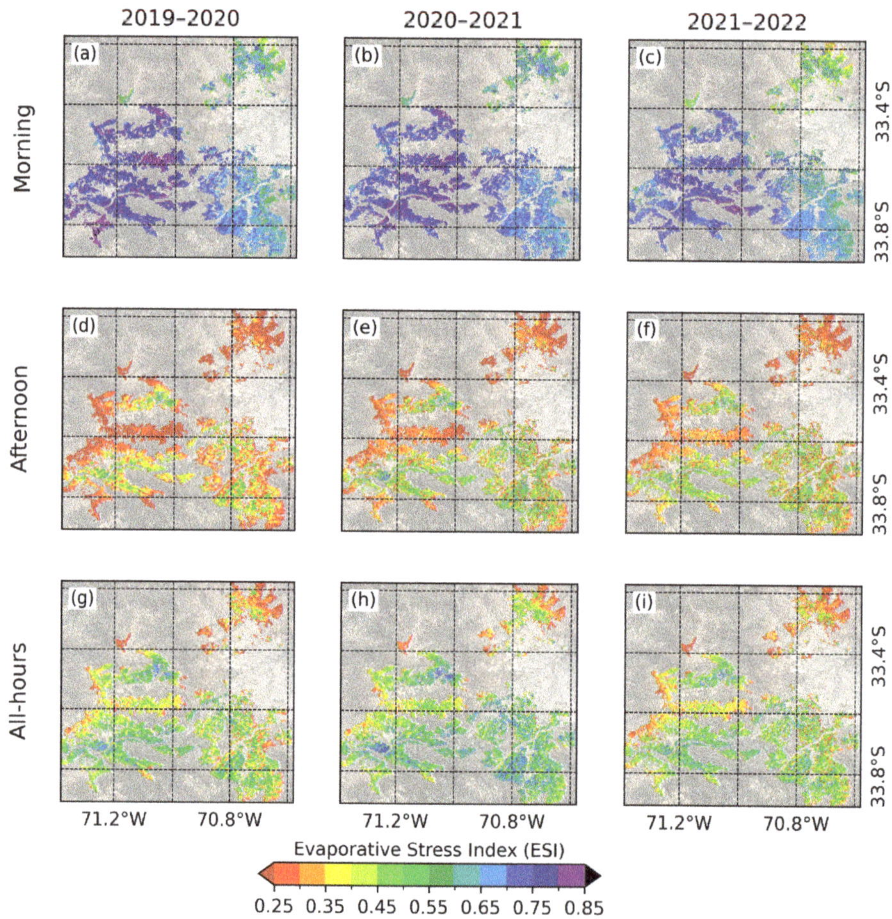

Figure 3. Variations in Evaporative Stress Index (ESI) from ECOSTRESS throughout the 126,000 ha of irrigated agricultural land of the Maipo River Basin across various times of day (vertically) and growing seasons (horizontally). Values of ESI approaching 1 (blue) indicate a close match between available water and atmospheric demand, whereas an ESI of 0 (red) represents maximum plant stress.

3.3. Spatial Correlation of ESI with Crop Health

When comparing different agricultural areas across the Maipo River Basin, we found some positive correlations between their ECOSTRESS ESI and Landsat 8 OLI NDVI (Figure 4). This means that some agricultural areas with higher ESI exhibited higher NDVI, and vice-versa. In addition, ESI appeared generally lower (i.e., higher stress) during the 2019–2020 growing cycle than in the following years, and this shift was reflected in lower NDVI over the same time period. The correlation between plant-water stress and crop health was consistently significant ($p < 0.001$) for the morning overpasses with Spearman's correlation coefficients ranging from 0.544 to 0.829 (Figure 4a–c). The relationship between the two variables also followed a weaker monotonic trend during some other times of day (Figure 4f,g,i).

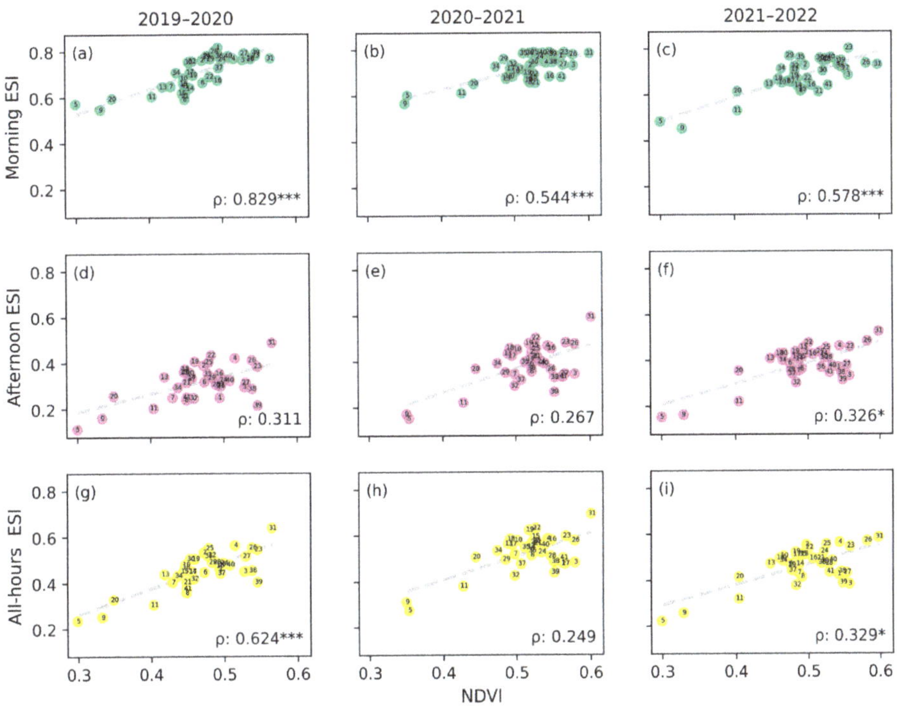

Figure 4. Trends in Evaporative Stress Index (ESI) from ECOSTRESS (y-axis) against Normalized Difference Vegetation Index (NDVI) from Landsat 8 (x-axis) of agricultural areas with different water allocations in the Maipo River Basin across various times of day (vertically) and growing seasons (horizontally). The positive correlation between the two variables is illustrated by a regression line (dashed) and its 95% confidence interval (shaded). Spearman's ρ coefficient is shown as a measure of the monotonic association between two variables. Significance levels are indicated based on the thresholds of $p < 0.05$ (*), $p < 0.01$ (**), and $p < 0.001$ (***).

3.4. Correlation of ESI with Water Rights

In exploring the link between crop-water stress and water rights, we found some positive correlations between ESI and the amounts of allocated water (Figure 5). Ranks in morning ESI were best aligned with those of documented surface- and ground-water allocations ($p < 0.001$ and ρ between 0.436 and 0.556). Based on this relationship, some areas such as Toma del Toro (ID#5) and Colina (ID#9), with relatively low allocated water, showed lower ESI. It should also be noted that the spatial variability in morning ESI was a fraction of the coefficient of variation obtained at other times of day (Figure S3). In addition, the spatial variability in the later hours of the day exhibited a significant correlation ($p < 0.05$) with the size of the agricultural areas, suggesting that dispersion around the mean might be further amplified across larger sample areas during the afternoon.

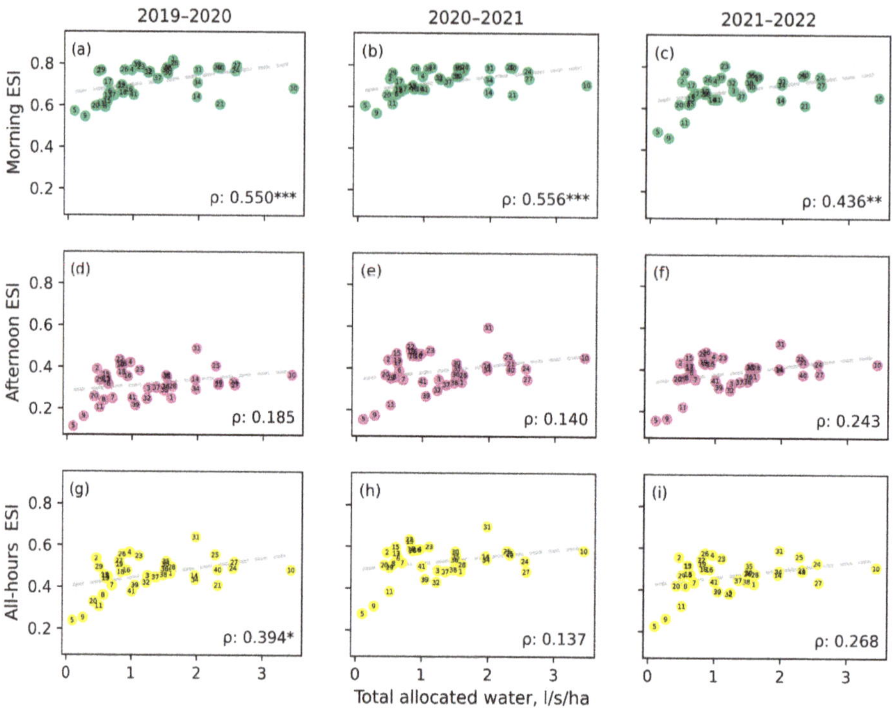

Figure 5. Trends in Evaporative Stress Index (ESI) from ECOSTRESS (y-axis) against total allocated water (x-axis) of agricultural areas with different water allocations in the Maipo River Basin across various times of day (vertically) and growing seasons (horizontally). The positive correlation between the two variables is illustrated by a regression line (dashed) and its 95% confidence interval (shaded). Spearman's ρ coefficient is shown as a measure of the monotonic association between two variables. Significance levels are indicated based on the thresholds of $p < 0.05$ (*), $p < 0.01$ (**), and $p < 0.001$ (***).

3.5. Variations in Crop Productions

When further examining the distribution of ESI, we found that it also varied across irrigated parcels based on crop types (Figure 6). The dispersion of ESI for each crop type is substantially larger for the afternoon and all hours combined, than it is in the morning time. Focusing on morning ESI only, we also observed that the stress imposed on avocado parcels is generally lower (i.e., higher ESI) than it is for grapevines (Figure 6a). Going further, morning ESI also appeared to shift down (i.e., further exacerbated) in the growing cycle 2021–2022 in comparison to the prior year (i.e., on average −5.0% for grapevine, −5.1% for walnut, −3.8% for maize, −3.9% for avocado).

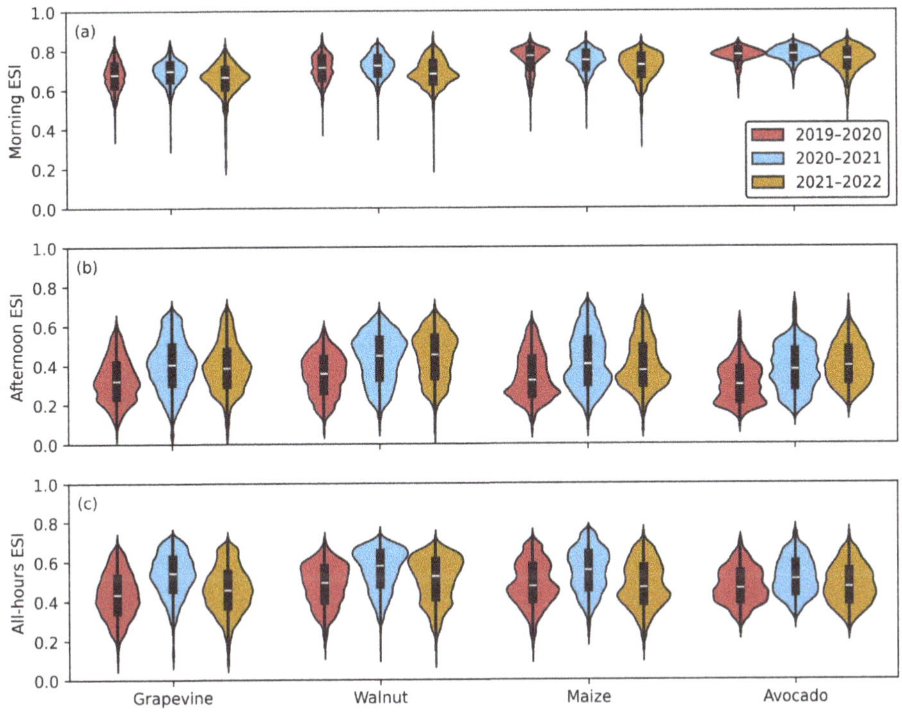

Figure 6. Comparison of the distributions in Evaporative Stress Index (ESI) from ECOSTRESS for the 4 major crop types in the Maipo River Basin (irrigated land only) across different times of day and growing seasons.

4. Discussion

4.1. Need for ECOSTRESS ESI

A rapid scale-up in the application of satellite data is needed to give new insight into drought conditions and limited water resources, specifically in the Andes region [92]. Previously, fluctuations in NDVI served as an indicator of crop coefficients and productivity [59,62]. However, there can be a lag in the greenness response of crops to rapid-onset drought events. NDVI can also be affected by other factors such as late frost, pests, or flooding. It is critical for drought indices to capture land-atmosphere interactions. The Palmer Drought Severity Index (PDSI) includes temperature (one of the many drivers of ET) but does not account for humidity or incoming radiation [4,93]. Other common indices of droughts attempt to model water availability based on precipitation, soil moisture, and/or ET [8,94–98]. Generally, drought indices fail to capture fine-scale heterogeneity [3,6]. Common indicators like the U.S. Drought Monitor are coarse in size (with a resolution of 0.025 degree) and lack the granularity necessary to support decision-making [99]. In contrast, ECOSTRESS measures TIR radiation at a 70-m spatial resolution to model evaporative stress relying on the principle that well-irrigated fields are cooler than water-stressed crops.

4.2. Limited Understanding of ECOSTRESS ESI

In general, our understanding of atmospheric drying (over the course of a day) and soil drying (over the course of a season) and their combined effects on ET remains limited, especially at the field scale [9]. Current ET retrievals based on PT-JPL showed considerable overestimation across heterogeneous landscapes and complex terrain [32,100]. As for the overpass time, some studies reported that the PT-JPL model performed particularly well in the early morning [31], produced lower biases later in the day [34], or yielded no significant differences at different revisit times [101]. Overall, there have been very few prior applications of any ECOSTRESS PT-JPL data to Mediterranean climates [22,34,35] or South America [22]. In fact, only one paper previously used the ESI variable of ECOSTRESS and differentiated observations before and after 12:00 p.m. [35].

4.3. New Insights into ECOSTRESS ESI

To the best of our knowledge, our study is the first to examine how instantaneous values of ESI varies as a function of the ECOSTRESS acquisition time. We found substantial variations in crop-water stress based on the revisit time over multiple fields and throughout several growing seasons. We also used a more conservative grouping of ECOSTRESS overpasses by differentiating ESI observations between early morning (i.e., 5:00 a.m. to 9:59 a.m.) and early afternoon (i.e., 12:00 p.m. to 16:59 p.m.). Despite a consistent megadrought, we also observed some shifts in ECOSTRESS ESI between the seasons studied. It is worth noting that these inter-seasonal changes looked substantially different across various times of day. A decrease in afternoon ESI (e.g., lower stress) between the 2019–2020 and 2020–2021 cycles could potentially be explained by lower seasonal temperatures (Table 1). To gain confidence in the ESI snapshots from different times of day, we examined how well ECOSTRESS observations agreed with patterns across irrigated land. We found that morning ESI systematically aligned best with crop health and water rights and exhibited lower spatial variability (even across large agricultural areas). Our results held consistent, whether we used Kendall's τ or Spearman's ρ coefficients (Tables S1 and S2). These findings contradict previous studies that found greater discrepancies in ECOSTRESS PT-JPL products during the morning hours [32–34]. However, we should note that our study focuses on relatively larger study areas (i.e., >900 ha.) and longer timescales (i.e., median over a season), which might not be as sensitive to shifts in micro-meteorological conditions.

4.4. Societal Implications

Our research showed how agricultural parcels with the same crop faced more or less stress over the course of a season, and how various crops also exhibited different stress responses from year to year. However, one of our most actionable findings identified irrigated areas, which appeared to be disproportionately affected by lower water allocations. As such, ECOSTRESS ESI proved helpful in highlighting areas with sub-optimal crop conditions. This kind of knowledge could help guide much-needed adaptation actions as irrigated crops are faced with reduced water availability in the future. Some of these measures might include early sowing or double cropping, the production of drought-resistant crops, the adoption of irrigation technologies, near-real-time monitoring tools, and the redesign of subsidy programs [44,102–106]. This type of information at the community level is even more relevant in the context of smallholder and subsistence agriculture [107]. Therefore, the unprecedented spatial resolution and diurnal sampling provided by ECOSTRESS ESI could offer additional insights to support sustainable water management (SDG #6) and reduce irrigated agriculture's vulnerability (SDG #2) and inequalities (SDG #10) in the midst of climate extremes (#SDG #13).

4.5. Future Work

Accessing information on referential water-use rights was critical to our analysis but proved difficult. In line with the recommendations of others, we suggest that these data be compiled by authorities at the national level and be made publicly available [92]. While documented values of water allocation were used in this analysis, it should be noted that referential water-use amounts were not necessarily utilized at their full volumetric rate during the timeframe studied. This is because water availability varies from year to year. Therefore, the installation of flow meters and the reporting of water withdrawals across the basin would provide a better measure of consumptive water use [52].

In addition, our research was possible because the study area was not severely impacted by cloud cover and had frequent ECOSTRESS observations with nominal quality or higher (Figures S2 and S3). Moving forward, ECOSTRESS will require additional testing, especially in complex terrain and humid regions with missing data [31,33]. It is also important to continue exploring the relationship between TIR and optical signal, and to what extent the former goes beyond the latter. ECOSTRESS serves as a precursor for upcoming TIR missions [108]. Gaining insights into where these remotely sensed data have discrepancies and why is critical to improving agricultural monitoring [37]. In the future, one could hope to monitor agricultural outputs and crop yield using this information from space [18,19]. But for data from ECOSTRESS or other missions to be used operationally, it should also be made available online on a near-real-time basis.

5. Conclusions

ECOSTRESS measures TIR radiation at a 70-m spatial resolution (about 50 times finer than MODIS). Its unique vantage point onboard the ISS provides observations of agricultural productions at different revisit times for critical regions, such as the Maipo River Basin. Current understanding of the effects of overpass time on PT-JPL retrievals is limited, especially for the ESI product. To the best of our knowledge, this study is the first to examine how ESI varies as a function of the ECOSTRESS overpass time. Throughout our analyses, we found substantial variations in ECOSTRESS ESI at different times of the day. Based on that, we recommend differentiating ECOSTRESS observations into two groups for early morning (i.e., 5:00 a.m. to 9:59 a.m.) and early afternoon (i.e., 12:00 p.m. to 16:59 p.m.). In parts of our study area, morning ESI values were up to four times higher (i.e., lower stress) than in the afternoon. In addition, we found contrasting spatial patterns in ESI based on the overpass time. Therefore, we analyzed how well the different ECOSTRESS snapshots agreed with patterns in the environment. To that end, we created an extensive compilation of surface- and ground-water allocations across a combined 126,000 hectares of irrigated land in the Maipo River Basin. We identified a positive correlation ($p < 0.001$) between the ranks in morning ESI and those in allocated water, meaning that areas with higher water allocations tended to exhibit higher morning ESI (i.e., lower stress). We also found that year after year, morning ESI was best aligned with Landsat 8 OLI NDVI (a proxy for crop health), yielding correlation coefficients between 0.544 and 0.829. Beyond these monotonic associations, ECOSTRESS ESI proved helpful in detecting shifts in evaporative stress and areas with sub-optimal crop conditions. This type of information could inform future decisions for sustainable water withdrawals and crop selection in different parts of the basin and beyond. More broadly, our work gave new insights into the diurnal sampling provided by ECOSTRESS and the effects of overpass time on TIR measurements, in support of future agricultural monitoring.

Supplementary Materials: The following supporting information can be downloaded at: https://www.mdpi.com/article/10.3390/rs16173174/s1, Figure S1: Number of ESI observations from ECOSTRESS; Figure S2: Number of NDVI observations from Landsat 8; Figure S3: Trends in the relative standard deviation of ESI against the size of agricultural areas; Table S1: Correlation coefficients between ESI and NDVI; Table S2: Correlation coefficients between ESI and Water-Use Rights.

Author Contributions: Conceptualization, B.D.G., C.C.C.-M. and F.N.-R.; methodology, B.D.G. and C.C.C.-M.; software, B.D.G. and C.C.C.-M.; validation, B.D.G.; formal analysis, B.D.G.; data curation, B.D.G. and C.C.C.-M.; writing—original draft preparation, B.D.G.; writing—review and editing, B.D.G., C.C.C.-M., F.N.-R., D.D.G. and V.L.; visualization, B.D.G.; supervision, V.L.; project administration, B.D.G.; funding acquisition, B.D.G., D.D.G. and V.L. All authors have read and agreed to the published version of the manuscript.

Funding: This project was made possible by an award from the Center for Global Health Equity at the University of Virginia. This work was also supported by funding from the University of Virginia School of Engineering and Applied Science and the Jefferson Scholars Foundation.

Data Availability Statement: The ECOSTRESS products discussed in this study are publicly available via the NASA Earthdata Search and the AppEEARS platforms at https://lpdaac.usgs.gov/products/eco4esiptjplv001/ (accessed on 11 June 2024). Surface reflectance data from Landsat 8 OLI (Level 2, Collection 2, Tier 1) can be acquired using the Google Earth Engine Data Catalog at https://earthengine.google.com/ (accessed on 23 July 2024). Our compilation of water-use rights in the Maipo River Basin is also made accessible online at https://github.com/ccalvocm/research/blob/main/data/arDAA.gpkg (accessed on 26 August 2024).

Acknowledgments: The authors would like to recognize the staff members of Chile's Centro de Información de Recursos Naturales (CIREN) for their guidance and feedback.

Conflicts of Interest: The authors declare no conflict of interest.

References

1. Sauer, T.; Havlík, P.; Schneider, U.A.; Schmid, E.; Kindermann, G.; Obersteiner, M. Agriculture and resource availability in a changing world: The role of irrigation. *Water Resour. Res.* **2010**, *46*, e2009WR007729. [CrossRef]
2. Howden, S.M.; Soussana, J.F.; Tubiello, F.N.; Chhetri, N.; Dunlop, M.; Meinke, H. Adapting agriculture to climate change. *Proc. Natl. Acad. Sci. USA* **2007**, *104*, 19691–19696. [CrossRef] [PubMed]
3. Heim, R.R., Jr. A review of twentieth-century drought indices used in the United States. *Bull. Am. Meteorol. Soc.* **2002**, *83*, 1149–1166. [CrossRef]
4. Fisher, J.B.; Whittaker, R.J.; Malhi, Y. ET come home: Potential evapotranspiration in geographical ecology. *Glob. Ecol. Biogeogr.* **2011**, *20*, 1–18. [CrossRef]
5. Begg, J.E.; Turner, N.C. Crop water deficits. *Adv. Agron.* **1976**, *28*, 161–217. [CrossRef]
6. Wang, K.; Dickinson, R.E. A review of global terrestrial evapotranspiration: Observation, modeling, climatology, and climatic variability. *Rev. Geophys.* **2012**, *50*, e2011RG000373. [CrossRef]
7. Lobell, D.B.; Gourdji, S.M. The influence of climate change on global crop productivity. *Plant Physiol.* **2012**, *160*, 1686–1697. [CrossRef]
8. Narasimhan, B.; Srinivasan, R. Development and evaluation of Soil Moisture Deficit Index (SMDI) and Evapotranspiration Deficit Index (ETDI) for agricultural drought monitoring. *Agric. For. Meteorol.* **2005**, *133*, 69–88. [CrossRef]
9. Koehler, T.; Wankmüller, F.J.; Sadok, W.; Carminati, A. Transpiration response to soil drying vs. increasing vapor pressure deficit in crops–physical and physiological mechanisms and key plant traits. *J. Exp. Bot.* **2023**, *74*, 4789–4807. [CrossRef]
10. Nelson, G.C.; Rosegrant, M.W.; Koo, J.; Robertson, R.; Sulser, T.; Zhu, T.; Ringler, C.; Msangi, S.; Palazzo, A.; Batka, M.; et al. *Climate Change: Impact on Agriculture and Costs of Adaptation*; International Food Policy Research Institute (IFPRI): Washington, DC, USA, 2009; Volume 21. Available online: https://ebrary.ifpri.org/utils/getfile/collection/p15738coll2/id/130648/filename/130821.pdf (accessed on 23 July 2024).
11. Dai, A. Drought under global warming: A review. *Wiley Interdiscip. Rev. Clim. Chang.* **2011**, *2*, 45–65. [CrossRef]
12. Fisher, J.B.; Melton, F.; Middleton, E.; Hain, C.; Anderson, M.; Allen, R.; McCabe, M.F.; Hook, S.; Baldocchi, D.; Townsend, P.A.; et al. The future of evapotranspiration: Global requirements for ecosystem functioning, carbon and climate feedbacks, agricultural management, and water resources. *Water Resour. Res.* **2017**, *53*, 2618–2626. [CrossRef]
13. Fisher, J.B. *Level-4 Evaporative Stress Index L4 (ESI_PT-JPL) Algorithm Theoretical Basis Document*; JPL: Pasadena, CA, USA, 2018. Available online: https://lpdaac.usgs.gov/documents/342/ECO4ESIPTJPL_ATBD_V1.pdf (accessed on 23 July 2024).
14. Anderson, M.C.; Hain, C.; Wardlow, B.; Pimstein, A.; Mecikalski, J.R.; Kustas, W.P. Evaluation of drought indices based on thermal remote sensing of evapotranspiration over the continental United States. *J. Clim.* **2011**, *24*, 2025–2044. [CrossRef]
15. Anderson, M.C.; Kustas, W.P.; Norman, J.M.; Hain, C.R.; Mecikalski, J.R.; Schultz, L.; González-Dugo, M.P.; Cammalleri, C.; D'Urso, G.; Pimstein, A.; et al. Mapping daily evapotranspiration at field to continental scales using geostationary and polar orbiting satellite imagery. *Hydrol. Earth Syst. Sci.* **2011**, *15*, 223–239. [CrossRef]
16. Anderson, M.C.; Hain, C.; Otkin, J.; Zhan, X.; Mo, K.; Svoboda, M.; Wardlow, B.; Pimstein, A. An intercomparison of drought indicators based on thermal remote sensing and NLDAS-2 simulations with US Drought Monitor classifications. *J. Hydrometeorol.* **2013**, *14*, 1035–1056. [CrossRef]
17. Otkin, J.A.; Anderson, M.C.; Hain, C.; Mladenova, I.E.; Basara, J.B.; Svoboda, M. Examining rapid onset drought development using the thermal infrared–based evaporative stress index. *J. Hydrometeorol.* **2013**, *14*, 1057–1074. [CrossRef]
18. Anderson, M.C.; Zolin, C.A.; Sentelhas, P.C.; Hain, C.R.; Semmens, K.; Yilmaz, M.T.; Gao, F.; Otkin, J.A.; Tetrault, R. The Evaporative Stress Index as an indicator of agricultural drought in Brazil: An assessment based on crop yield impacts. *Remote Sens. Environ.* **2016**, *174*, 82–99. [CrossRef]
19. Otkin, J.A.; Anderson, M.C.; Hain, C.; Svoboda, M.; Johnson, D.; Mueller, R.; Tadesse, T.; Wardlow, B.; Brown, J. Assessing the evolution of soil moisture and vegetation conditions during the 2012 United States flash drought. *Agric. For. Meteorol.* **2016**, *218*, 230–242. [CrossRef]
20. Nguyen, H.; Wheeler, M.C.; Otkin, J.A.; Cowan, T.; Frost, A.; Stone, R. Using the evaporative stress index to monitor flash drought in Australia. *Environ. Res. Lett.* **2019**, *14*, 064016. [CrossRef]

21. Anderson, M.C.; Norman, J.M.; Mecikalski, J.R.; Otkin, J.A.; Kustas, W.P. A climatological study of evapotranspiration and moisture stress across the continental United States based on thermal remote sensing: 2. Surface moisture climatology. *J. Geophys. Res. Atmos.* **2007**, *112*, e2006JD007507. [CrossRef]
22. Fisher, J.B.; Lee, B.; Purdy, A.J.; Halverson, G.H.; Dohlen, M.B.; Cawse-Nicholson, K.; Wang, A.; Anderson, R.G.; Aragon, B.; Arain, M.A.; et al. ECOSTRESS: NASA's next generation mission to measure evapotranspiration from the international space station. *Water Resour. Res.* **2020**, *56*, e2019WR026058. [CrossRef]
23. Hulley, G.; Hook, S.; Fisher, J.; Lee, C. ECOSTRESS, A NASA Earth-Ventures Instrument for studying links between the water cycle and plant health over the diurnal cycle. In Proceedings of the 2017 IEEE International Geoscience and Remote Sensing Symposium (IGARSS), Fort Worth, TX, USA, 23–28 July 2017; pp. 5494–5496. [CrossRef]
24. Fisher, J.B.; Tu, K.P.; Baldocchi, D.D. Global estimates of the land–atmosphere water flux based on monthly AVHRR and ISLSCP-II data, validated at 16 FLUXNET sites. *Remote Sens. Environ.* **2008**, *112*, 901–919. [CrossRef]
25. Fisher, J.B. *Level-3 Evapotranspiration L3 (ET_PT-JPL) Algorithm Theoretical Basis Document*; JPL: Pasadena, CA, CA, USA, 2018. Available online: https://lpdaac.usgs.gov/documents/335/ECO3ETPTJPL_ATBD_V1.pdf (accessed on 23 July 2024).
26. Xiao, J.; Fisher, J.B.; Hashimoto, H.; Ichii, K.; Parazoo, N.C. Emerging satellite observations for diurnal cycling of ecosystem processes. *Nat. Plants* **2021**, *7*, 877–887. [CrossRef] [PubMed]
27. Chen, Y.; Xia, J.; Liang, S.; Feng, J.; Fisher, J.B.; Li, X.; Li, X.; Liu, S.; Ma, Z.; Miyata, A.; et al. Comparison of satellite-based evapotranspiration models over terrestrial ecosystems in China. *Remote Sens. Environ.* **2014**, *140*, 279–293. [CrossRef]
28. Ershadi, A.; McCabe, M.F.; Evans, J.P.; Chaney, N.W.; Wood, E.F. Multi-site evaluation of terrestrial evaporation models using FLUXNET data. *Agric. For. Meteorol.* **2014**, *187*, 46–61. [CrossRef]
29. McCabe, M.F.; Ershadi, A.; Jimenez, C.; Miralles, D.G.; Michel, D.; Wood, E.F. The GEWEX LandFlux project: Evaluation of model evaporation using tower-based and globally gridded forcing data. *Geosci. Model Dev.* **2016**, *9*, 283–305. [CrossRef]
30. Michel, D.; Jiménez, C.; Miralles, D.G.; Jung, M.; Hirschi, M.; Ershadi, A.; Martens, B.; McCabe, M.F.; Fisher, J.B.; Mu, Q.; et al. The WACMOS-ET project–Part 1: Tower-scale evaluation of four remote-sensing-based evapotranspiration algorithms. *Hydrol. Earth Syst. Sci.* **2016**, *20*, 803–822. [CrossRef]
31. Liu, N.; Oishi, A.C.; Miniat, C.F.; Bolstad, P. An evaluation of ECOSTRESS products of a temperate montane humid forest in a complex terrain environment. *Remote Sens. Environ.* **2021**, *265*, 112662. [CrossRef]
32. Wu, J.; Feng, Y.; Liang, L.; He, X.; Zeng, Z. Assessing evapotranspiration observed from ECOSTRESS using flux measurements in agroecosystems. *Agric. Water Manag.* **2022**, *269*, 107706. [CrossRef]
33. Liang, L.; Feng, Y.; Wu, J.; He, X.; Liang, S.; Jiang, X.; de Oliveira, G.; Qiu, J.; Zeng, Z. Evaluation of ECOSTRESS evapotranspiration estimates over heterogeneous landscapes in the continental US. *J. Hydrol.* **2022**, *613*, 128470. [CrossRef]
34. Hu, T.; Mallick, K.; Hitzelberger, P.; Didry, Y.; Boulet, G.; Szantoi, Z.; Koetz, B.; Alonso, I.; Pascolini-Campbell, M.; Halverson, G.; et al. Evaluating European ECOSTRESS Hub Evapotranspiration Products Across a Range of Soil-Atmospheric Aridity and Biomes Over Europe. *Water Resour. Res.* **2023**, *59*, e2022WR034132. [CrossRef]
35. Pascolini-Campbell, M.; Lee, C.; Stavros, N.; Fisher, J.B. ECOSTRESS reveals pre-fire vegetation controls on burn severity for Southern California wildfires of 2020. *Glob. Ecol. Biogeogr.* **2022**, *31*, 1976–1989. [CrossRef]
36. Goffin, B.D.; Thakur, R.; Carlos, S.D.C.; Srsic, D. Maipo River Valley Agriculture: Determining Crop Coefficients Using Remote Sensing for the Maipo River Valley Basin in Chile. NASA DEVELOP National Program: Virginia—LaRC, 2022. [Unpublished Manuscript]. Available online: https://ntrs.nasa.gov/api/citations/20220013971/downloads/2022Sum_PUP_MaipoRiverValleyAg_TechPaper_FD_v2.docx.pdf (accessed on 23 July 2024).
37. Fritz, S.; See, L.; Bayas, J.C.L.; Waldner, F.; Jacques, D.; Becker-Reshef, I.; Whitcraft, A.; Baruth, B.; Bonifacio, R.; Crutchfield, J.; et al. A comparison of global agricultural monitoring systems and current gaps. *Agric. Syst.* **2019**, *168*, 258–272. [CrossRef]
38. Dirección General de Aguas (DGA); Centro de Información de Recursos Naturales (CIREN). *Redefinición de la Clasificación Red Hidrográfica a Nivel Nacional*; Dirección General de Aguas (DGA): Santiago, Chile, 2014. Available online: https://snia.mop.gob.cl/repositoriodga/handle/20.500.13000/6786 (accessed on 1 December 2023).
39. Centro de Información de Recursos Naturales (CIREN). *Progama de Actualización de Cobertura y Uso de Suelo Agrícola: Región de Valparaíso*; CIREN: Santiago, Chile, 2020.
40. Centro de Información de Recursos Naturales (CIREN). *Diagnóstico Estudio Monitoreo Territorial Hortícola de la Región Metropolitana de Santiago*; CIREN: Santiago, Chile, 2021; Publicación N°225. Available online: https://bibliotecadigital.ciren.cl/handle/20.500.13082/147767 (accessed on 23 July 2024).
41. Centro de Información de Recursos Naturales (CIREN). *Diagnóstico Territorial de la Situación Hortícola Región de O'Higgins*; CIREN: Santiago, Chile, 2017; Publicación N°200. Available online: https://bibliotecadigital.ciren.cl/handle/20.500.13082/26496 (accessed on 23 July 2024).
42. Peña-Guerrero, M.D.; Nauditt, A.; Muñoz-Robles, C.; Ribbe, L.; Meza, F. Drought impacts on water quality and potential implications for agricultural production in the Maipo River Basin, Central Chile. *Hydrol. Sci. J.* **2020**, *65*, 1005–1021. [CrossRef]
43. Meza, F.J.; Vicuña, S.; Jelinek, M.; Bustos, E.; Bonelli, S. Assessing water demands and coverage sensitivity to climate change in the urban and rural sectors in central Chile. *J. Water Clim. Chang.* **2014**, *5*, 192–203. [CrossRef]
44. Meza, F.J.; Silva, D. Dynamic adaptation of maize and wheat production to climate change. *Clim. Chang.* **2009**, *94*, 143–156. [CrossRef]

45. Falvey, M.; Garreaud, R.D. Regional cooling in a warming world: Recent temperature trends in the southeast Pacific and along the west coast of subtropical South America (1979–2006). *J. Geophys. Res. Atmos.* **2009**, *114*, e2008JD010519. [CrossRef]
46. Garreaud, R.D.; Alvarez-Garreton, C.; Barichivich, J.; Boisier, J.P.; Christie, D.; Galleguillos, M.; LeQuesne, C.; McPhee, J.; Zambrano-Bigiarini, M. The 2010–2015 megadrought in central Chile: Impacts on regional hydroclimate and vegetation. *Hydrol. Earth Syst. Sci.* **2017**, *21*, 6307–6327. [CrossRef]
47. Garreaud, R.D.; Boisier, J.P.; Rondanelli, R.; Montecinos, A.; Sepúlveda, H.H.; Veloso-Aguila, D. The central Chile mega drought (2010–2018): A climate dynamics perspective. *Int. J. Climatol.* **2020**, *40*, 421–439. [CrossRef]
48. Meza, F.J.; Wilks, D.S.; Gurovich, L.; Bambach, N. Impacts of climate change on irrigated agriculture in the Maipo Basin, Chile: Reliability of water rights and changes in the demand for irrigation. *J. Water Resour. Plan. Manag.* **2012**, *138*, 421–430. [CrossRef]
49. Aldunce, P.; Araya, D.; Sapiain, R.; Ramos, I.; Lillo, G.; Urquiza, A.; Garreaud, R. Local perception of drought impacts in a changing climate: The mega-drought in central Chile. *Sustainability* **2017**, *9*, 2053. [CrossRef]
50. Cai, W.; McPhaden, M.J.; Grimm, A.M.; Rodrigues, R.R.; Taschetto, A.S.; Garreaud, R.D.; Dewitte, B.; Poveda, G.; Ham, Y.-G.; Santoso, A.; et al. Climate impacts of the El Niño–southern oscillation on South America. *Nat. Rev. Earth Environ.* **2020**, *1*, 215–231. [CrossRef]
51. Donoso, G. Management of water resources in agriculture in Chile and its challenges. *Cienc. Investig. Agrar. Rev. Latinoam. Cienc. Agric.* **2021**, *48*, 171–185. [CrossRef]
52. Melo, O.; Báez, N.; Acuña, D. Towards Sustainable Agriculture in Chile, Reflections on the Role of Public Policy. *Cienc. Investig. Agrar. Rev. Latinoam. Cienc. Agric.* **2021**, *48*, 186–209. [CrossRef]
53. Fernández, F.J.; Vásquez-Lavín, F.; Ponce, R.D.; Garreaud, R.; Hernández, F.; Link, O.; Zambrano, F.; Hanemann, M. The economics impacts of long-run droughts: Challenges, gaps, and way forward. *J. Environ. Manag.* **2023**, *344*, 118726. [CrossRef]
54. Taucare, M.; Viguier, B.; Figueroa, R.; Daniele, L. The alarming state of Central Chile's groundwater resources: A paradigmatic case of a lasting overexploitation. *Sci. Total Environ.* **2024**, *906*, 167723. [CrossRef]
55. Hook, S.; Fisher, J. ECOSTRESS Evaporative Stress Index PT-JPL Daily L4 Global 70 m V001 [Data Set]. NASA EOSDIS Land Processes Distributed Active Archive Center: Sioux Falls, SD, USA, 2019. Available online: https://lpdaac.usgs.gov/products/eco4esiptjplv001/ (accessed on 11 June 2024).
56. Halverson, G.; Fisher, J.B.; Lee, C.M. *Level-4 Evaporative Stress Index Priestley-Taylor Jet Propulsion Laboratory (PT-JPL) Data User Guide*; JPL: Pasadena, CA, USA, 2019. Available online: https://lpdaac.usgs.gov/documents/425/ECO4ESIPTJPL_User_Guide_V1.pdf (accessed on 23 July 2024).
57. Masek, J.G.; Vermote, E.F.; Saleous, N.E.; Wolfe, R.; Hall, F.G.; Huemmrich, K.F.; Gao, F.; Kutler, J.; Lim, T.K. A Landsat surface reflectance dataset for North America, 1990–2000. *IEEE Geosci. Remote Sens. Lett.* **2006**, *3*, 68–72. [CrossRef]
58. Vermote, E.; Justice, C.; Claverie, M.; Franch, B. Preliminary analysis of the performance of the Landsat 8/OLI land surface reflectance product. *Remote Sens. Environ.* **2016**, *185*, 46–56. [CrossRef]
59. Zambrano, F.; Vrieling, A.; Nelson, A.; Meroni, M.; Tadesse, T. Prediction of drought-induced reduction of agricultural productivity in Chile from MODIS, rainfall estimates, and climate oscillation indices. *Remote Sens. Environ.* **2018**, *219*, 15–30. [CrossRef]
60. Tucker, C.J. Red and photographic infrared linear combinations for monitoring vegetation. *Remote Sens. Environ.* **1979**, *8*, 127–150. [CrossRef]
61. Oumarou Abdoulaye, A.; Lu, H.; Zhu, Y.; Alhaj Hamoud, Y.; Sheteiwy, M. The global trend of the net irrigation water requirement of maize from 1960 to 2050. *Climate* **2019**, *7*, 124. [CrossRef]
62. Goffin, B.D.; Thakur, R.; Carlos, S.D.C.; Srsic, D.; Williams, C.; Ross, K.; Neira-Román, F.; Cortés-Monroy, C.C.; Lakshmi, V. Leveraging remotely-sensed vegetation indices to evaluate crop coefficients and actual irrigation requirements in the water-stressed Maipo River Basin of Central Chile. *Sustain. Horiz.* **2022**, *4*, 100039. [CrossRef]
63. NASA JPL. *NASA Shuttle Radar Topography Mission Global 1 Arc Second [Data Set]*; NASA EOSDIS Land Processes Distributed Active Archive Center: Sioux Falls, SD, USA, 2013. Available online: https://lpdaac.usgs.gov/products/srtmgl1v003/ (accessed on 9 March 2024).
64. Funk, C.C.; Peterson, P.J.; Landsfeld, M.F.; Pedreros, D.H.; Verdin, J.P.; Rowland, J.D.; Romero, B.E.; Husak, G.J.; Michaelsen, J.C.; Verdin, A.P. *A Quasi-Global Precipitation Time Series for Drought Monitoring: Data Series 832*; U.S. Geological Survey: Reston, VA, USA, 2014. [CrossRef]
65. Hulley, G.; Hook, S. MODIS/Terra Land Surface Temperature/3-Band Emissivity 8-Day L3 Global 1km SIN Grid V061 [Data Set]. NASA EOSDIS Land Processes Distributed Active Archive Center: Sioux Falls, SD, USA, 2021. Available online: https://lpdaac.usgs.gov/products/mod21a2v061/ (accessed on 22 July 2024).
66. Duran-Llacer, I.; Munizaga, J.; Arumí, J.L.; Ruybal, C.; Aguayo, M.; Sáez-Carrillo, K.; Arriagada, L.; Rojas, O. Lessons to be learned: Groundwater depletion in Chile's Ligua and Petorca watersheds through an Interdisciplinary approach. *Water* **2020**, *12*, 2446. [CrossRef]
67. Muñoz, A.A.; Klock-Barría, K.; Alvarez-Garreton, C.; Aguilera-Betti, I.; González-Reyes, Á.; Lastra, J.A.; Chávez, R.O.; Barría, P.; Christie, D.; Rojas-Badilla, M.; et al. Water crisis in Petorca Basin, Chile: The combined effects of a mega-drought and water management. *Water* **2020**, *12*, 648. [CrossRef]
68. Olivera-Guerra, L.; Quintanilla, M.; Moletto-Lobos, I.; Pichuante, E.; Zamorano-Elgueta, C.; Mattar, C. Water dynamics over a Western Patagonian watershed: Land surface changes and human factors. *Sci. Total Environ.* **2022**, *804*, 150221. [CrossRef]

69. Dirección General de Aguas (DGA). *Plan Estratégico de Gestión Hídrica en la Cuenca del Maipo*; S.I.T. N° 471; Realizado por ICASS SpA; DGA: Santiago, Chile, 2021.
70. Comisión Nacional de Riego (CNR). *Estudio Integral de Optimización del Regadío de la 3a Sección del Río Maipo y Valles del Yali y Alhué*; Realizado por Geofun Ltda. Agosto; CNR: Santiago, Chile, 2001; Volume 1.
71. Dirección General de Aguas (DGA). *Diagnóstico, Disponbilidad y Requerimientos de Agua en la Región Metropolitana*; Departamento de Estudios S.I.T. N°10. Realizado por IPLA Ltda.; DGA: Santiago, Chile, 1993.
72. Comisión Nacional de Riego (CNR). *Anexo 4. Hidrología Río Angostura*; CNR: Santiago, Chile, 2016.
73. Comisión Nacional de Riego (CNR). *Anexo 4. Hidrología Canal Hospital*; CNR: Santiago, Chile, 2016.
74. Sociedad del Canal de Maipo. Available online: https://www.scmaipo.cl/canalistas (accessed on 27 October 2023).
75. Dirección General de Aguas (DGA). *Evaluación de los Recursos Hídricos Superficiales en la Cuenca del Río Maipo*; Informe Técnico; Realizado por: Departamento de Administración de Recursos Hídricos; S.D.T. N°145; DGA: Santiago, Chile, 2003.
76. Asociación de Canales Unidos de Buin. *Memoria Balance*; Asociación de Canales Unidos de Buin: Buin, Chile, 2013.
77. Asociacion Canal Huidobro. Caudales Historicos. Available online: https://www.canalhuidobro.cl/2021/03/09/caudales-historicos/ (accessed on 27 October 2023).
78. Asociación de Canalistas Canal de Pirque. Caudales y Estadísticas. Available online: https://canaldepirque.cl/caudales-2/ (accessed on 27 October 2023).
79. Comisión Nacional de Riego (CNR). *Programa de Capacitación Organizacional Piloto en la Tercera Sección del Río Maipo. Región Metropolitana*; CNR: Santiago, Chile, 2007.
80. CEPIA Ingenieros Consultores. *Anexo 9.4.1 Análisis Hidrológico. Construcción de Obra de Distribución en Canales Santa Cruz, Romero y Castillo*; CEPIA Ingenieros Consultores: Talca, Chile, 2019.
81. Dirección General de Aguas (DGA). *Catastro de Usuarios de Aguas de la Subcuenca del Río Mapocho Región Metropolitana. Informe Final*; Realizado por IPLA Ingenieros Consultores; DGA: Santiago, Chile, 1989.
82. Asociación del Canal de Las Mercedes. *Estatutos Refundidos de la Asociación del Canal de Las Mercedes*; Asociación del Canal de Las Mercedes: Santiago, Chile, 2023.
83. Comisión Nacional de Riego (CNR). *Estudio Para el Desarrollo Agrícola y Manejo de Aguas del Area Metropolitana. Informe Principal*; CNR: Santiago, Chile, 1999; Volume 1.
84. Aquasys Ingenieros Consultores. *Anexo 4 Hidrología Río Peuco*; Aquasys Ingenieros Consultores: Santiago, Chile, 2016.
85. Dirección General de Aguas (DGA). *Recopilación y Sistematización de Información de Derechos de Agua Otorgados por el SAG*; Realizado por CIREN; DGA: Santiago, Chile, 2013.
86. Dirección General de Aguas (DGA). *Diagnóstico Nacional de Organizaciones de Usuarios: Informe Final*; Realizado por Laboratorio de Análisis Territorial, Facultad de Ciencias Agronómicas, Universidad de Chile; DGA: Santiago, Chile, 2018.
87. Dirección General de Aguas (DGA). *Derechos de Aprovechamiento de Aguas Registrados en DGA*; DGA: Santiago, Chile, 2023.
88. Lehmann, E.L.; D'Abrera, H.J. *Nonparametrics: Statistical Methods Based on Ranks*; Springer: New York, NY, USA, 2006; Volume 464.
89. Sneyers, R. *On the Statistical Analysis of Series of Observations*; Technical Note No. 143 WMO No. 415; World Meteorological Organization: Geneva, Switzerland, 1990; p. 192.
90. Kendall, M. *Rank Correlation Measures*; Charles Griffin: London, UK, 1975; p. 15.
91. Yue, S.; Pilon, P.; Cavadias, G. Power of the Mann–Kendall and Spearman's rho tests for detecting monotonic trends in hydrological series. *J. Hydrol.* **2002**, *259*, 254–271. [CrossRef]
92. Altemus-Cullen, K. A review of applications of remote sensing for drought studies in the Andes region. *J. Hydrol. Reg. Stud.* **2023**, *49*, 101483. [CrossRef]
93. Palmer, W.C. *Meteorological Drought*; United States Weather Bureau: Washington, DC, USA, 1965; pp. 1–56.
94. Selyaninov, G.T. Methods of agricultural climatology. *Agric. Meteorol* **1930**. (In Russian)
95. McKee, T.B.; Doesken, N.J.; Kleist, J. The relationship of drought frequency and duration to time scales. In Proceedings of the 8th Conference on Applied Climatology, Anaheim, CA, USA, 17–22 January 1993; Volime 17; pp. 179–183. Available online: https://climate.colostate.edu/pdfs/relationshipofdroughtfrequency.pdf (accessed on 23 July 2024).
96. Mu, Q.; Zhao, M.; Kimball, J.S.; McDowell, N.G.; Running, S.W. A remotely sensed global terrestrial drought severity index. *Bull. Am. Meteorol. Soc.* **2013**, *94*, 83–98. [CrossRef]
97. Dabrowska-Zielinska, K.; Malinska, A.; Bochenek, Z.; Bartold, M.; Gurdak, R.; Paradowski, K.; Lagiewska, M. Drought model diss based on the fusion of satellite and meteorological data under variable climatic conditions. *Remote Sens.* **2020**, *12*, 2944. [CrossRef]
98. Oertel, M.; Meza, F.J.; Gironás, J. Multivariate Standardized Drought Indices to Identify Drought Events: Application in the Maipo River Basin. In *Towards Water Secure Societies: Coping with Water Scarcity and Quality Challenges*; Springer: Berlin/Heidelberg, Germany, 2021; pp. 141–160. [CrossRef]
99. Steinemann, A. Drought information for improving preparedness in the western states. *Bull. Am. Meteorol. Soc.* **2014**, *95*, 843–847. [CrossRef]
100. Wen, J.; Fisher, J.B.; Parazoo, N.C.; Hu, L.; Litvak, M.E.; Sun, Y. Resolve the Clear-Sky Continuous Diurnal Cycle of High-Resolution ECOSTRESS Evapotranspiration and Land Surface Temperature. *Water Resour. Res.* **2022**, *58*, e2022WR032227. [CrossRef]

101. Guillevic, P.C.; Olioso, A.; Hook, S.J.; Fisher, J.B.; Lagouarde, J.P.; Vermote, E.F. Impact of the revisit of thermal infrared remote sensing observations on evapotranspiration uncertainty—A sensitivity study using AmeriFlux Data. *Remote Sens.* **2019**, *11*, 573. [CrossRef]
102. Meza, F.J.; Silva, D.; Vigil, H. Climate change impacts on irrigated maize in Mediterranean climates: Evaluation of double cropping as an emerging adaptation alternative. *Agric. Syst.* **2008**, *98*, 21–30. [CrossRef]
103. Donoso, G.; Blanco, E.; Franco, G.; Lira, J. Water footprints and irrigated agricultural sustainability: The case of Chile. *Int. J. Water Resour. Dev.* **2016**, *32*, 738–748. [CrossRef]
104. Jordan, C.; Donoso, G.; Speelman, S. Measuring the effect of improved irrigation technologies on irrigated agriculture. A study case in Central Chile. *Agric. Water Manag.* **2021**, *257*, 107160. [CrossRef]
105. Jordan, C.; Donoso, G.; Speelman, S. Irrigation subsidy policy in Chile: Lessons from the allocation, uneven distribution and water resources implications. *Int. J. Water Resour. Dev.* **2023**, *39*, 133–154. [CrossRef]
106. Moletto-Lobos, I.; Mattar, C.; Barichivich, J. Performance of satellite-based evapotranspiration models in temperate pastures of Southern Chile. *Water* **2020**, *12*, 3587. [CrossRef]
107. Morton, J.F. The impact of climate change on smallholder and subsistence agriculture. *Proc. Natl. Acad. Sci. USA* **2007**, *104*, 19680–19685. [CrossRef]
108. Cawse-Nicholson, K.; Townsend, P.A.; Schimel, D.; Assiri, A.M.; Blake, P.L., Buongiorno, M.F.; Campbell, P.; Carmon, N.; Casey, K.A.; Correa-Pabón, R.E.; et al. NASA's surface biology and geology designated observable: A perspective on surface imaging algorithms. *Remote Sens. Environ.* **2021**, *257*, 112349. [CrossRef]

remote sensing

Article

Spatial Distribution of Cropping Systems in South Asia Using Time-Series Satellite Data Enriched with Ground Data

Murali Krishna Gumma [1,*], Pranay Panjala [1], Sunil K. Dubey [2], Deepak K. Ray [3], C. S. Murthy [2], Dakshina Murthy Kadiyala [4], Ismail Mohammed [1] and Yamano Takashi [5]

1 International Crops Research Institute for the Semi-Arid Tropics (ICRISAT), Patancheru 502324, India; pranay.panjala@icrisat.org (P.P.)
2 Mahalanobis National Crop Forecast Centre (MNCFC), New Delhi 110012, India
3 Institute on the Environment, University of Minnesota, St. Paul, MN 55108, USA
4 Department of Agronomy, Acharya NG Ranga Agricultural University, Guntur 522034, India
5 Asian Development Bank, Dhaka 1207, Bangladesh
* Correspondence: muralikrishna.gumma@icrisat.org or gummamk@gmail.com

Abstract: A cropping system practice is the sequential cultivation of crops in different crop seasons of a year. Cropping system practices determine the land productivity and sustainability of agriculture in regions and, therefore, information on cropping systems of different regions in the form of maps and statistics form critical inputs in crop planning for optimal use of resources. Although satellite-based crop mapping is widely practiced, deriving cropping systems maps using satellites is less reported. Here, we developed moderate-resolution maps of the major cropping systems of South Asia for the year 2014–2015 using multi-temporal satellite data together with a spectral matching technique (SMT) developed with an extensive set of field observation data supplemented with expert-identified crops in high-resolution satellite images. We identified and mapped 27 major cropping systems of South Asia at 250 m spatial resolution. The rice-wheat cropping system is the dominant system, followed by millet-wheat and soybean-wheat. The map showing the cropping system practices of regions opens up many use cases related to the agriculture performance of the regions. Comparison of such maps of different time periods offers insights on sensitive regions and analysis of such maps in conjunction with resources maps such as climate, soil, etc., enables optimization of resources vis-à-vis enhancing land productivity. Thus, the current study offers new opportunities to revisit the cropping system practices and redesign the same to meet the challenges of food security and climate resilient agriculture.

Keywords: cropping systems; South Asia; crop type mapping; time-series analysis; crop phenology detection

1. Introduction

A cropping system indicates the crop types and sequences practiced in a region over the crop seasons of a year. Generally, a crop year consists of multiple crop seasons of roughly defined duration. The crops grown during these seasons comprise the cropping system of that region, characterizing the agricultural practices followed in that area as a whole rather than in an individual parcel of land. In India, there are three crop seasons: *kharif* (June to November), *rabi* (November to March) and *zaid* (March to June). Crops are grown in one, two or all three of these seasons, with variations from region to region. For instance, if rice is grown in the initial season followed by wheat in the second season and no crop in the third, then it is called a rice-wheat cropping system. The agroecological and agrometeorological conditions of a region have a bearing on the cropping system practiced there.

Cropping systems play a role in determining the land productivity of a region by facilitating optimal utilization of resources to maximize crop yields [1]. However, resource-intensive cropping systems tend to impact soil health, ground and surface-water hydrology, environmental sustainability, climate resilience of agriculture and income security of farmers. Sustainable agriculture is closely linked to cropping system practices [2,3]. Therefore, monitoring cropping systems is of critical importance to sustainable agriculture and food production. In particular, cropping system maps can provide important information on crop types, the number of crops grown and the management practices followed in irrigated or rainfed systems during a crop year at a particular location. Given the growing population and rapid climate change, there is increasingly a need to monitor and reduce emissions in the quest for sustainable agriculture development.

Cereals and legumes, especially short-duration crops, are crucial for food and nutrition security in Asia and Africa. They have the potential to improve livelihoods by enhancing incomes and providing stable employment. However, increasing cropping intensity through better irrigation technology, fertilizer use and mechanization has led to growing pressure on croplands and other environmental concerns. The need to allocate water to other competing sectors (urban and industrial development) and reduced water availability in a highly varying climate inhibit the expansion of irrigated cropland area [4,5]. Additionally, the adverse impacts of accelerated irrigation development and fertilizer use have been manifest in increasing salinity in croplands, leading to loss of soil fertility.

Given that context, mapping cropping systems can provide a crucial input into decision-making to improve cropping systems, explore opportunities to increase production and raise farmers' incomes under conditions where there can be no further expansion of agricultural area. Additionally, accurate crop information from the lowest administrative level is necessary to estimate country-level crop-wise acreage [6–8], which is a basic input that feeds into agricultural policies. Since dryland cropping systems are highly mutable and are influenced by biophysical, social and economic factors [9–12], accurate estimation of cropping patterns is essential for planning. In such a scenario, geospatial products can provide input parameters for food security studies, and also for developing seasonal cropping patterns of cereals followed by short-duration legumes [13,14].

There is a need to revisit the cropping system practices of different regions in view of the growing importance of sustainable agriculture and food systems. With climate change leading to increased exposure of agriculture to various risks, and fast-changing food habits altering the demand and supply scenario of food, there is increasing recognition of the importance of carbon-neutral economies and also reassessment of the global trade practices. Mapping and inventory taking of croplands following different cropping systems are basic information products that can aid in the development of more suitable cropping systems in different regions. Spatial information on current cropping systems integrated with layers of information on resources such as soil, weather and water will enable a holistic assessment of the sustainability of current agricultural systems.

There have been numerous studies on mapping cropping patterns using satellite data, but they are limited to the subdistrict and plot levels [15–17] and single or major crops at large scale such as rice, wheat and sugarcane [18–22]. Currently, the various data sources available for use include regional and subnational statistical data on cultivated areas, coarse/medium-scale land-use/land-cover (LULC) maps and paddy maps from the 1980s to the 2010s [23–25], maps of rice areas using medium spatial resolution data [19,26–28] and a combination of various other sources [29,30].

Modern remote sensing is characterized by frequent imaging of the earth's surface to produce large streams of spectral data in high to moderate resolutions and making available analysis-ready data and biophysical products in the public domain to map agriculture systems. These developments in remote sensing offer numerous opportunities to produce more accurate and customized information on crop cover and its changes.

Information on spatial distribution of different cropping systems is important for spatial analysis to identify areas appropriate for sustainable food production, for a better understanding of where croplands are available to introduce/grow short-duration legumes across winter fallows, and to overcome current farmers' income uncertainties. Analysis of near-real-time satellite imagery can be a substitute for ground surveys to estimate the acreage of major crop types; moreover, it is relatively quick, cheap and independent [31–35]. Some studies have used satellite data to map agricultural areas at various temporal and spatial resolutions [36–41]; moderate resolution imaging spectrometer (MODIS) normalized difference vegetation index (MODIS NDVI) time-series data to map croplands, cropping intensity at lower administrative levels and for river basins [42–44]; and land surface water index (LSWI) data to monitor natural vegetation- and soil-related wetness at a regional scale [26,45–48].

However, to our knowledge, none of these studies have dealt with entire cropping systems, i.e., crops grown in a sequence over the entire year. This study maps major cropping systems across South Asia with the help of MODIS time-series data and spectral matching techniques (SMTs) to provide precise information on cropping sequences with the objective of enabling improvement of crop water productivity and proper natural resource management.

2. Materials and Methods

2.1. Study Area

South Asia is located between 5°38′ and 36°54′N latitudes and 61°05′ and 97°14′E longitudes, covering nearly 477 million ha. The region comprises Bangladesh, Bhutan, India, Nepal, Pakistan and Sri Lanka (Figure 1), and is divided, based on similar climatic conditions, into six agroecological zones (AEZs) [49] in which agriculture is strongly influenced by rainfall and seasonal winds. South Asia is home to 65% of the underprivileged people who live in rural and remote areas and depend on agriculture for their livelihood [50]. People living in 70% of the countryside depend on coastal fisheries and land. South Asia has nine major river basins: Brahmaputra, Godavari, Indus, Ganges, Tapti, Krishna, Kaveri, Narmada and Mahanadi. Irrigation projects in this region serve a command area of nearly 133 million ha [51]. Rice and wheat are the staple foods in the region; the former crop is grown throughout the year in some areas.

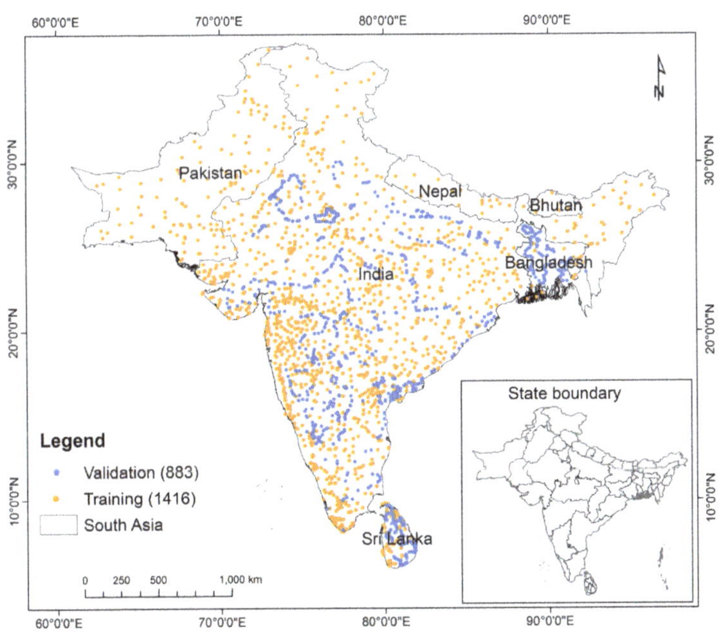

Figure 1. Spatial distribution of ground data collected for the crop year 2014–2015.

The major cropping systems in South Asia vary by country (Table 1). India has diversified cropping systems, but rice-wheat is dominant in northern India and rice-rice in southern India. In Sri Lanka, rice-rice, rice-fallow and other crops-fallow are the dominant cropping systems. In Bangladesh, rice-rice, rice-pulses-rice, rice-fallow-rice and jute-pulses-rice are popular; Bhutan favors rice-fallow and mixed crops. The dominant cropping systems in Nepal are rice-wheat, maize-wheat and maize-rice, whereas in Pakistan they are rice-wheat, rice-rice, rice-pulses and cotton. Thus, rice-rice and rice-wheat are the common major cropping systems across the region.

Table 1. Country-wise major cropping systems in South Asia.

Country	Total Geographical Area ('000 ha)	Total Gross Planted Area ('000 ha)	Major Cropping Systems
Bangladesh	14,804	15,002	Rice-rice, rice-pulses-rice, rice-fallow-rice, jute-pulses-rice
Bhutan	4365	121	Rice-fallow, mixed crops
India	345,623	184,443	Diversified cropping systems (Table 2)
Nepal	16,210	4208	Rice-wheat, maize-wheat, maize-rice
Pakistan	89,167	22,817	Rice-wheat, rice-rice, rice-pulses, cotton
Sri Lanka	6453	2076	Rice-rice, rice-fallow, other crops-fallow
Total	476,622	228,668	

Note: The gross planted area indicates the total sown area within a crop year, including multiple cropping cycles where applicable.

Table 2. Field samples used for training and validation, and national statistics on cropping systems in India.

Classified Data	Training Samples	Validation Samples	Cropping Systems in India (M ha)
01. Rice-wheat	42	46	14.8
02. Rice-rice	15	88	2.4
03. Rice-pulses	18	51	4
04. Pulses/rice-rice	13	107	4.5
05. Soybean-wheat	18	33	8.3
06. Pulses-wheat	17	22	9.2
07. Maize-wheat	35	24	NA
08. Millet-wheat	53	38	10.1
09. Maize-wheat	3	13	2.5
10. Maize-chickpea	16	14	4
11. Millet-mustard	10	20	3.8
12. Pulses-maize	9	10	2
13. Sugarcane	22	17	5.1
14. Groundnut-pulses	7	15	4
15. Sorghum-fallow	9	18	1.3
16. Rice-fallow	23	58	12.6
17. Pigeonpea-fallow	28	27	5.5
18. Groundnut/cotton	9	15	NA
19. Cotton-fallow	77	43	15.3
20. Millet-fallow	18	13	3.8
21. Sorghum-fallow	19	16	3.2
22. Pulses-fallow	9	11	1.3
23. Fallow-chickpea	4	16	1.4
24. Groundnut-fallow	18	14	9.7
25. Mixed crops	94	55	NA
26. Other LULC	773	61	NA
27. Rice-fallow/mixed crops	57	38	NA
Total samples	1416	883	

2.2. Ground Reference Data

Ground survey information with geographical coordinates (Table 2) was collected broadly throughout South Asia for the crop year 2014–2015 for identification of classes of cropping systems together with an accuracy assessment exercise (Figure 1). Ground data collection was conducted using two distinct approaches. For the training data, homogeneous patches were selected, and comprehensive crop-related information was collected to generate ideal spectra signatures. In contrast, the validation data were gathered from various random locations, focusing primarily on land use, land cover, crop name rather than detailed information. A total of 2203 field samples were collected, of which 1416 points were used for training to develop an ideal spectral data bank (ISDB) and 883 for class validation (Table 2). Land-use land-cover information was collected for each sample (at least 250 m × 250 m) along with data on crop type, cropping intensity and irrigation water source. During collection of ground data, the area around the sample was categorized into one of three classes: small (≤10 ha); medium (10–15 ha); and large (≥15 ha) to enable acreage estimation. In some areas, farmers were interviewed to glean information on sowing dates, irrigation practices and cropping patterns to aid class identification. For areas that could not be visited due to lack of proper roads, information was obtained from secondary sources, including local agriculture experts and records. Class names were allotted in the field using a labeling protocol [31,52]. Data limitations in some areas led to classes being identified on the basis of prior experience with crop signatures.

Data on India's agriculture were obtained from the official websites of the Department of Agriculture of the Government of India; other countries' data were obtained from the respective national agricultural/statistics departments.

As per Indian national statistics, the dominant rainfed cropping systems are cotton-fallow (15.3 M ha) followed by rice-fallow (12.6 M ha) and groundnut-fallow (9.7 M ha). The major irrigated cropping systems are rice-wheat (14.8 M ha) followed by millet-wheat (10.1 M ha). The other major cropping systems are pulses-wheat (9.2 M ha) and soybean-wheat (8.3 M ha) (Table 2).

3. Methodology

The workflow of this study consisted of three major steps: (1) preparation of satellite data including mosaicking and stacking; (2) running the machine learning algorithm, i.e., unsupervised classification (K-means) on Google Earth Engine (GEE) and identification of classes and labeling with the help of SMTs; and (3) accuracy assessment using validation data and national statistics (Figure 2).

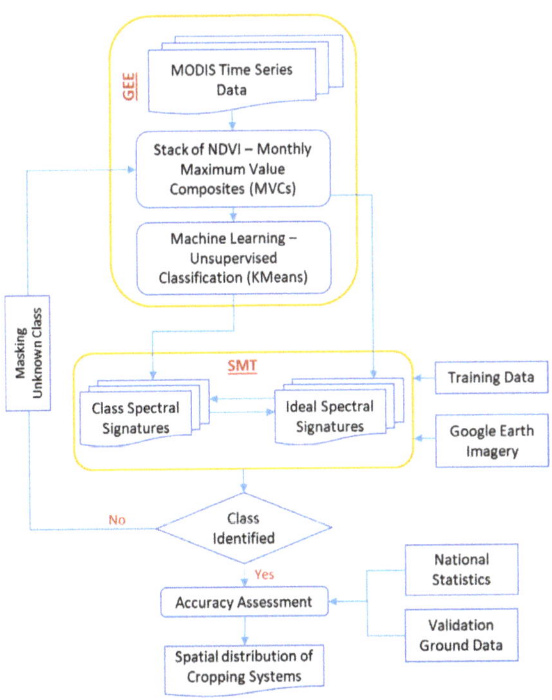

Figure 2. The steps to map crop dominance classification.

3.1. *Preparation of Satellite Data and Running Machine Learning Algorithm*

In GEE, the readily available MOD13Q1 V6.1 product provides preprocessed NDVI data for every 16 days on a per-pixel basis, masked for water, clouds, heavy aerosols and cloud shadows. Twelve MODIS tiles covering the study area for the study year June 2014–May 2015 were considered for our analysis.

The monthly maximum NDVI imagery was calculated using Equation (1).

$$NDVIMVC_i = Max(NDVI_{i1}, NDVI_{i2}) \tag{1}$$

where, MVC_i = Month maximum composite of ith month (e.g., "i" is June–May) and i1, i2 = 16-day images in a month.

The images were first stacked as monthly maximum-value composites, which means that the maximum NDVI values for every month were stacked for the study year 2014–15, making an analysis-ready data (ARD) cube of 12 months, one for each month, thus providing the NDVI variation in a crop year and overall coverage of the whole study area. MODIS data has temporal resolution of one day. Since this study considered monthly maximum images, there was less chance of missing pixel values; in an exceptional case, linear regression interpolation was done to fill the gap.

The final ARD cube was set into a K-means unsupervised machine learning algorithm with 100 clusters for every agroecological zone. The machine learning algorithm classifies the ARD cube into 100 clusters with 100 class spectral signatures.

3.2. *Spectral Matching Techniques for Class Identification and Labeling*

3.2.1. Ideal Spectral Signatures (ISS)

Ideal spectral signatures for cropping systems were generated based on independent training data drawn from extensive field survey information using time-series data. The ground samples were grouped based on their distinctive characteristics and later classified into classes of unique cropping systems. Figure 3 shows the ISS of various classes.

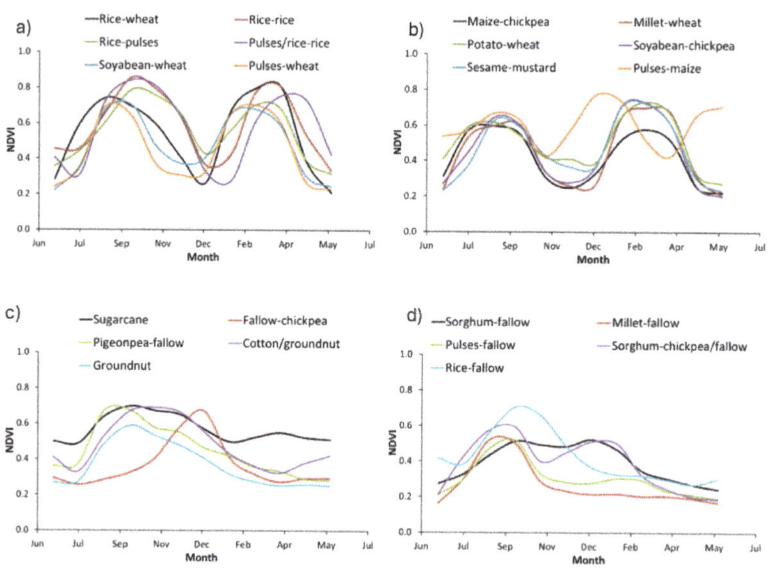

Figure 3. (**a**,**b**) Ideal Spectra Signatures of double crops (**c**,**d**) Ideal Spectra Signatures of single crop.

The spectral signature curves explain cropping systems behavior over time. Regularly, NDVI values are low during initial crop development, high through peak growth and low again at harvesting stage, due to low to high reflectance value of near infrared region band during the crop development. In Figure 3a,b, double crop (rice-rice, rice-wheat systems, etc.) classes exhibit two peaks in a single curve, signifying that the first peak is in the start season and the second peak is in the later season. Figure 3c,d shows only a single crowning, which means that there was only a single crop (e.g., groundnut-fallow, sorghum-fallow, etc.) in a crop year. Depending on the NDVI values and ground data, we labeled classes as a combination of crop intensity and crop names.

3.2.2. Class Spectra Signatures (CSS)

Class spectra signatures (Figure 4a) were generated for the 100 classes (clusters) from unsupervised classification (K-means) of the 16-day NDVI temporal data. The signatures were generated for every individual class by taking the NDVI mean of all classified pixels of each cluster for every month.

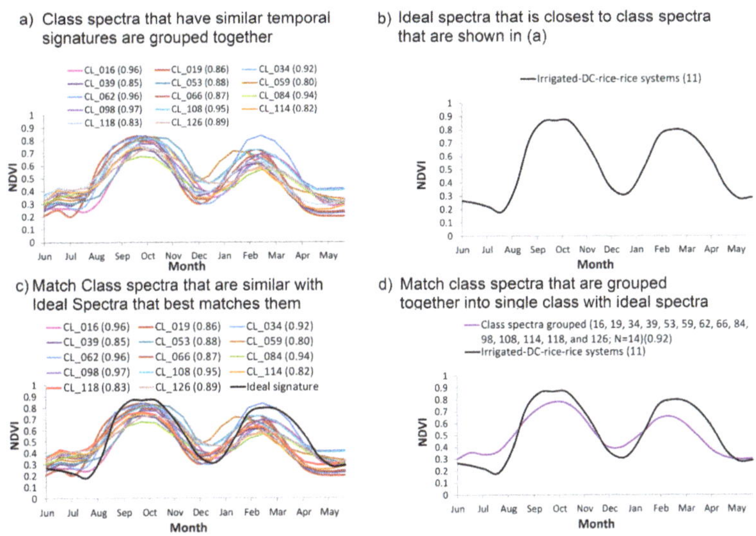

Figure 4. Matching CSS with ISS and grouping of similar classes.

3.2.3. Matching CSS with ISS to Group Classes Using SMTs

The initial 100 CSS signatures were grouped using quantitative spectral matching techniques (QSMTs), i.e., based on similar spectral signatures [53,54] (Figure 4a). The CSS (Figure 4c) signatures are then matched with the ISS (Figure 4b) and a match is determined by QSMTs with spectral correlation similarity (SCS) R-square value of 0.70 or higher are grouped and labeled as a class (Figure 4d). Classes with good visual matches both in terms of magnitude and shape of the spectral signatures are considered for class labeling.

The preliminary labeling of classes was validated, mainly by using field data, Google Earth and other secondary sources. The process was repeated to classify and label all 100 classes, leading to a finalization of classes. Wherever there was limited ground data, high-resolution data from Google Earth was used for reference. In the event of any ambiguity in the labeling of classes—e.g., some classes may not correlate with spectral signatures—such classes were reclassified using the process above [52,54].

3.3. Accuracy Assessment

Accuracy assessment was then carried out with 883 independent field points that were not used in the classification process using an error matrix. The overall classification accuracy of cropping systems and users' and producers' accuracies were calculated for each cropping system.

3.4. Sub-Pixel Areas

Highly precise area estimates from the coarse resolution imagery are only possible through sub-pixel areas (SPAs) [13,52]. This study used MOD13Q1 in which every pixel covers 250 m × 250 m, i.e., an area of about 6.25 ha. Full pixel areas (FPAs) are not an accurate representation of actual areas, as they may include areas falling under other LULC classes or mixed classes. On the other hand, sub-pixel areas (SPAs) involving actual area calculation, include the percentage of crop area. Therefore, areas based on SPA provide a near approximation of the cropped area.

Within the cropland class, there is a high possibility of discrepancies in the form of other LULC results. However, some classes have >50% area cropped and the proportions can vary widely. The percentage of various classes (water, cropland, etc.) is decided based on observation of ground data, photographs from ground data collection and by visualizing the field in Google Earth imagery. For example, in a rice-rice cropping system, a rice area sub pixel area (SPA), i.e., actual area is calculated based on the crop land area fraction (CAF), i.e., percent of the rice area in 250 m × 250 m area of random rice field observation samples in that class. This process was carried out for each cropping system obtained. To get the actual areas, the FPA was multiplied by the CAF [55]. Since major cropping systems were mapped, the areas obtained are nearly equivalent to SPAs, with a few exceptions. Each class is evaluated for its actual area as follows:

$$\text{SPAs} = \text{FPAs} \times \text{CAFs} \tag{2}$$

4. Results

4.1. Cropping Systems

We find there are 27 major cropping systems in South Asia (Figure 5 and Table 3). Non-croplands are found in the extreme northern and western parts of South Asia and the Himalayas in the north, and are ~55% of the land cover of South Asia. The non-croplands consist mainly of deserts, mountains, glaciers and hills. The croplands are in the river valleys and vegetated plains. Around 45 percent of South Asia comprises croplands.

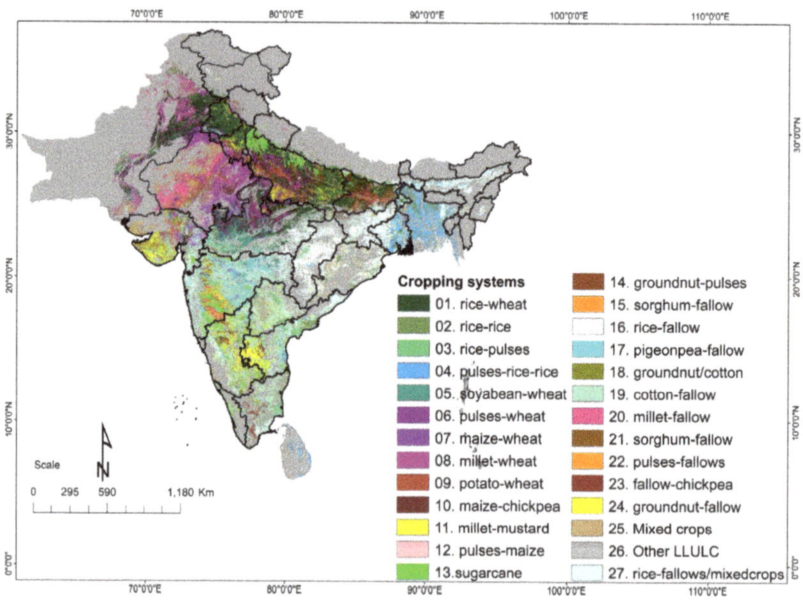

Figure 5. The distribution of 27 major cropping systems across South Asia.

Table 3. The area covered under major cropping systems of South Asia.

Cropping Systems	Sub-Pixel Area (SPA) Fractions							Full Pixel Area FPA ('000 ha)
	Trees	Shrubs	Water	Grasses	Orchards	Other LULC	Crop Area	
01. Rice-wheat	1.9	6.2	3.9	0.3	0.5	3.9	83.4	23,884
02. Rice-rice	1.7	2.5	1.5	0.4	0.0	2.9	91.1	5608
03. Rice-pulses	2.1	2.2	1.7	0.1	0.0	5.6	88.4	5967
04. Pulses/rice-rice	0.7	2.1	2.4	0.0	0.7	4.0	90.0	8537
05. Soybean-wheat	2.0	3.1	0.2	0.0	0.0	3.1	91.7	7378
06. Pulses-wheat	1.6	2.3	0.0	0.0	0.3	4.7	91.2	5007
07. Maize-wheat	1.6	2.8	0.6	0.0	1.1	4.4	89.4	6877
08. Millet-wheat	1.4	0.8	0.3	0.0	0.5	3.9	93.3	13,469
09. Maize-wheat	1.6	0.8	0.9	0.0	1.9	3.9	91.0	3193
10. Maize-chickpea	1.1	8.3	0.8	0.0	0.6	3.0	86.2	6342
11. Millet-mustard	1.7	0.6	0.0	0.0	0.4	3.8	93.4	4259
12. Pulses-maize	3.0	9.3	1.7	0.0	5.0	2.3	78.7	2740
13. Sugarcane	1.6	0.0	0.1	0.0	0.3	2.4	95.6	5509
14. Groundnut-pulses	2.1	2.7	0.9	0.0	0.0	4.4	89.9	4699
15. Sorghum-fallow	3.0	5.8	0.7	0.0	0.8	4.7	85.0	1667
16. Rice-fallow	4.8	1.6	0.4	0.0	0.0	1.8	91.4	13,414
17. Pigeon pea-fallow	3.9	10.2	0.4	0.2	4.5	11.7	69.2	10,035
18. Groundnut/cotton	1.4	1.0	0.4	0.2	0.6	2.4	94.0	4241
19. Cotton-fallow	1.9	3.9	1.3	0.5	0.0	5.0	87.3	19,045
20. Millet-fallow	2.9	2.4	0.1	0.0	0.0	2.9	91.7	4284
21. Sorghum-fallow	2.6	5.7	1.3	0.4	0.0	7.1	82.9	4991
22. Pulses-fallows	1.8	13.4	1.1	0.1	2.0	7.2	74.5	2538
23. Fallow-chickpea	0.7	22.3	0.1	0.0	4.0	1.7	71.3	1999
24. Groundnut-fallow	1.2	1.9	1.0	0.0	4.0	2.3	89.6	6251
25. Mixed crops	4.0	10.8	3.4	0.5	0.0	1.3	80.0	31,596
26. Other LULC	0.5	52.4	0.5	0.0	0.0	31.0	15.6	214,625
27. Rice-fallows/mixed crops	0.9	22.9	2.7	0.1	3.8	9.4	60.2	24,360

There are 14 double cropped cropping systems and 11 single crops cropping systems, 1 mixed crop cropping system, and 1 other land use class, i.e., non-croplands in South Asia. Each cropping system has unique crop or crop-combinations. For example, there are several cropping systems with rice as one of the crops such as rice-rice double crop (i.e., kharif season rice followed by rabi season rice), or rice-fallow system in which there is only a single rice crop (i.e., kharif season rice followed by fallow winter season rice). Some other major classes were rice-wheat, rice-pulses, sorghum-fallow, millet-fallow and mixed crops.

4.2. Spatial Distribution of Each Cropping System

There is, in general, a dominating major crop of the rainy (kharif) season such as rice together with a secondary crop in the rabi and/or zaid summer season (not common). Below, we describe our findings of fifteen major cropping systems of South Asia in greater details (Figure 6):

Rice-wheat: This is the dominant cropping system in the South Asia covering nearly 19.9 Mha stretching from Pakistan (along the Indus River basin) across into northern India (Indo-Gangetic plains) covering the states of Punjab in the west to Bihar in the east. In this system, rice crop is planted at the start of the rainy season (kharif) in June (analysis of spectral signature—Figure 3) and harvested at the end of summer monsoon season starting in August and lasting through September. As harvest progresses, NDVI reduces more gradually till it reaches a minimum and then around November wheat crop starts to emerge starting first with Bihar and ending last with those in Pakistan. Finally, wheat harvests start in March–April and they proceed much faster than rice harvests.

Rice-rice: Rice followed by a second rice crop can be observed in eastern India down to the south of India, Pakistan and in entire Bangladesh covering nearly 5.1 Mha of area. This cropping system is prevalent in irrigated lands and coastal areas. Unfortunately, the practice of mono-cropping in these regions has led to soil degradation and a deficiency in essential micronutrients. The first rice crop is in the kharif season, and sown during the monsoon months, typically between June and July, and is ready for harvest from September to October. After harvesting the first rice crop, the same fields are used for the second rice crop, which is planted during the dry winter months of the rabi season, usually between November and December. The second rice crop harvest starts from February to March as seen in their spectral signature.

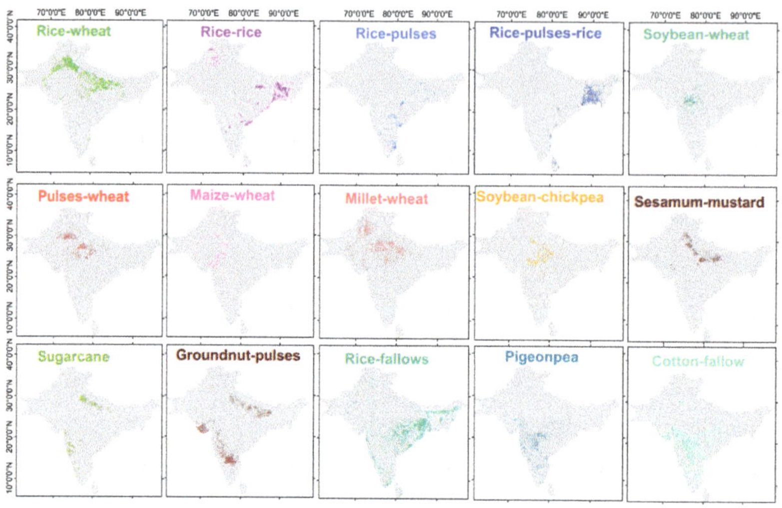

Figure 6. Maps of spatial distribution of individual major cropping systems.

Rice-pulses: Rice cultivation in rainy season and followed by pulses cultivation is observed in east to south India covering about 5.2 Mha of area. This cropping system is practiced in parts of Chhattisgarh, Odisha, Tamil Nadu and Bihar states in India. It has led to high yields of both rice and pulses. However, the challenge lies in the irregular availability of irrigation in these regions. Rice cultivation takes place during the kharif season, where the sowing of rice seeds begins from June to July and harvesting between the months of September and October. Our spectral signature shows that subsequently, pulses are sown between November and December in this cropping system regions with the matured pulses harvested between February and March. Nearly 0.3 Mha is observed in Nepal and Bangladesh and no significant area of this cropping system is present in Pakistan, Bhutan and Sri Lanka.

Rice-pulses-rice: This is the cropping system of three crops in a single crop year in which rice is followed by pulses and then by another rice crop. This cropping system is observed mostly in Bangladesh, but also in a few parts of Sri Lanka and India where irrigation coverage is excellent. The first rice crop is sown between June and July and is typically harvested from September to October. Subsequently, pulses are sown between November and December and then harvested from February to March. After the pulse harvest, preparations are made for the second rice crop to be sown in April to May, followed by the harvesting of the second rice crop from June to July. It is this latter zaid season crop that is irrigated.

Soybean-wheat: The cultivation of soybean followed by wheat is seen majorly in Madhya Pradesh state of India covering nearly 6.6 Mha of area. This system is also found in Rajasthan, and certain regions of southern Maharashtra. Soybean cultivation in India primarily occurs during the kharif season, with our spectral signature showing the sowing of soybean seeds between June and July and harvest taking place from September to October. In the subsequent rabi season, after the soybean harvest, typically occurring in October to November, fields are prepared for wheat cultivation. Wheat is sown from November to December, and the mature wheat crop is harvested between March and April. Approximately 0.1 Mha of this cropping system is observed in Pakistan, whereas there is no significant area in other countries.

Pulses–wheat: Unlike the rice-pulse-rice system, this system is concentrated in northern India and into neighboring Pakistan, spanning an area of approximately 4.5 million hectares (Table 1). It is prevalent in several states, especially northern Indian regions, including Uttar Pradesh, Madhya Pradesh, Rajasthan, Haryana and Punjab states. The usual pattern involves planting kharif pulses like chickpeas, pigeon peas and lentils during the monsoon season, followed by the rabi wheat crop. Pulses cultivation follows a specific timeline, with pulses being sown from June to August and harvested between October and November. Wheat is sown from November to December, and the mature wheat crop is ready for harvesting between March and April.

Maize-wheat: Maize followed by wheat cultivation is observed in western parts of India and Pakistan, covering nearly 6.1 Mha of area. In this cropping system, maize is the primary crop during the kharif season, while wheat takes center stage in the rabi season. This system is widely practiced in states like Uttar Pradesh, Rajasthan, Madhya Pradesh, Bihar and Punjab. A significant portion of the maize-wheat system is rainfed, which means it heavily relies on rainfall, posing a major challenge due to its unpredictability. In the kharif season of India, maize sowing is carried out between June and July, and the maize harvest occurs during September to October. Subsequently, in the rabi season that follows, wheat is sown between November and December, and the mature wheat crop is usually ready for harvesting between March and April. Approximately 0.1 Mha of this cropping system is observed in Pakistan, whereas there is no significant area in other countries.

Millet-wheat: The cultivation of millet followed by wheat is observed in northern parts of India and also in Pakistan, covering 12.5 Mha of area. These are significant cropping systems in the western part of India, particularly in the semi-arid regions of Gujarat, Rajasthan and Haryana. Wheat is cultivated as a rabi crop, while Millet is grown during the kharif season. Millet cultivation is carried out between June and July, with millet ready for harvesting in September to October. Subsequently, in the following rabi season, wheat is sown from November to December, and the mature wheat crop is typically harvested between March and April.

Soybean-chickpea: This type of cropping system is observed in the central parts of India covering nearly 5.4 Mha of area, especially in the areas of Madhya Pradesh and some areas of Uttar Pradesh states. During the kharif season in India, soybean sowing starts between June and July, with the soybean harvest typically taking place from September to October. In the subsequent rabi season, chickpeas are sown from November to December and are ready for harvesting between February and March.

Sesame-mustard: The cultivation of sesame followed by mustard is observed in Uttar Pradesh state of India covering nearly 3.9 Mha of area. The sesame-mustard cropping system is adopted in various states across India, including Rajasthan, Gujarat, Madhya Pradesh, Uttar Pradesh, Bihar and portions of Haryana and Punjab. In the kharif season, sesame cultivation starts with sowing between June and July, and the sesame crop is typically harvested from September to October. In the subsequent rabi season, Mustard is sown from November to December and is usually ready for harvesting between February and March.

Sugarcane: Sugarcane is a whole-year crop, i.e., cultivated throughout the crop year mostly in Uttar Pradesh and Maharashtra states. Sugarcane is grown in several states across India, with Uttar Pradesh, Maharashtra, Karnataka, Tamil Nadu, Andhra Pradesh and Bihar being the major sugarcane-producing states. Sugarcane is usually planted in the pre-monsoon or early monsoon season. In northern India, it is planted from March to June, while in southern India, planting can even extend into July.

Rice-fallows: Rice fallows are predominantly found in both rainfed and irrigated areas of the south Asia. The significant rainfed rice-fallows area found in India of about 11.4 Mha and mixed irrigated rice-fallows of nearly 13.3 Mha, whereas Bangladesh and Pakistan have nearly 0.5 Mha.

Pigeon pea-fallows: The pigeon pea-fallow cropping system is mainly practiced in the semi-arid and arid regions of India, including areas in Rajasthan, Gujarat, Maharashtra, Karnataka and Madhya Pradesh, covering about 6.7 Mha. Generally, depending upon the rainfall, the sowing of pigeon pea will be done and cultivation period ranges from 6 months to more than one year in some areas.

Cotton-fallow: Cotton-fallows areas are also significantly found in India, which is about 16.2 Mha of area. This cropping systems is predominantly in states of Karnataka, Maharashtra and some parts of Gujarat. Cotton sowing happens in June-July, and harvesting takes place from October to January.

The remote sensing-based data on full-pixel crop areas of major cropping systems (Table 3) was further refined as the actual area statistics which were calculated using SPA by taking random samples of each class and identified the percentage of crop found in respective class pixels. The SPA was calculated given the possibility of non-crop pixels in crop pixels due to coarser resolution.

Sub-pixel areas in each cropping system were calculated. It was observed that on an average, 10 to 20 percent of each class contains other LULC. We calculated the sub-pixel areas for each class to get the actual area by eliminating 10–20 percent accordingly.

4.3. Country-Wise Cropping Systems

Country-wise cropping systems and their respective sub-pixel areas shows the dominant cropping sequence followed in each country (Table 4). In Pakistan, the dominant cropping systems are rice-wheat (4.3 M ha) followed by millet-wheat (3.4 M ha) and rice-rice (1.4 M ha). In Nepal, the major cropping systems are rice-wheat (0.4 M ha) and rice-pulses (0.36 M ha). In Bhutan, the cropping systems are rice-fallow (0.012 M ha) and pulses-rice-rice (0.007 M ha). In Bangladesh, the dominant cropping systems are pulses-rice-rice (3.8 M ha) followed by rice-rice (1.1 M ha), whereas in Sri Lanka, the dominant cropping system is pulses-rice-rice (0.49 M ha). In India, there are highly diversified cropping systems such as rice-wheat (15.1 M ha), rice-fallow (11.4 M ha), cotton-fallow (16.2 M ha) and rice-fallow/mixed crops (13.3 M ha), followed by a significant shared area with different cropping systems. The dominance of rice-fallow cropping systems provides opportunities for intensification with short-duration pulse crops.

Table 4. Country-wise net crop area under different cropping systems.

Cropping Systems	Net Area ('000 ha)					
	Pakistan	Nepal	Bhutan	Bangladesh	Sri Lanka	India
01. Rice-wheat	4328	420	0	2	0	15,170
02. Rice-rice	1412	4	0	1126	3	2564
03. Rice-pulses	50	365	1	399	4	4456
04. Pulses-rice-rice	30	39	7	3816	497	3293
05. Soybean-wheat	145	0	0	0	0	6620
06. Pulses-wheat	610	1	0	0	0	3955
07. Maize-wheat	826	0	0	0	0	5324
08. Millet-wheat	3406	7	0	0	53	9094
09. Potato-wheat	0	100	0	0	0	2806
10. Soybean-chickpea	123	1	0	0	0	5341
11. Sesamum-mustard	112	10	0	0	2	3854
12. Pulses-maize	0	0	0	0	0	2156
13. Sugarcane	0	76	0	0	0	5189
14. Groundnut-pulses	0	19	0	0	0	4204
15. Sorghum-fallow	0	0	0	0	0	1417
16. Rice-fallow	95	125	2	633	1	11,403
17. Pigeonpea-fallow	181	0	0	0	4	6758
18. Groundnut/cotton	13	0	0	1	1	3971
19. Cotton-fallow	393	5	1	1	2	16,231
20. Millet-fallow	740	6	1	2	6	3175
21. Sorghum-fallow	472	2	1	0	0	3661
22. Pulses-fallows	264	0	0	0	1	1625
23. Fallow-chickpea	39	0	0	0	7	1379
24. Groundnut-fallow	0	0	0	0	0	5601
25. Mixed crops	1619	403	28	864	739	21,623
26. Other LULC	9605	1940	611	843	681	19,800
27. Rice-fallows/mixed crops	406	258	12	407	239	13,349
Net cropped area	24,871	3781	664	8096	2240	184,020

4.4. Comparison of Remote Sensing-Derived Crop Area Statistics to Survey Based National Statistics

Data on area by cropping system was extracted from a remote sensing-derived cropping systems map and compared with National Agriculture Statistics (NAS) for the year 2014–15 obtained from the Department of Agriculture. The analysis reveals significant discrepancies between the reported areas in NAS and those derived from the remote sensing map for various crop combinations. Crop combinations like pigeon pea-fallow (77.56%), maize-chickpea (54.95%) and pulses-fallow (67.82%) show substantial positive differences. Conversely, soybean-wheat (−13.02%), millets-fallow (−8.91%) and millets-wheat (−3.44%) exhibit negative differences. Some combinations, such as rice-fallow (−0.99%) and sugarcane (6.46%), show minimal differences, indicating a varying level of alignment between reporting and mapping methods (Figure 7).

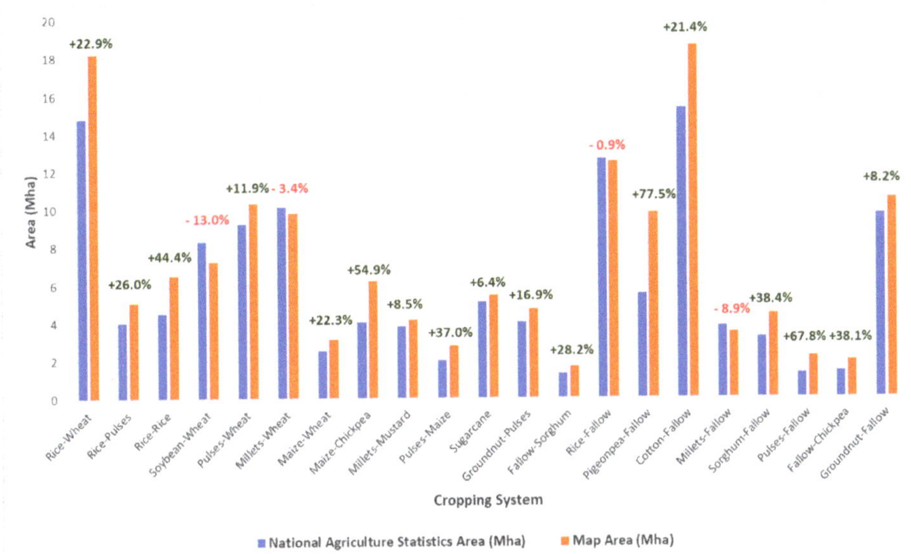

Figure 7. Comparison of remote sensing-derived crop areas with that from national statistics.

The accuracy assessment was calculated for identified cropping systems with 883 independent ground points. Overall classification accuracy was observed to be 76.59 percent, with kappa coefficient of 0.75. User and producer accuracies of major cropping systems ranged from 70 per cent to 90 percent, whereas mixed cropping system showed less accuracy ranging from 50 percent to 70 percent.

The major rice-wheat cropping system showed a user accuracy of 89 percent and producer accuracy of 72 percent, and pulses-wheat systems revealed a user accuracy of 92 percent and producer accuracy of 100 percent. Likewise, each cropping system was validated using validation points, their individual class accuracies and kappa coefficient determined (Table 5). Due to domination of small-holding farms and mixed crops in major areas of south Asia, the accuracy of some of the minor cropping systems was lower. The mapping accuracy of such cropping systems can be increased by utilizing high resolution satellite data along with more extensive ground data collection to build better models.

Table 5. Accuracy assessment of the croplands product by error matrix method.

Classified Data	Row Total	Reference Total	Classified Total	Number Correct	Producer Accuracy (%)	User Accuracy (%)	Kappa
01. Rice-wheat	46	46	37	33	72	89	0.89
02. Rice-rice	88	88	57	50	57	88	0.86
03. Rice-pulses	51	51	52	35	69	67	0.65
04. Pulses/rice-rice	107	107	106	95	89	90	0.88
05. Soybean-wheat	33	33	35	29	88	83	0.82
06. Pulses-wheat	22	22	24	22	100	92	0.91
07. Maize-wheat	24	24	24	22	92	92	0.91
08. Millet-wheat	38	38	39	36	95	92	0.92
09. Potato-wheat	13	13	14	11	85	79	0.75
10. Maize-chickpea	14	14	15	13	93	87	0.86
11. Millet mustard	20	20	11	11	55	100	1.00
12. Pulses-maize	10	10	12	10	100	83	0.67
13. Sugarcane	17	17	12	11	65	92	0.92
14. Groundnut-pulses	15	15	20	12	80	60	0.59
15. Sorghum-fallow	18	18	15	14	78	93	0.93
16. Rice-fallow	58	58	41	26	45	63%	0.61
17. Pigeonpea-fallow	27	27	38	22	81	58	0.57
18. Groundnut/cotton	15	15	15	8	53	53	0.53
19. Cotton-fallow	43	43	48	33	77	69	0.67
20. Millet-fallow	13	13	13	13	100	100	1.00
21. Sorghum-fallow	16	16	14	11	69	79	0.78
22. Pulses-fallows	11	11	10	10	91	100	1.00
23. Fallow-chickpea	16	16	17	16	100	94	0.94
24. Groundnut-fallow	14	14	18	12	86	67	0.66
25. Mixed crops	55	55	50	41	75	82	0.81
26. Other LULC	61	61	91	52	85	57	0.54
27. Rice-fallow/mixed crops	38	38	55	38	100	69	0.68
Total	883	883	883	686	Overall accuracy = 76.59%;	Kappa = 0.7545	

5. Discussion

While the classification of irrigated, homogenous crops such as rice and wheat were relatively easy throughout South Asia, identification of rainfed crops like groundnut and pulses was difficult. Further, cropping patterns in most regions of south Asia vary widely, with a high possibility of mixed crops. In India and Bangladesh, rice is the major irrigated crop and cultivated in large homogenous extent, whereas in Pakistan, wheat is the major irrigated crop, comprising more than 25 percent of all crops. Rice and wheat cropping system areas could be identified and classified more correctly. The identification of other crops mainly depends upon extensive ground data, their ideal signatures and crop phenology. Western South Asia comprising Afghanistan and Pakistan has large tracts of uncultivated area, whereas eastern South Asia, i.e., northeastern India, contains hill areas and the north part of south Asia contains ice glaciers. Crop identification in hilly regions was affected by shadows and different cropping patterns. Nearly 26 percent of South Asian cropland consists of double crops (i.e., rainy and winter season crops). Double crops in irrigated areas were identified using spectral curves and ground data. Only 45 percent of the total area in south Asia was under cultivation, with other LULC comprising mainly deserts, glaciers and hilly areas.

Knowledge of regional cropping systems is essential to maximize land productivity and improve crop production towards the objective of sustainable agriculture. Of late, high-intensity cropping practices and overexploitation of resources have resulted in unsustainable agriculture. The 27 cropping systems identified in this study cover all major and dominant agricultural practices in the region. These novel information products can serve as baseline data for developing sustainable agriculture solutions. When integrated with thematic layers such as weather, soil, groundwater availability and land degradation in GIS environments, they can unlock opportunities to develop action-oriented solutions to improve the performance of agriculture systems. It is also possible to produce maps with finer granularity using Sentinel data of 10 m resolution once more ground level data are collected. Sustainable agriculture and food system solutions need such innovative information products, which are potentially the low-hanging benefits from modern remote sensing technology.

Policy Implications: Cropping system maps and statistics form basic inputs for formulating sustainable agriculture plans to optimize land productivity. The major challenges of agriculture in South Asia are food security, income security and climate resilience, which can be addressed only through systematic efforts to optimize crop patterns and resource consumption in producing crops. The maps of cropping systems can now be superimposed on resources and climatology data layers to offer deeper insights to decision-makers on the efficiency of the current agricultural systems and to prioritize critical areas for immediate interventions. The design and implementation of such interventions needs a strong base of data and information on cropping systems and related analytics that our maps will now provide. These interventions will involve both management and structural measures at the grass root level, and may involve certain incentives and investments, but cannot proceed without understanding the entire cropping system of the local area. Specifically, the government of India's flagship projects on water use efficiency in agriculture, the National Food Security Mission (NFSM) and Agroforestry development mission will become more effective when cropping systems information is integrated within their design and implementation plans. The cropping system determines multiple ecosystem services, pest predation and water quality. For example, the development of agroforestry plans in a region should take into account the existing cropping system. Carbon sequestration and carbon emission potentials of croplands are directly related to the cropping systems followed in a region. Therefore, any effort towards achieving carbon neutral agroecosystems will fundamentally get guided by cropping system practices.

Policymakers across South Asia will find our cropping system maps to be quite useful, since they provide a thorough understanding of the geographical distribution of various crops within an area. These maps will help with all governmental planning of resource allocation, making investments in pest management, irrigation and crop-specific subsidies more effective. Furthermore, they are crucial in guaranteeing food security since they enable policymakers to evaluate the yield of vital food crops and create plans for agricultural output diversification. By highlighting regions with significant production potential for particular crops, cropping sequence type maps will also aid in market growth and strengthen local and regional economies. They also aid in the identification of vulnerable regions that require protection, the evaluation of the effects of farming on natural resources and the development of sustainable agricultural policy. All things considered, these maps are invaluable resources that support the development of evidence-based policies and well-informed decisions about a variety of aspects of land use and agriculture.

6. Conclusions

Knowledge of regional cropping systems is essential to maximize land productivity and improve crop production towards the objective of sustainable agriculture. Of late, high-intensity cropping practices and overexploitation of resources have resulted in unsustainable agriculture.

This study has developed a new geospatial application in the agriculture sector to address crop planning and sustainable agriculture practices.

Remote sensing imagery from MODIS 250 m was used to generate maps and statistics on different cropping systems practiced in South Asia. The 27 cropping systems identified in this study cover all major and dominant agricultural practices in the region. These novel information products can serve as baseline data for developing sustainable agriculture solutions. When integrated with thematic layers such as weather, soil, groundwater availability and land degradation in GIS environments, they can unlock numerous opportunities to develop action-oriented solutions to improve the performance of agriculture systems. It is also possible to produce maps with finer granularity using Sentinel data of 10 m resolution. Sustainable agriculture and food system solutions need such innovative information products, which are low-hanging benefits from modern remote sensing technology.

Author Contributions: M.K.G., C.S.M. and D.K.R.: conceptualization, methodology, investigation, developing algorithms, image analysis, ground data collection, writing—reviewing and editing; P.P.: validation, writing, testing machine learning algorithms; C.S.M., Y.T. and D.M.K.: investigation, formal analysis, accuracy assessment and editing; I.M.: ground data collection; S.K.D.: agricultural statistics; Y.T.: review and editing. All authors have read and agreed to the published version of the manuscript.

Funding: This study was funded by the Japan Fund for Prosperous and Resilient Asia and the Pacific financed by the Government of Japan through the Asian Development Bank (ADB) and Mahalano-bis National Crop Forecasting Centre (MNCFC). DKR was funded with support from the Institute on the Environment and WRI Land & Carbon Lab grant number G3275. Land & Carbon Lab is convened by World Resources Institute and the Bezos Earth Fund.

Data Availability Statement: The classification of cropping systems in this study was performed on the local computer and Google Earth Engine. The code utilized is available at https://github.com/ICRISAT-GBDS/South-Asia-Cropping-Systems.git (accessed on 17 May 2024).

Acknowledgments: Our thanks go to Prasad S. Thenkabail, Anthony Whitbread, ML Jat and Partha Saradhi Teluguntla for their valuable comments and support.

Conflicts of Interest: The authors declare no conflicts of interest.

References

1. Connor, D.J.; Loomis, R.S.; Cassman, K.G. *Crop Ecology: Productivity and Management in Agricultural Systems*; Cambridge University Press: Cambridge, UK, 2011.
2. Yang, T.; Siddique, K.H.; Liu, K. Cropping systems in agriculture and their impact on soil health—A review. *Glob. Ecol. Conserv.* **2020**, *23*, e01118. [CrossRef]
3. Mahlayeye, M.; Darvishzadeh, R.; Nelson, A. Cropping Patterns of Annual Crops: A Remote Sensing Review. *Remote Sens.* **2022**, *14*, 2404. [CrossRef]
4. Garnett, T.; Appleby, M.; Balmford, A.; Bateman, I.; Benton, T.; Bloomer, P.; Burlingame, B.; Dawkins, M.; Dolan, L.; Fraser, D. Sustainable intensification in agriculture: Premises and policies. *Science* **2013**, *341*, 33–34. [CrossRef] [PubMed]
5. Gray, J.; Friedl, M.; Frolking, S.; Ramankutty, N.; Nelson, A.; Gumma, M. Mapping Asian Cropping Intensity with MODIS. *IEEE J. Sel. Top. Appl. Earth Obs. Remote Sens.* **2014**, *7*, 3373–3379. [CrossRef]
6. Boryan, C.; Yang, Z.; Mueller, R.; Craig, M. Monitoring US agriculture: The US Department of Agriculture, National Agricultural Statistics Service, Cropland Data Layer Program. *Geocarto Int.* **2011**, *26*, 341–358. [CrossRef]
7. Khan, A.; Hansen, M.C.; Potapov, P.; Stehman, S.V.; Chatta, A.A. Landsat-based wheat mapping in the heterogeneous cropping system of Punjab, Pakistan. *Int. J. Remote Sens.* **2016**, *37*, 1391–1410. [CrossRef]
8. Skakun, S.; Franch, B.; Vermote, E.; Roger, J.-C.; Becker-Reshef, I.; Justice, C.; Kussul, N. Early season large-area winter crop mapping using MODIS NDVI data, growing degree days information and a Gaussian mixture model. *Remote Sens. Environ.* **2017**, *195*, 244–258. [CrossRef]
9. Seinn, S.; Ahmad, M.; Thapa, G.; Shrestha, R. Farmers Adaptation to Rainfall Variability and Salinity through Agronomic Practices in Lower Ayeyarwady Delta, Myanmar. *J. Earth Sci. Clim. Chang.* **2015**, *6*, 2157–7617.
10. Parr, J.; Stewart, B.; Hornick, S.; Singh, R. Improving the sustainability of dryland farming systems: A global perspective. In *Advances in Soil Science*; Springer: Berlin/Heidelberg, Germany, 1990; pp. 1–8.
11. Salvati, L.; Kosmas, C.; Kairis, O.; Karavitis, C.; Acikalin, S.; Belgacem, A.; Solé-Benet, A.; Chaker, M.; Fassouli, V.; Gokceoglu, C. Unveiling soil degradation and desertification risk in the Mediterranean basin: A data mining analysis of the relationships between biophysical and socioeconomic factors in agro-forest landscapes. *J. Environ. Plan. Manag.* **2015**, *58*, 1789–1803. [CrossRef]
12. Feyisa, G.L.; Palao, L.K.; Nelson, A.; Gumma, M.K.; Paliwal, A.; Win, K.T.; Nge, K.H.; Johnson, D.E. Characterizing and mapping cropping patterns in a complex agro-ecosystem: An iterative participatory mapping procedure using machine learning algorithms and MODIS vegetation indices. *Comput. Electron. Agric.* **2020**, *175*, 105595. [CrossRef]

13. Gumma, M.K.; Thenkabail, P.S.; Teluguntla, P.; Rao, M.N.; Mohammed, I.A.; Whitbread, A.M. Mapping rice-fallow cropland areas for short-season grain legumes intensification in South Asia using MODIS 250 m time-series data. *Int. J. Digit. Earth* **2016**, *9*, 981–1003. [CrossRef]
14. Thenkabail, P.S.; Hanjra, M.A.; Dheeravath, V.; Gumma, M. A holistic view of global croplands and their water use for ensuring global food security in the 21st century through advanced remote sensing and non-remote sensing approaches. *Remote Sens.* **2010**, *2*, 211–261. [CrossRef]
15. Mao, Z.-H.; Deng, L.; Duan, F.-Z.; Li, X.-J.; Qiao, D.-Y. Angle effects of vegetation indices and the influence on prediction of SPAD values in soybean and maize. *Int. J. Appl. Earth Obs. Geoinf.* **2020**, *93*, 102198. [CrossRef]
16. Xu, J.; Zhu, Y.; Zhong, R.; Lin, Z.; Xu, J.; Jiang, H.; Huang, J.; Li, H.; Lin, T. DeepCropMapping: A multi-temporal deep learning approach with improved spatial generalizability for dynamic corn and soybean mapping. *Remote Sens. Environ.* **2020**, *247*, 111946. [CrossRef]
17. Wang, M.; Liu, Z.; Ali Baig, M.H.; Wang, Y.; Li, Y.; Chen, Y. Mapping sugarcane in complex landscapes by integrating multi-temporal Sentinel-2 images and machine learning algorithms. *Land Use Policy* **2019**, *88*, 104190. [CrossRef]
18. Dong, J.; Xiao, X. Evolution of regional to global paddy rice mapping methods: A review. *ISPRS J. Photogramm. Remote Sens.* **2016**, *119*, 214–227. [CrossRef]
19. Xiao, X.; Boles, S.; Liu, J.; Zhuang, D.; Frolking, S.; Li, C.; Salas, W.; Moore III, B. Mapping paddy rice agriculture in southern China using multi-temporal MODIS images. *Remote Sens. Environ.* **2005**, *95*, 480–492. [CrossRef]
20. Zhang, G.; Xiao, X.; Biradar, C.M.; Dong, J.; Qin, Y.; Menarguez, M.A.; Zhou, Y.; Zhang, Y.; Jin, C.; Wang, J. Spatiotemporal patterns of paddy rice croplands in China and India from 2000 to 2015. *Sci. Total Environ.* **2017**, *579*, 82–92. [CrossRef]
21. Gumma, M.K.; Nelson, A.; Thenkabail, P.S.; Singh, A.N. Mapping rice areas of South Asia using MODIS multitemporal data. *J. Appl. Remote Sens.* **2011**, *5*, 053547. [CrossRef]
22. Sun, H.; Xu, A.; Lin, H.; Zhang, L.; Mei, Y. Winter wheat mapping using temporal signatures of MODIS vegetation index data. *Int. J. Remote Sens.* **2012**, *33*, 5026–5042. [CrossRef]
23. Huke, R.E. *Rice Area by Type of Culture: South, Southeast, and East Asia*; International Rice Research Institute: Los Banos, CA, USA; Laguna, Philippines, 1982.
24. Huke, R.E.; Huke, E.H. *Rice Area by Type of Culture: South, Southeast, and East Asia, A Revised and Updated Data Base*; International Rice Research Institute: Los Banos, CA, USA; Laguna, Philippines, 1997.
25. Aselman, I.; Crutzen, P.J. Global distribution of natural freshwater wetlands and rice paddies, their net primary productivity, seasonality and possible methane emissions. *J. Atmos. Chem.* **1989**, *8*, 307–358. [CrossRef]
26. Xiao, X.; Boles, S.; Frolking, S.; Li, C.; Babu, J.Y.; Salas, W.; Moore Iii, B. Mapping paddy rice agriculture in South and Southeast Asia using multi-temporal MODIS images. *Remote Sens. Environ.* **2006**, *100*, 95–113. [CrossRef]
27. Takeuchi, W.; Yasuoka, Y. Sub-pixel mapping of rice paddy fields over Asia using MODIS time series. In *Remote Sensing of Global Croplands for Food Security (Remote Sensing Applications Series)*; Thenkabail, P., Lyon, J.G., Turral, H., Biradar, C., Thenkabail, P., Eds.; CRC Press: Boca Raton, FL, USA, 2009.
28. Sun, H.; Huang, J.; Huete, A.R.; Peng, D.; Zhang, F. Mapping paddy rice with multi-date moderate-resolution imaging spectroradiometer (MODIS) data in China. *J. Zhejiang Univ. Sci. A* **2009**, *10*, 1509–1522. [CrossRef]
29. Frolking, S.; Yeluripati, J.B.; Douglas, E. New district-level maps of rice cropping in India: A foundation for scientific input into policy assessment. *Field Crops Res.* **2006**, *98*, 164–177. [CrossRef]
30. Frolking, S.; Qiu, J.; Boles, S.; Xiao, X.; Liu, J.; Zhuang, Y.; Li, C.; Qin, X. Combining remote sensing and ground census data to develop new maps of the distribution of rice agriculture in China. *Glob. Biogeochem. Cycles* **2002**, *16*, 3801–3810. [CrossRef]
31. Thenkabail, P.S.; Biradar, C.M.; Noojipady, P.; Dheeravath, V.; Li, Y.; Velpuri, M.; Gumma, M.; Gangalakunta, O.R.P.; Turral, H.; Cai, X.; et al. Global irrigated area map (GIAM), derived from remote sensing, for the end of the last millennium. *Int. J. Remote Sens.* **2009**, *30*, 3679–3733. [CrossRef]
32. Badhwar, G.D. Automatic corn-soybean classification using landsat MSS data. I. Near-harvest crop proportion estimation. *Remote Sens. Environ.* **1984**, *14*, 15–29. [CrossRef]
33. Lobell, D.B.; Asner, G.P.; Ortiz-Monasterio, J.I.; Benning, T.L. Remote sensing of regional crop production in the Yaqui Valley, Mexico: Estimates and uncertainties. *Agric. Ecosyst. Environ.* **2003**, *94*, 205–220. [CrossRef]
34. Thiruvengadachari, S.; Sakthivadivel, R. *Satellite Remote Sensing for Assessment of Irrigation System Performance: A Case Study in India*; Research Report 9; International Irrigation Management Institute: Colombo, Sri Lanka, 1997.
35. Thenkabail, P.S. Global Croplands and their Importance for Water and Food Security in the Twenty-first Century: Towards an Ever Green Revolution that Combines a Second Green Revolution with a Blue Revolution. *Remote Sens.* **2010**, *2*, 2305–2312. [CrossRef]
36. Dheeravath, V.; Thenkabail, P.S.; Chandrakantha, G.; Noojipady, P.; Reddy, G.P.O.; Biradar, C.M.; Gumma, M.K.; Velpuri, M. Irrigated areas of India derived using MODIS 500 m time series for the years 2001–2003. *ISPRS J. Photogramm. Remote Sens.* **2010**, *65*, 42–59. [CrossRef]
37. Goetz, S.J.; Varlyguin, D.; Smith, A.J.; Wright, R.K.; Prince, S.D.; Mazzacato, M.E.; Tringe, J.; Jantz, C.; Melchoir, B. Application of Multitemporal Landsat Data to Map and Monitor Land Cover and Land Use Change in the Chesapeake Bay Watershed. In *Analysis of Multi-Temporal Remote Sensing Images*; Smits, P., Bruzzone, L., Eds.; World Scientific Publishers: Singapore, 2024; pp. 223–232.

38. Thenkabail, P.S.; Schull, M.; Turral, H. Ganges and Indus river basin land use/land cover (LULC) and irrigated area mapping using continuous streams of MODIS data. *Remote Sens. Environ.* **2005**, *95*, 317–341. [CrossRef]
39. Velpuri, N.M.; Thenkabail, P.S.; Gumma, M.K.; Biradar, C.B.; Noojipady, P.; Dheeravath, V.; Yuanjie, L. Influence of Resolution in Irrigated Area Mapping and Area Estimations. *Photogramm. Eng. Remote Sens.* **2009**, *75*, 1383–1395. [CrossRef]
40. Varlyguin, D.; Wright, R.; Goetz, S.J.; Prince, S.D. Advances in Land Cover Classification: A Case Study from the Mid-Atlantic Region. In Proceedings of the American Society for Photogrammetry and Remote Sensing, St. Louis, MO, USA, 23–27 April 2001; Volume 7. Available online: https://citeseerx.ist.psu.edu/document?repid=rep1&type=pdf&doi=278fb1793b69fd19ffad89f70f846cf39dfdbf6f (accessed on 11 February 2024).
41. Knight, J.F.; Lunetta, R.L.; Ediriwickrema, J.; Khorram, S. Regional Scale Land-Cover Characterization using MODIS-NDVI 250 m Multi-Temporal Imagery: A Phenology Based Approach. *GISci. Remote Sens.* **2006**, *43*, 1–23. [CrossRef]
42. Biggs, T.W.; Thenkabail, P.S.; Gumma, M.K.; Scott, C.A.; Parthasaradhi, G.R.; Turral, H.N. Irrigated area mapping in heterogeneous landscapes with MODIS time series, ground truth and census data, Krishna Basin, India. *Int. J. Remote Sens.* **2006**, *27*, 4245–4266. [CrossRef]
43. Gaur, A.; Biggs, T.W.; Gumma, M.K.; Parthasaradhi, G.; Turral, H. Water scarcity effects on equitable water distribution and land use in Major Irrigation Project—A Case study in India. *J. Irrig. Drain. Eng.* **2008**, *134*, 26–35. [CrossRef]
44. Gumma, M.K.; Thenkabail, P.S.; Muralikrishna, I.V.; Velpuri, M.N.; Gangadhararao, P.T.; Dheeravath, V.; Biradar, C.M.; Acharya Nalan, S.; Gaur, A. Changes in agricultural cropland areas between a water-surplus year and a water-deficit year impacting food security, determined using MODIS 250 m time-series data and spectral matching techniques, in the Krishna River basin (India). *Int. J. Remote Sens.* **2011**, *32*, 3495–3520. [CrossRef]
45. Sakamoto, T.; Van Nguyen, N.; Kotera, A.; Ohno, H.; Ishitsuka, N.; Yokozawa, M. Detecting temporal changes in the extent of annual flooding within the Cambodia and the Vietnamese Mekong Delta from MODIS time-series imagery. *Remote Sens. Environ.* **2007**, *109*, 295–313. [CrossRef]
46. Ratana, P.; Huete, A.R.; Yin, Y.; Jacobson, A. Interrelationship among among MODIS vegetation products across an Amazon Eco-climatic gradient. In Proceedings of the Geoscience and Remote Sensing Symposium, 2005, IGARSS'05, Proceedings. Seoul, Republic of Korea, 29 July 2005; IEEE International: Piscataway, NJ, USA, 2005; pp. 3009–3012.
47. Becerra, J.; Alvalá, R.; Berlato, M.; Molion, L.; Berri, G.; Bertossa, G.; Bertagnolli, C.; Peres, R.; Ferreira, N.; Schuch, N. Detection of Tropical Savannah (Cerrado) Physiognomies in the Legal Amazon by the Application of the Vegetation and Moisture Indices with MODIS Time Series Data. In Proceedings of the 8th International Conference on Southern Hemisphere Meteorology and Oceanography (ICSHMO), Foz de Iguaçu, Brazil, 24–28 April 2006; pp. 861–868.
48. Chandrasekar, K.; Sesha Sai, M.V.R.; Roy, P.S.; Dwevedi, R.S. Land Surface Water Index (LSWI) response to rainfall and NDVI using the MODIS Vegetation Index product. *Int. J. Remote Sens.* **2010**, *31*, 3987–4005. [CrossRef]
49. Choice, H. Agroecological Zones. 2009. Available online: http://harvestchoice.org/production/biophysical/agroecology (accessed on 1 May 2010).
50. WorldBank. Rural Population (% of Total Population)—South Asia. 2021. Available online: https://data.worldbank.org/indicator/SP.RUR.TOTL.ZS?locations=8S (accessed on 25 March 2023).
51. Thenkabail, P.S.; Biradar, C.M.; Noojipady, P.; Dheeravath, V.; Li, Y.J.; Velpuri, M.; Reddy, G.; Cai, X.; Gumma, M.; Turral, H. *A Global Irrigated Area Map (GIAM) Using Remote Sensing at the End of the Last Millennium*; International Water Management Institute: Colombo, Sri Lanka, 2008.
52. Gumma, M.K.; Thenkabail, P.S.; Maunahan, A.; Islam, S.; Nelson, A. Mapping seasonal rice cropland extent and area in the high cropping intensity environment of Bangladesh using MODIS 500m data for the year 2010. *ISPRS J. Photogramm. Remote Sens.* **2014**, *91*, 98–113. [CrossRef]
53. Homayouni, S.; Roux, M. Material Mapping from Hyperspectral Images using Spectral Matching in Urban Area. In Proceedings of the IEEE Workshop in Honour of Prof. Landgrebe, Washington, DC, USA, 6 October 2003.
54. Thenkabail, P.; GangadharaRao, P.; Biggs, T.; Gumma, M.; Turral, H. Spectral matching techniques to determine historical land-use/land-cover (LULC) and irrigated areas using time-series 0.1-degree AVHRR Pathfinder datasets. *Photogramm. Eng. Remote Sens.* **2007**, *73*, 1029–1040.
55. Thenkabail, P.S.; Biradar, C.M.; Noojipady, P.; Cai, X.; Dheeravath, V.; Li, Y.; Velpuri, M.; Gumma, M.; Pandey, S. Sub-pixel area calculation methods for estimating irrigated areas. *Sensors* **2007**, *7*, 2519–2538. [CrossRef]

remote sensing MDPI

Article

Early Season Forecasting of Corn Yield at Field Level from Multi-Source Satellite Time Series Data

Johann Desloires [1,2,*], Dino Ienco [1,3] and Antoine Botrel [2]

1 TETIS, Université de Montpellier, INRAE, 500 Rue Jean François Breton, 34000 Montpellier, France; dino.ienco@inrae.fr
2 Syngenta France SA, 1228 Chem. de l'Hobit, 31790 Saint-Sauveur, France; antoine.botrel@syngenta.com
3 INRIA, 860 Rue de St-Priest, 34090 Montpellier, France
* Correspondence: johann.desloires@gmail.com

Abstract: Crop yield forecasting during an ongoing season is crucial to ensure food security and commodity markets. For this reason, here, a scalable approach to forecast corn yields at the field-level using machine learning and satellite imagery from Sentinel-2 and Landsat missions is proposed. The model, evaluated on 1319 corn fields in the U.S. Corn Belt from 2017 to 2022, integrates biophysical parameters from Sentinel-2, Land Surface Temperature (LST) from Landsat, and agroclimatic data from ERA5 reanalysis dataset. Resampling the time series over thermal time significantly enhances predictive performance. The addition of LST to our model further improves in-season yield forecasting, through its capacity to detect early drought, which is not immediately visible to optical sensors such as the Sentinel-2. Finally, we propose a new two-stage machine learning strategy to mitigate early season partially available data. It consists in extending the current time series on the basis of complete historical data and adapting the model inference according to the crop progress.

Keywords: yield forecasting; machine learning; thermal time; Sentinel-2; land surface temperature; early season forecasting

1. Introduction

Monitoring and predicting crop phenology, growth, and yield is crucial for global food security, market dynamics, policy making, and decision making [1]. Accurate early season estimations of crop yield provides farmers with an estimate of their production, enabling them to assess risks, determine insurance premiums, and evaluate input costs [2]. In addition to supporting individual farmers, it contributes to a broader understanding of the complex interplay between environmental factors and management practices in agriculture [3]. This understanding facilitates the development of more effective and flexible within-season management strategies [4,5] and enables the anticipation of market demand by forecasting supply [6].

Traditional methods such as manual sampling and field campaigns are labor-intensive, and provide limited insights into the spatial variability of crop yield. These limitations have led to the development of alternative approaches for estimating yields during the growing season [7]. Using advanced technologies and data-driven methods, these yield forecasts not only reduce the work and time spent on measurements, but also improve the spatial coverage and accuracy of the obtained information [1]. In recent years, there have been remarkable advances in yield forecasting methods that use Earth Observation (EO) data, satellite remote sensing imagery, acquired via modern missions. This technology has become a valuable source of information for monitoring agricultural practices. By exploiting EO data, within-season (or early season) crop yield forecasts can now be provided, enabling farmers and stakeholders to make informed decisions leading to optimal production [8]. This advancement primarily stems from the availability of near real-time data (NRT) EO datasets from open sources with optical instruments on satellites like SPOT,

MODIS, PROBA-V, and Landsat which have played a crucial role in such operational crop monitoring [9]. These instruments offer several benefits, including daily revisit cycles, global coverage, long-term data archives, and low- or no-cost accessibility. However, there is a quest for generic time series analysis methods for crop mapping and monitoring that can be deployed at a large scale, taking advantage of the global coverage, with high spatial and temporal resolution, provided by modern Earth observation missions.

The launch of Sentinel-2A (S2A) in late 2015 and Sentinel-2B (S2B) in early 2017 has significantly enhanced crop monitoring capabilities. S2A provides a 10-day revisit time period over Europe and Africa, and 20 days elsewhere, while S2B ensures a 5-day revisit time period worldwide since February 2018 [10]. This unprecedented revisit time is particularly suitable for in-season crop monitoring. Unlike previous Earth Observation (EO) missions, Sentinel-2 enables the derivation of red-edge-based vegetation indices, which exhibit stronger correlations with agronomic parameters compared to red-based indices [11]. The combination of Copernicus' free and open access policy with the high resolution of Sentinel-2 images allows for the construction of dense and consistent time series throughout the crop growth cycle in most regions of the world [12]. Consequently, Sentinel-2 satellite imagery, with its spectral bands (visible, near infrared, red-edge, and short-wave infrared) and spatial resolutions (10 m, 20 m, and 60 m), has been successfully exploited in recent years for modeling crop grain yield at field and within-field scales [1,13,14].

Before the advent of Sentinel-2, Landsat satellite data played a crucial role in accurately mapping crop types and predicting yields at the field level in agricultural landscapes, worldwide [15]. With a temporal resolution of 8 days, Landsat 7 and 8 (before 2022) and Landsat 8 and 9 (after 2022) images offer a tangible advantage in crop monitoring when coupled with Sentinel-2, fully enriching the latter with the information provided by Landsat thermal bands [16]. The Landsat missions are currently the only constellation equipped with thermal bands, provided at an adequate spatial resolution (100 meters), for precise monitoring at the scale of individual fields. Land surface temperature (LST), derived from these thermal bands, is used to monitor heat stress and drought, which can explain some of the variability in yields between years [2,17]. Indeed, relying solely on early season optical remote sensing data can make it difficult to detect the onset of drought, which mainly captures information on the upper canopy. This is because drought symptoms tend to appear earlier in the lower leaves, potentially underestimating the negative effects of drought on yield [18]. At the county scale, studies have shown a negative correlation between MODIS diurnal LST and mid-summer corn yield forecasts [2]. However, in satellite-based agricultural modeling, studies mainly focus on vegetation indices in the visible and near-infrared part of the electromagnetic spectrum, and potential data related to Land Surface Temperature are often neglected [19]. We therefore considered it important to evaluate LST as an input variable in our task of early season crop yield prediction.

When it comes to yield estimation methods, mechanistic crop growth models are the standard choice, as they are designed and calibrated to simulate yield formation processes using soil information, climate and farm management practices. They enable crop yields to be predicted at any time and in any place [5], but the need for extensive (and often costly) data on field-specific biotic and abiotic factors limits the large-scale deployment of these approaches over the ongoing season [2,15,20]. In contrast, machine learning (ML) algorithms can handle complex relationships between predictors and the target variables, leading to their increased use in the agriculture domain [21–23].

A key limitation of ML methods for crop yield prediction is their dependence on data acquired under specific local conditions, which may result in inaccurate forecasts when confronted with data acquired under unseen conditions not included in the model's training data [24]. This may be partly explained by possible differences in crop progress and climate/environmental condition changes from season to season, or from location to location, where these seasonal changes in phenology are primarily influenced by temperature and water regimes [25]. To face such generalization issues affecting modern ML approaches, in the context of yield prediction, a possible solution is to resample remote sensing satellite image data over periods calculated over thermal time, i.e., the number of growing degrees–days accumulated from the sowing date, rather than calendar time [26], with the aim of mitigating possible shifts in crop phenology at a given date, due to different temperature regimes.

The research is founded upon a cut-off date of mid-August, approximately two months prior to the corn harvest. This timeframe corresponds to the initial stages of maize reproduction, including heading and pollination, with regional variations influenced by planting dates [27]. At this juncture, grain filling commences, during which kernel size begins to develop, while the kernel count has already been established in prior stages. Consequently, pre-stage crop conditions and weather patterns can significantly impact yield potential. The objective of this study is to contribute to the comprehension of utilizing remote sensing data for estimating in-season corn yield through the application of machine learning methodologies.

From a methodological point of view, our approach is based on a real-life deployment scenario of a machine learning framework. In this framework, a model is trained using reference data collected during previous seasons. With regard to the selection and processing of predictors, our methodology includes (1) the incorporation of thermal time to account for the various phenological advances of crops, which depend on temperature regimes rather than a fixed number of days, (2) the integration of Land Surface Temperature (LST) and agroclimatic stress indicators to address the limitation of optical imagery which does not reach its full potential during the growing season, and (3) the exploitation of comprehensive historical information covering the whole season to improve predictions for the current early season forecasting task.

In this study, a multi-year proprietary dataset is utilized, comprising data from 1319 corn production fields predominantly situated across 29 counties within the U.S. Corn Belt spanning the years 2017 to 2022. The dataset includes the average yield of each field, and for evaluation purposes, each year is assessed independently, with the remaining years serving as the training dataset for the model.

The rest of this paper is organized as follows. Section 2 introduces the data available on the study site, and Section 3 the proposed framework associated with experimental settings. The results are reported and discussed in Section 4. Finally, Section 5 concludes.

2. Materials

2.1. Corn Yield Dataset

The research site is primarily located in the U.S. Corn Belt, the largest corn-producing states in the United States. A total of 1319 corn fields are available, of which 849 fields were located in Nebraska, 423 in Iowa, 29 in Illinois, and 21 in Wisconsin.

The yield dataset utilized in this study, provided by Syngenta (Syngenta is an international leading science-based agtech company https://www.syngenta.com/en/company (accessed on 10 October 2023)), comprises direct measurements of corn yield obtained from seed production fields. These fields are utilized for crossbreeding two corn varieties, consisting of male and female plants. During harvesting, only the female plants are collected, as the male plants are sterile. Therefore, the yield data is obtained solely from the harvested female acres. The measurements are derived from the weight of harvested corn, adjusted for moisture content to 15.5% and measured in green bushels per female acre (GB/FA).

The size of the fields ranged from 3.9 to 276 female acres. Female parental lines were on average planted on day of the year 137 ($\pm$9.3) and harvested on day 263 ($\pm$9.8), and the average length of the growing was 126 days ($\pm$9.1). The distribution of yield across different years is depicted in Figure 1b. In Figure 2, the average yield data at the county level across all historical records is depicted to assess spatial heterogeneity. For data confidentiality, the yield values were scaled between 0 and 1 using min–max normalization across all fields.

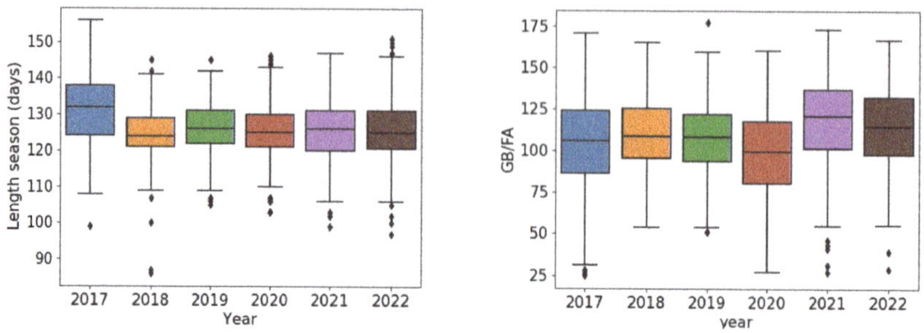

(**a**) Distribution of growing season length in days by year (**b**) Yield distribution by year

Figure 1. The boxplots illustrate the strong disparities in distribution observed between years, indicating significant heterogeneity in our data. Seasonal durations (**a**) ranged from 110 to 150 days from sowing to harvest, with a median duration of around 125 days. As for crop yields (**b**), they ranged from 25 to 175 bushels per acre, highlighting significant year-to-year variability. For example, the median yield in 2020 was 100 bushels, while it was 125 bushels in 2021. All data points, including those outside the boxplot whiskers, are retained in this study as they represent genuine measurements.

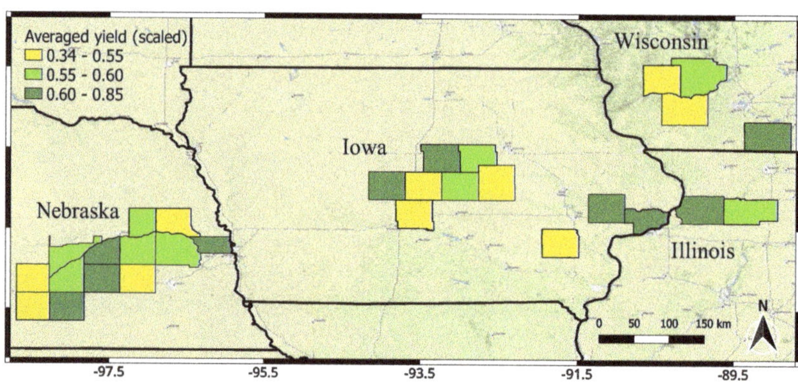

Figure 2. Seed production fields are spatially distributed over 29 counties in eastern Nebraska, Iowa, Illinois and Wisconsin.

In this study, the machine learning model utilized some in situ features, namely the corn cultivar duration (FAO maturity groups) of the female parental lines, whether the field was irrigated or not, and the geographic coordinates (expressed in decimal degrees and rounded off by a factor of 1), to integrate information from the local environment and the farming practices.

2.2. Satellite Data

2.2.1. Acquisition

Raw data from the Sentinel-2 and Landsat missions were collected using the open-source eo-learn Python library developed by Sinergise (https://github.com/sentinel-hub/eo-learn, accessed on 10 October 2023). This framework provides a Python client for downloading and pre-processing level 2A atmosphere-corrected data from the SentinelHub cloud platform. The raw band pixels were resampled to a 10-meter reference grid using nearest neighbor interpolation. Within each image, pixels identified as saturated, shaded, or cloudy using the scene classification map (SLC) obtained from sen2cor [28] were excluded from subsequent analyses. Then, the percentage of pixels previously identified as invalid relative to the total number of pixels within the field bounding box was calculated. Only acquisition dates with more than 50% valid pixels were retained for the analysis.

To ensure the model receives adequate data for accurate predictions during the growing season, specific rules were established to select observations from the whole set of data. More precisely, a minimum of four Sentinel-2 and Landsat 7/8/9 images was required between the beginning of May and the end of July, with at least one image needed in the first half of August. Challenges arose from limited image availability, particularly in 2017, where only Sentinel-2A was accessible, averaging eight images between May and September (Table 1). Additionally, in 2019, a technical issue with the Landsat-8 satellite temporarily reduced the number of exploitable Landsat images. Termed as a "degraded image quality" event, this issue occurred in May 2019, resulting in a decreased number of exploitable images for that season.

Table 1. Summary statistics in-situ data. The values reported are mean ± standard deviation.

Year	# Fields	# Valid Sentinel-2 Images	# Valid Landsat Images
2017	250	8.1 ± 2.6	15.9 ± 5.9
2018	189	17.8 ± 7.5	14.8 ± 5.2
2019	185	15.8 ± 8.1	8.1 ± 4.9
2020	226	19.0 ± 7.8	14.1 ± 3.3
2021	221	17.3 ± 6.1	14.9 ± 3.9
2022	248	19.2 ± 7.6	14.9 ± 6.4
Summary	1319	16.2 ± 6.6	13.8 ± 4.9

Figure 3a,b depict the distribution of the number of Sentinel-2 and Landsat 7/8/9 images from April to September, aggregated over all considered years. August emerges with the highest number of associated images, with a median of four Sentinel-2 images and three Landsat images. However, in May, a maximum of only two available Sentinel-2 images was observed, indicating potential systematic cloud coverage over our study area during this period.

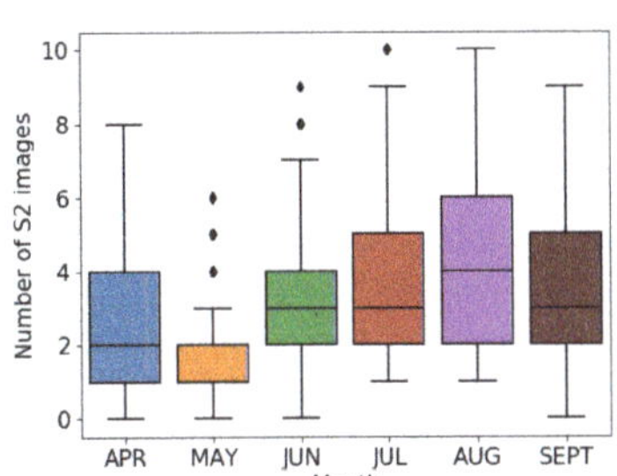

(**a**) Distribution of the number of images per month for Sentinel-2.

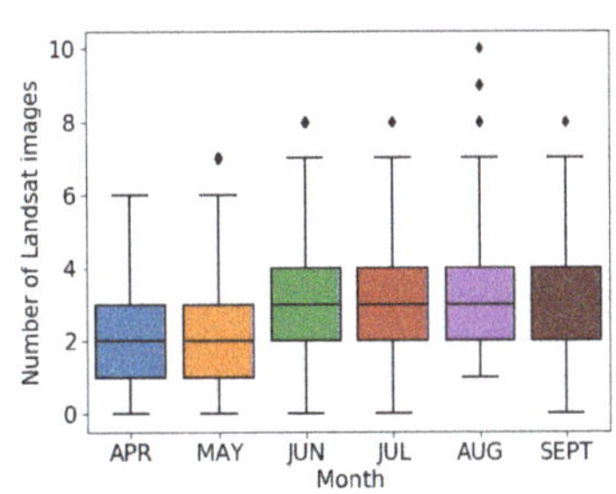

(**b**) Distribution of the number of images per month for Landsat 7/8/9.

Figure 3. Distribution of the number of images per month for Sentinel-2 and Landsat 7/8/9 data between 2017 and 2022. Points outside the boxplot whiskers correspond to fields lying on two relative orbits.

2.2.2. Biophysical Parameters

The estimation of Leaf Area Index (LAI) and Leaf Chlorophyll Content (C_{ab}) was facilitated using an Artificial Neural Network (ANN) pre-trained on the raw Sentinel-2 bands accessible within the Sentinel Application Platform (SNAP) [29]. These biophysical parameters are of particular interest due to their ability to enhance the signal-to-noise ratio compared to raw bands and/or vegetation indices when predicting maize yield for a new season using Sentinel-2 data [26].

The time series were linearly interpolated and smoothed using the Savitzky–Golay filter, which is commonly used to smooth noisy signals [30,31]. A half-width size of 15 was used for the smooth window, and a third-degree polynomial was used to set the weighting coefficient.

2.2.3. Land Surface Temperature

Land Surface Temperature (LST) was derived by applying the single-channel algorithm [32] to Landsat 8 Band 10 and Landsat 7 Band 6. For consistency, all raw band pixels were resampled to a 50-meter reference grid using nearest neighbor interpolation. Additionally, a binary erosion of radius 1 was applied to eliminate border pixels around the field. In subsequent analyses, pixels identified as cloud and shadow masks, based on thresholding the quality assessment (QA) band, were discarded. The percentage of identified invalid pixels was then calculated relative to the total number of pixels within the field bounding box. Only acquisition dates with more than 50% valid pixels were retained for the analysis.

The time series underwent linear interpolation and smoothing using the Savitzky–Golay filter [30]. Outliers were masked out initially by applying an empirical threshold of 10 °C between day of the year 100 and day of the year 300.

2.3. Agroclimatic Data

In addition to Sentinel-2 optical imagery to monitor seasonal changes and Land Surface Temperature (LST) for drought assessment, our objective was to develop early stress indicators impacting future yield potential. This involved assessing extreme weather conditions like drought, hail, floods, frost, high temperatures, and strong winds, known to have a considerable impact on corn productivity [33].

Previous research has highlighted precipitation and air temperature during the growing season, particularly during the late vegetative and early reproductive stages, as possible influence factors for corn yield deviations [34]. Excessive heat can negatively impact physiological processes, including water stress, root growth, flowering, and lead to premature maturity and senescence [35]. Temperature was found to be more influential than rainfall in estimating corn yield, with rainfall during May and August being relatively less important [36]. Moreover, Johnson [2] showed that precipitations have no correlation with corn yields regardless of the seasonality for midsummer corn yield forecasting at the county level. Given these contradictions and the inclusion of both irrigated and rainfed fields in the study, integration of information related to crop water requirements was avoided.

In order to comprehensively capture temperature dynamics throughout the study period, additional agrometeorological information was incorporated. This included daily means, minimums and maximums, as well as the number of days with temperatures above 30 °C (86 °F) and the accumulated number of Growing Degree Days (GDD). These temperature-related variables were selected because of their proven influence on maize growth and development [25].

ERA5, a high-resolution reanalysis dataset (https://www.sciencedirect.com/topics/engineering/reanalysis, accessed on 10 October 2023) produced by ECMWF (European Centre for Medium-Range Weather Forecasts), was used to obtain historical weather data. However, these data has a time lag of approximately two months and is not available in real-time. To ensure timely forecasts and to place themselves in a operational scenario, we used ERA5T, which is an initial release data available with a time lag of around 5 days.

Table 2 presents a summary of the variables considered to address the early season forecasting of corn yield at the field-level. These variables were derived from daily time series data, which were aggregated into periods, as detailed in Section 3.1.

Table 2. Variables used in crop yield forecasting during 2017–2022.

Category	Abbreviation	Definition	Source
Satellite data	C_{ab}	Leaf Chlorophyll Content	Sentinel-2
	$C_{ab_{MAX}}$	Maximum value of C_{ab}	
	LAI	Leaf Area Index	
	$C_{ab_{CUM}}$	Accumulated daily C_{ab}	
	LST	Land Surface Temperature	Landsat
Agroclimatic	T_{max}	Daily maximum temperature	ERA5T
	T_{min}	Daily minimum temperature	
	GDD	Accumulated Growing Degree Days	
	GDD_{Peak}	Accumulated GDD when LAI is maximum	
	$Heat_{Days}$	# Days $T_{max} > 30$ °C	
	WG_{Days}	# Days Wind Gust > 60 km/h	
In-situ	ReM	Relative maturity group	Syngenta P&S
	Lon	Longitude	
	Lat	Latitude	
	Irr	Irrigation (yes/no)	

3. Methods

In this section, the established early season yield prediction strategy is presented. Yield forecasting is performed for an ongoing season in which harvest data are not yet available, i.e., extrapolating yield to test samples using machine learning methods trained on data from other years.

Section 3.1 briefly introduces the temporal resampling of the time series used in this study, based on thermal rather than calendar time. In Section 3.2.1, the proposal is made to adapt the model inference according to the number of periods available, rather than imputing or deleting missing observations as a baseline. Then, it is suggested to enrich the partially available early season dataset from the full dataset, in terms of time period, available from other years (e.g. historical data). To this end, Section 3.2.2 introduces the workflow to predict an extra period for each field, which will then be used as an observation for the inference phase.

3.1. Temporal Resampling

To resample the time series data based on different temporal criteria, two approaches were adopted. Firstly, for calendar time, the data was divided into fixed 9-day periods starting from the crop fields' sowing date.

Thermal time offers the advantage of mitigating potential temporal shifts across seasons or locations [37]. Measured in growing-degree days (GDDs), thermal time accumulates mean daily air temperatures at 2 meters above the ground, surpassing a crop-specific threshold. While the accumulation of GDDs over the growing season can vary by location and year, the thermal time required for corn plants to reach specific developmental stages remains relatively constant [38]. This computation can be expressed as follows:

$$GDD^{(t)} = \sum_{i=1}^{t} \max\left(\frac{T_{min}^{(i)} + T_{max}^{(i)}}{2} - T_{base}, 0\right) \quad (1)$$

where $T_{min}^{(i)}$ and $T_{max}^{(i)}$ denote the minimum and maximum daily temperatures for a given day i, respectively, and T_{base} represents the base temperature for corn, conventionally set at 10 °C to account for limited growth below this threshold. The upper threshold temperature was typically fixed at 30 °C, assuming a relatively consistent development rate between these two thresholds.

The time series data was aggregated at intervals corresponding to the accumulation of 120 GDDs, as suggested in [26]. Figure 4 displays the distribution of elapsed days and accumulated Growing Degree Days (GDDs) since planting. Generally, the variability in time units between sowing and the cut-off date (August 15th) was less pronounced than that of GDDs.

The final period was retained if it covered at least 80% of the chosen time unit before mid-August. This translated to a minimum duration of 7 days and an accumulation of 95 Growing Degree Days (GDDs) for both calendar and thermal time analyses, respectively. This completeness rule applied solely to fields in the current season for which yield forecasting was conducted, as the entire season's data were available from the training years. Figure 5a,b illustrate the percentage of fields in the dataset that reached the final periods for both calendar and thermal time analyses, corresponding to periods 7 through 10.

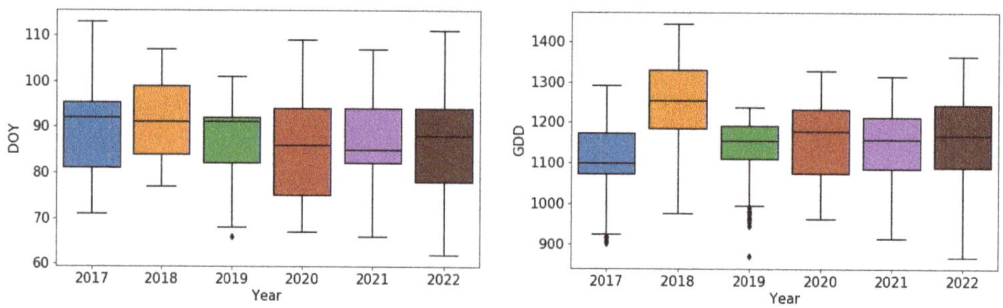

(**a**) (**b**)

Figure 4. Distribution of the number of days (calendar or thermal) (**a**) and accumulated Growing Degree Days (GDDs) between the planting date and the 15th of August across different years (**b**). These boxplots illustrate the variability in the dataset distribution throughout the growing season, providing insights into the temporal patterns of crop development.

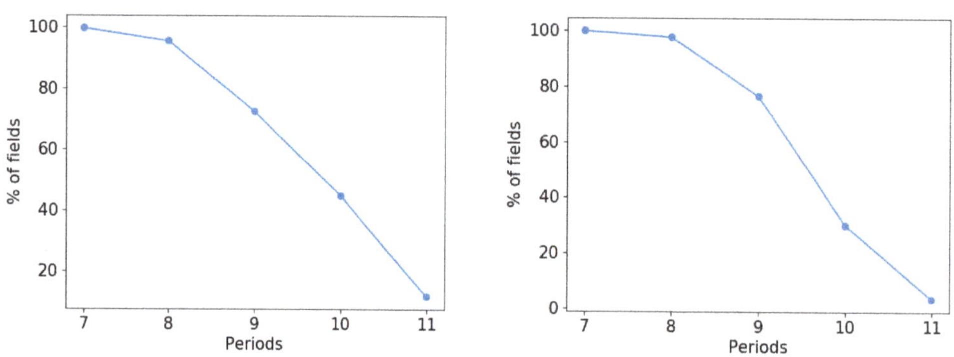

(**a**) Field completion rate considering calendar time-derived periods up to the final period. *t*

(**b**) Field completion rate considering thermal time-derived periods up to the final period. *t*

Figure 5. Percentage of fields reaching a certain period after resampling of time series (calendar (**a**) or thermal time (**b**)) at 15 August.

3.2. Yield Estimation Methods

3.2.1. Adapting to Field Progress with Multiple Random Forest Models

In the previous subsection, the issue of missing periods for early season forecasting was addressed, stemming from later sowing dates (calendar time) or lower temperature regimes (thermal time). A straightforward solution would involve deleting the periods for which values are missing. However, this approach would discard part of the available information. To maximize the use of the available data, we propose a strategy to accommodate the number of periods for each field.

Regarding the choice of the machine learning algorithm, Random Forest (RF) [39] has demonstrated considerable successful in predicting corn yield using Sentinel-2 data [13] and Landsat data [5,24]. The scikit-learn package [40] was employed for implementation, with all models empirically parameterized with 500 trees and a maximum depth of 7, while all other parameters were set to their default values. Figure 6 depicts our early season inference strategy, wherein Random Forest (RF) machine learning methods trained on data from previous years are tested. We evaluated early season yield forecasting for each year independently, training the model on data from the years excluding the one being predicted, thereby treating each season as an independent entity.

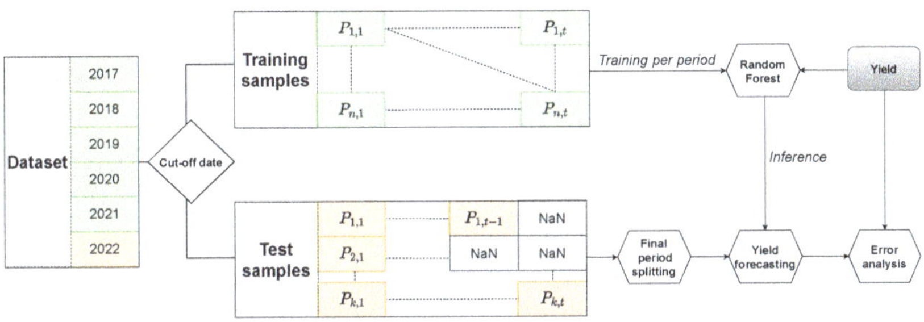

Figure 6. Training and inference procedure during a new season (e.g., 2022). A cut-off date is applied to the test samples, resulting in missing periods P as each field progresses. We retain all periods in the training samples that correspond to the most advanced period observed in the test samples. Individual models are then calibrated for each of the last periods observed in the test samples (e.g., $P_{.,t}$). Training years are shown in green, while the test year is shown in orange.

3.2.2. Forecasting Seasonal Trajectories Using Comprehensive Historical Data

The forecasting of C_{ab} was proposed as it has been shown to be the best yield predictor derived from Sentinel-2 data [26]. All available information prior to the cut-off date, derived from Sentinel-2 (S2), crop stress indices (CS) from agroclimatic data, and Land Surface Temperature (LST) from Landsat, was considered. This multivariate approach was preferred over a univariate method that would neglect potential stress factors that could later influence the temporal trajectories of C_{ab}, such as earlier and more abrupt senescence. In previous work, statistical models based on climatic factors such as precipitation and temperature have been applied to predict the seasonal trajectories of the NDVI vegetation index [41–43].

A Deep Neural Network (DNN), integrating a Long Short-Term Memory (LSTM) layer [44] to model the inputs with a time dimension was proposed. LSTM layers are designed to model information from long time periods and effectively mitigate the issue of vanishing gradients. Additionally, a data augmentation strategy was considered, in which time series profiles of the 15% and 85% quantiles derived from Sentinel-2 data aggregated to the field level were included, together with the median values previously obtained. This approach aims to leverage additional factors and predict a potential accelerated decrease in the C_{ab} trajectory after the reproductive stage, capturing dynamic changes induced by potential stressors that the model can get profit on.

The output of the DNN was specifically designed to predict the M periods ahead that can be forecasted along the C_{ab} trajectory at period t. These predicted periods are denoted by x_t. Our objective was to capture the contextual information of the seasonal trajectory more effectively through their temporal dependencies by considering a wider forecast horizon than the available one. This would allow the model to gain a deeper understanding of the underlying dynamics, resulting in more accurate and reliable forecasts for the initially desired forecast horizon. The forecast value for (x_t), denoted as $y_t^{(p)}$, can be expressed as follows:

$$y_t^{(p)} = \left(x_t^{(p)}, x_{t+1}^{(p)}, \ldots, x_{t+M-1}^{(p)}\right)^T$$

In the DNN architecture (Figure 7), "static" data are incorporated into the training dataset, referring to data with fixed values throughout the season. This includes important information such as the relative maturity of the variety, which can impact the observed phenology during the season, characterized indirectly by the time series of C_{ab}. To incorporate these data, a two-branch approach was employed. One branch consists of *Dense* (or *Fully Connected*) layers dedicated to processing the static inputs, while the other branch utilizes an LSTM layer to handle the time-dependent inputs. Merging or combining the internal representations of the two data types in subsequent dense layers via concatenation allows static and temporal information to be integrated, enabling the model to leverage the full set of available information to forecast the seasonal trajectory.

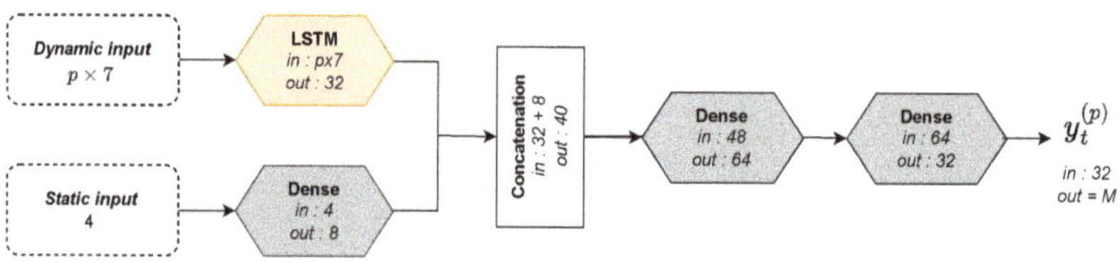

Figure 7. DNN architecture with LSTM and dense layers. Concatenation of dynamic and static paths. Node dimensions indicated. Each dense hidden layer is followed by batch normalization and ReLU activation function with a dropout rate of 0.5. DNN was trained over 100 epochs with a batch size of 32, learning rate of 10e^{-4} using the Adam optimizer. Implemented through the Tensorflow library [45].

The Huber loss function was chosen [46]. This loss is characterized by a threshold (the δ threshold) to allow a transition between mean absolute error (MAE) and mean squared error (MSE), with the aim to handle both outliers and small errors in the prediction model. If $y_t = (x_t, x_{t+1}, \ldots, x_{t+M-1})^T$ denotes the corresponding real value, the loss can be defined as follows:

$$L(y^{(p)}, y) = \frac{1}{M} \sum_t \begin{cases} \frac{1}{2}(y_t^{(p)} - y_t)^2, & \text{if } |y_t^{(p)} - y_t| \leq \delta \\ \delta |y_t^{(p)} - y_t| - \frac{1}{2}\delta^2, & \text{otherwise} \end{cases} \quad (2)$$

4. Results and Discussion

In this section, the results of our different preprocessing techniques and early season yield estimation strategies, constituting our contribution as outlined in Section 3.2, are presented. Two evaluation metrics, R-squared (R^2) and MAPE (Mean Absolute Percentage Error), were utilized to assess the performance of the methods.

In Section 4.1, an evaluation of the two temporal resampling techniques for time series observations (calendar and thermal time) is presented using the $\text{RF}_{PERIODS}$ method. Then, in Section 4.2, the impact of different data sources on the evaluation metrics of the early season model was examined. Then, in Section 4.3, the results of the yield prediction model are presented, where the first step involves predicting one period of C_{ab} ahead for the test samples, which are then used in model inference during the season. In Section 4.4, detailed results are provided for predicting the yield for each season independently in mid-August with the most effective strategy we have designed. Finally, in Section 4.5.2, we draw possible future perspectives, taking into account the conclusions drawn from the analyses described earlier.

4.1. Comparison between Calendar and Thermal Time Temporal Resampling

For accurate results with the early season strategy outlined in Section 3.2.1, the $RF_{PERIODS}$ model was employed. This model trains a separate RF model for each period and utilizes the last observed period as a rule for inference in the test samples. Resampling the time series over thermal time has led to a reduction in the mean absolute percentage error from 17.02% to 16.25% and an increase in the proportion of yield variability explained by the model (R^2) from 0.25 to 0.30 (Table 3).

Table 3. MAPE and R^2 evaluation metrics around the 15 August. The values reported are mean ± standard deviation.

Metric	Resampling	$RF_{PERIODS}$
MAPE	Calendar	17.02 ± 2.23
	Thermal	**16.25 ± 1.99**
R^2	Calendar	0.25 ± 0.07
	Thermal	**0.30 ± 0.08**

4.2. Contribution of Different Data Sources to Early Season Forecasting

In Table 2, various data sources were compared: biophysical variables from Sentinel-2 (S2), a combination of Sentinel-2 data with Land Surface Temperature (LST) (S2 + LST), and the integration of these sources with crop stress indices (CS) from agroclimatic data (S2 + LST + CS). Additionally, end-of-season metrics ($RF_{P=end}$) were presented, utilizing a total of 14 periods for both training and testing. This allowed for the assessment of the early season forecasting strategy against predictions made with comprehensive data. The aim was to determine if the identified variables were beneficial solely for early season crop yield forecasting or had broader applicability for strategies informed at the end of the season.

Tables 4 and 5 provide the average and standard deviations of the MAPE and R^2 scores, respectively. Considering Land Surface Temperature (LST) as a data source resulted in a reduction of the MAPE error for the $RF_{PERIODS}$ model from 16.54% to 16.39% and a significant increase in the proportion of variance explained, from 0.26 to 0.29. Subsequent addition of agroclimatic crop stress indices (CS) further improved the results, leading to a MAPE of 16.25%. These findings confirm the effectiveness of LST and CS in detecting early stress during the growing season. However, there was no observed improvement in the performance of the $RF_{P=end}$ model with the inclusion of these factors, suggesting that the optical data alone were sufficient for end-of-season forecasting.

Table 4. MAPE evaluation metric per data source. The values reported are mean ± standard deviation.

Source	$RF_{PERIODS}$	$RF_{P=end}$
S2	16.54 ± 1.94	15.26 ± 1.78
S2 + LST	16.39 ± 1.86	15.32 ± 0.96
S2 + LST + CS	**16.25 ± 1.99**	15.36 ± 1.08

Table 5. R^2 evaluation metric per data source. The values reported are mean ± standard deviation.

Source	$RF_{PERIODS}$	$RF_{P=end}$
S2	0.26 ± 0.07	0.39 ± 0.08
S2 + LST	0.29 ± 0.07	0.38 ± 0.07
S2 + LST + CS	**0.30 ± 0.08**	0.39 ± 0.07

4.3. *Leverage Out-of-Year Data to Refine Early Season Yield Forecasts*

Chlorophyll-a and b (C*ab*) values were forecasted one period ahead for each test sample using the DNN method. These forecasts were then integrated into corn yield forecasting with $RF_{PERIODS}$, aiming to expand the early season time series. During inference, RF models were determined based on the last predicted period. Evaluation measures for this approach, labeled DNN, are presented in Table 6.

Our strategy, that consists in the use of the DNN to impute the next period of a test sample from a given year, showed the most promising performance among all the evaluated methods. It achieved an MAPE of 16.09%. These results confirm the potential of using out-of-year (e.g. historical data) after the cut-off date to extend data acquired during the ongoing season for the early season yield forecasting task. Figure 8 presents a bar chart of the MAPE and R^2 evaluation metrics by year. It is evident that a systematic improvement, across all seasons, exists thus, resulting in increased R^2 values or decreased MAPE, except for 2019 where we observed a slight degradation in results. This could be attributed, in part, to the limited availability of Landsat images used for estimating LST during that season.

Table 6. R^2 and MAPE evaluation metrics when using $RF_{PERIODS}$ whose period breakdown depends on the last period observed at the cut-off date plus a period predicted by DNN. The values reported are mean ± standard deviation.

Method	MAPE	R^2
$RF_{PERIODS}$	16.25 ± 1.99	0.30 ± 0.08
DNN	**16.09 ± 1.96**	**0.31 ± 0.08**

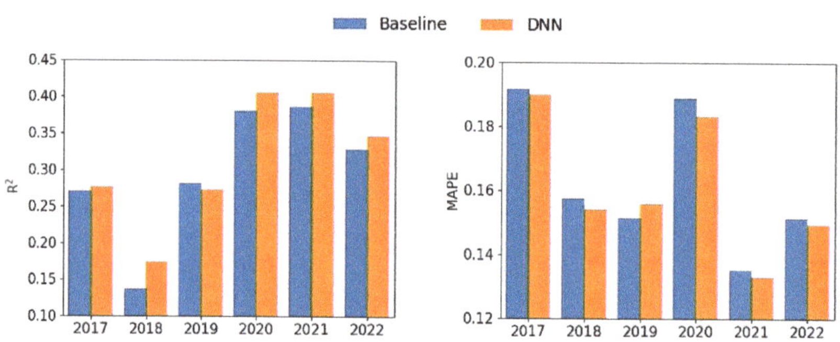

Figure 8. Model performance evaluation for early season yield forecasting in an independent test year, illustrated by R-squared (**left**) and MAPE (**right**). The baseline method (blue) is compared with $RF_{PERIODS}$, incorporating an additional period predicted by DNN, for period breakdown (orange).

4.4. *Fine Evaluation of the Best Early Season Strategy*

4.4.1. Impact of Available Images on Accuracy

The correlation between the number of available images and the mean absolute error (in percentage) for in-season yield forecasts, generated at the cut-off date of 15 August using evaluation data from years not included in model training, was investigated. In Figure 9, the number of S2 images between May and July was categorized into four bins, and the mean error was calculated, with the standard deviation illustrating the variability across years. A clear correlation between the number of available images and model performance was observed. Specifically, a significant improvement was noted when the number of images exceeded 12 during this period, indicating an average of at least one image per week.

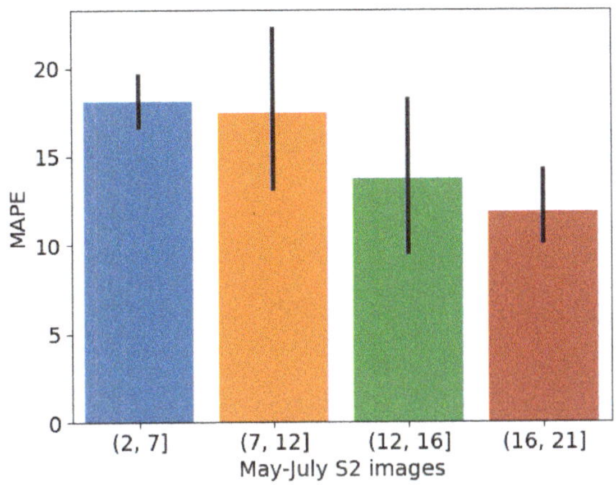

Figure 9. Mean Absolute Percentage Error (MAPE) computed across different total ranges of Sentinel-2 images available from May to August. The standard deviation, calculated across multiple years, is represented by the bars.

4.4.2. Visual Inspection

To quantitatively and qualitatively assess the yield forecasts, the extent to which the predicted and observed yields aligned with the 45° line passing through the origin (representing a perfect correspondence between predicted and observed) was examined. In Figure 10, scatter plots of observed corn yields against predicted yields are presented. The alignment of predicted and observed yields on the 45° line through the origin was satisfactory for 2020, 2021, and 2022. However, for 2017 and 2018, the predicted values were positively correlated but struggled to explain much of the variance around the observed mean values of corn yields. These results suggest the ability to predict a trend in maize yields averaged over at field scale.

4.5. Limitations and Perspectives

4.5.1. Limitations

Satellite-derived data covers only a portion of the spectrum of actual crop yield variation. Optical data primarily reflects the photosynthetically active biomass within fields but may not fully capture grain yield, which depends on multiple factors [47]. Despite this limitation, our approach to early season yield prediction using freely available remote sensing data offers valuable advantages for monitoring crop growth and development throughout the season at the field scale. Emphasizing the use of imagery acquired up to the first half of August, the scalable nature of our framework enables extensive monitoring at field scale, thus potentially providing benefits to farmers and other stakeholders involved in farm management.

Another inherent limitation of our approach concerns data availability. Although our method assumes the presence of at least one image during the first half of August, practical implementation can be affected by variations in image availability, which can affect forecast accuracy. Data limitation also applies to the use of meteorological data from ERA5T, which only provides weather information up to five days before the current date. This time constraint may lead to delays in providing larger-scale forecasts, as our approach requires a lead time of around five days.

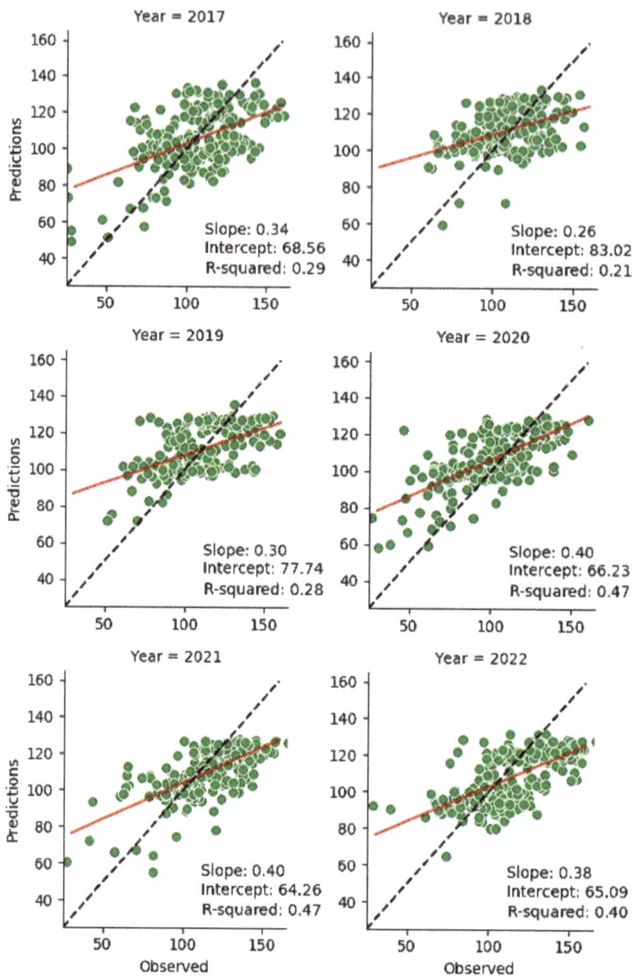

Figure 10. These scatter plots depict the correlation between observed (x-axis) and predicted (y-axis) yield values obtained using DNN (Deep Neural Network) models. The black dashed line represents the ideal 45° line of perfect prediction, while the red line indicates the linear regression fit of the scatter plot, accompanied by the fitting metric reported.

4.5.2. Future Work

There are several ways to further advance the forecasting of maize yield. One approach is to incorporate soil surface temperature (LST) into the Soil Surface Energy Balance Algorithm (SEBAL) [48] for more effective early season yield prediction and drought monitoring. LST data provides insights into energy flows and water stress in agricultural systems, enhancing estimates of evapotranspiration and crop water requirements. Integrating indicators of cumulative drought [5] or crop water requirements into the forecasting models would compensate for the limitations of optical satellite data, which primarily measure canopy surface potential and may underestimate the negative impacts of drought on yield [18]. This approach would be the most appropriate if we had precise information on irrigation dates and doses throughout the season. Additionally, utilizing finer-scale weather data, such as PRISM (https://prism.oregonstate.edu/, accessed on 10 October

2023), tailored to specific regions in the United States, could improve the accuracy and localization of yield predictions.

Secondly, from a methodological point of view, investigations of alternative machine learning methods can bring some additional values. In [26], the authors concluded that no single machine learning regression model emerged as a clear winner in predicting yield values for a new season, and performance can vary depending on the year. The authors suggested using a Stacked Averaging Ensemble (SAE) that combines multiple models could potentially outperform individual models, necessitating the optimization of hyperparameters using a leave-one-year-out validation strategy. This is particularly important for sensitive models like Support Vector Machines (SVM) or Extreme Gradient Boosting (XGBoost).

5. Conclusions

This study analyzed and compared the performance of data-driven methods for early season corn yield forecasting (mid-August) from ground-based data in the U.S. using Sentinel-2, Landsat 7/8/9, agroclimatic and in-situ data. The capacity of estimating yield based on out-of-year corn yield seeds production dataset was shown to be influenced by various aspects, namely: (1) the way in which temporal dynamic is considered (thermal or calendar); (2) the way we take into account the progress of the field based on thermal time by inferring a different model regarding the last period available; (3) how we could exploit the full seasonal trajectory information derived from Sentinel-2 data available in the training years to better infer the current season by forecasting the next C_{ab} period for the test samples. By resampling the time series over thermal time, building a per-period model based on expected field advancement, and predicting the short-term seasonal trend of C_{ab} using a DNN, our early season yield estimation approach explained 31% of the yield variation, with an associated MAPE of 16.1%. This early season result (estimated around mid-August) remains more than comparable to the MAPE achieved at the end of the season (Mid-October) that is estimated to 15.2%. This result suggests that valuable yield crop estimation (comparable to end of season) can be obtained early in the season, possibly providing farmers with an estimate of their production two months before the harvest time.

Author Contributions: J.D.: Conceptualization, Methodology, Software, Validation, Formal analysis, Investigation, Data Curation, Writing—Original Draft, Writing—Review and Editing, Visualization. D.I.: Conceptualization, Methodology, Writing—Review and Editing, Supervision, Project Administration, Resources, Funding Acquisition. A.B.: Validation, Supervision, Project Administration, Resources, Funding Acquisition. All authors have read and agreed to the published version of the manuscript.

Funding: This research was funded by the French National Association of Research and Technology through the Convention Industrielle de Formation par la REcherche (CIFRE refered as 2019/1993) Ph.D. Grant.

Data Availability Statement: The original contributions presented in the study are included in the article, further inquiries can be directed to the corresponding author.

Acknowledgments: The authors would like to thank the European Space Agency (ESA), France for sponsoring the SentinelHub account as part of the Network of Resources (NoR) program, and Syngenta Production and Supply for making its agronomic data available.

Conflicts of Interest: The authors J.D. and A.B were employed by the company Syngenta. The remaining author D.I. declares that the research was conducted in the absence of any commercial or financial relationships that could be construed as a potential conflict of interest.

Abbreviations

The following abbreviations are used in this manuscript:

ANN	Artficial Neural Network
C_{ab}	Leaf Chlorophyll Content (μg/cm^{-2})
DNN	Deep Neural Network
ERA5	ECMWF ReAnalysis 5
GDD	Growing Degree Days
LST	Land Surface Temperature
LSTM	Long Short Term Memory
MAPE	Mean Absolute Percentage Error
NDVI	Normalized Difference Vegetation Index
RF	Random Forest
S2	Sentinel-2
SNAP	Sentinel Application Platform

References

1. Zhao, Y.; Potgieter, A.B.; Zhang, M.; Wu, B.; Hammer, G.L. Predicting wheat yield at the field scale by combining high-resolution Sentinel-2 satellite imagery and crop modelling. *Remote Sens.* **2020**, *12*, 1024. [CrossRef]
2. Johnson, D.M. An assessment of pre-and within-season remotely sensed variables for forecasting corn and soybean yields in the United States. *Remote Sens. Environ.* **2014**, *141*, 116–128. [CrossRef]
3. Beres, B.L.; Hatfield, J.L.; Kirkegaard, J.A.; Eigenbrode, S.D.; Pan, W.L.; Lollato, R.P.; Hunt, J.R.; Strydhorst, S.; Porker, K.; Lyon, D.; et al. Toward a Better Understanding of Genotype × Environment × Management Interactions—A Global Wheat Initiative Agronomic Research Strategy. *Front. Plant Sci.* **2020**, *11, 89–93*. [CrossRef] [PubMed]
4. Tewes, A.; Hoffmann, H.; Krauss, G.; Schäfer, F.; Kerkhoff, C.; Gaiser, T. New approaches for the assimilation of LAI measurements into a crop model ensemble to improve wheat biomass estimations. *Agronomy* **2020**, *10*, 446. [CrossRef]
5. Shuai, G.; Basso, B. Subfield maize yield prediction improves when in-season crop water deficit is included in remote sensing imagery-based models. *Remote Sens. Environ.* **2022**, *272*, 112938. [CrossRef]
6. Carletto, C.; Jolliffe, D.; Banerjee, R. From tragedy to renaissance: Improving agricultural data for better policies. *J. Dev. Stud.* **2015**, *51*, 133–148. [CrossRef]
7. Van Klompenburg, T.; Kassahun, A.; Catal, C. Crop yield prediction using machine learning: A systematic literature review. *Comput. Electron. Agric.* **2020**, *177*, 105709. [CrossRef]
8. FAO. *Handbook for Defining and Setting up a Food Security Information and Early Warning System (FSIEWS)*; FAO: Rome, Italy, 2000.
9. Chipanshi, A.; Zhang, Y.; Kouadio, L.; Newlands, N.; Davidson, A.; Hill, H.; Warren, R.; Qian, B.; Daneshfar, B.; Bedard, F.; et al. Evaluation of the Integrated Canadian Crop Yield Forecaster (ICCYF) model for in-season prediction of crop yield across the Canadian agricultural landscape. *Agric. For. Meteorol.* **2015**, *206*, 137–150. [CrossRef]
10. Johnson, D.M.; Mueller, R. Pre-and within-season crop type classification trained with archival land cover information. *Remote Sens. Environ.* **2021**, *264*, 112576. [CrossRef]
11. Kanke, Y.; Tubana, B.; Dalen, M.; Harrell, D. Evaluation of red and red-edge reflectance-based vegetation indices for rice biomass and grain yield prediction models in paddy fields. *Precis. Agric.* **2016**, *17*, 507–530. [CrossRef]
12. Gascon, F.; Bouzinac, C.; Thépaut, O.; Jung, M.; Francesconi, B.; Louis, J.; Lonjou, V.; Lafrance, B.; Massera, S.; Gaudel-Vacaresse, A.; et al. Copernicus Sentinel-2A calibration and products validation status. *Remote Sens.* **2017**, *9*, 584. [CrossRef]
13. Kayad, A.; Sozzi, M.; Gatto, S.; Marinello, F.; Pirotti, F. Monitoring Within-Field Variability of Corn Yield using Sentinel-2 and Machine Learning Techniques. *Remote Sens.* **2019**, *11*, 2873. [CrossRef]
14. Li, F.; Miao, Y.; Chen, X.; Sun, Z.; Stueve, K.; Yuan, F. In-Season Prediction of Corn Grain Yield through PlanetScope and Sentinel-2 Images. *Agronomy* **2022**, *12*, 3176. [CrossRef]
15. Kang, Y.; Özdoğan, M. Field-level crop yield mapping with Landsat using a hierarchical data assimilation approach. *Remote Sens. Environ.* **2019**, *228*, 144–163. [CrossRef]
16. Wang, Q.; Blackburn, G.A.; Onojeghuo, A.O.; Dash, J.; Zhou, L.; Zhang, Y.; Atkinson, P.M. Fusion of Landsat 8 OLI and Sentinel-2 MSI data. *IEEE Trans. Geosci. Remote Sens.* **2017**, *55*, 3885–3899. [CrossRef]
17. Khanal, S.; Fulton, J.; Shearer, S. An overview of current and potential applications of thermal remote sensing in precision agriculture. *Comput. Electron. Agric.* **2017**, *139*, 22–32. [CrossRef]
18. Thelen, K. Assessing Drought Stress Effects on Corn Yield. *Field Crop Advisory Team Alert Newsletter. Michigan State Univ.* 2007. Available online: http://msue.anr.msu.edu/news/assessing (accessed on 10 October 2023).
19. Pede, T.; Mountrakis, G.; Shaw, S.B. Improving corn yield prediction across the US Corn Belt by replacing air temperature with daily MODIS land surface temperature. *Agric. For. Meteorol.* **2019**, *276*, 107615. [CrossRef]

20. Palosuo, T.; Kersebaum, K.C.; Angulo, C.; Hlavinka, P.; Moriondo, M.; Olesen, J.E.; Patil, R.H.; Ruget, F.; Rumbaur, C.; Takáč, J.; et al. Simulation of winter wheat yield and its variability in different climates of Europe: A comparison of eight crop growth models. *Eur. J. Agron.* **2011**, *35*, 103–114. [CrossRef]
21. Behmann, J.; Mahlein, A.K.; Rumpf, T.; Römer, C.; Plümer, L. A review of advanced machine learning methods for the detection of biotic stress in precision crop protection. *Precis. Agric.* **2015**, *16*, 239–260. [CrossRef]
22. Schwalbert, R.; Amado, T.; Nieto, L.; Corassa, G.; Rice, C.; Peralta, N.; Schauberger, B.; Gornott, C.; Ciampitti, I. Mid-season county-level corn yield forecast for US Corn Belt integrating satellite imagery and weather variables. *Crop Sci.* **2020**, *60*, 739–750. [CrossRef]
23. Ma, Y.; Zhang, Z.; Yang, H.L.; Yang, Z. An adaptive adversarial domain adaptation approach for corn yield prediction. *Comput. Electron. Agric.* **2021**, *187*, 106314. [CrossRef]
24. Deines, J.M.; Patel, R.; Liang, S.Z.; Dado, W.; Lobell, D.B. A million kernels of truth: Insights into scalable satellite maize yield mapping and yield gap analysis from an extensive ground dataset in the US Corn Belt. *Remote Sens. Environ.* **2021**, *253*, 112174. [CrossRef]
25. Duveiller, G.; Frederic, B.; Defourny, P. Using Thermal Time and Pixel Purity for Enhancing Biophysical Variable Time Series: An Interproduct Comparison. *IEEE Trans. Geosci. Remote Sens.* **2013**, *51*, 2119–2127. [CrossRef]
26. Desloires, J.; Ienco, D.; Botrel, A. Out-of-year corn yield prediction at field-scale using Sentinel-2 satellite imagery and machine learning methods. *Comput. Electron. Agric.* **2023**, *209*, 107807. [CrossRef]
27. Van Donk, S.J.; Petersen, J.L.; Davison, D.R. Effect of amount and timing of subsurface drip irrigation on corn yield. *Irrig. Sci.* **2013**, *31*, 599–609. [CrossRef]
28. Main-Knorn, M.; Pflug, B.; Louis, J.; Debaecker, V.; Müller-Wilm, U.; Gascon, F. Sen2Cor for Sentinel-2. In Proceedings of the Image and Signal Processing for Remote Sensing XXIII, Warsaw, Poland, 11–13 September 2017; 10427, p. 3.
29. Weiss, M.; Baret, F. *S2ToolBox Level 2 Products: LAI, FAPAR, FCOVER*; Institut National de la Recherche Agronomique (INRA): Avignon, France, 2016.
30. Savitzky, A.; Golay, M.J.E. Smoothing and Differentiation of Data by Simplified Least Squares Procedures. *Anal. Chem.* **1964**, *36*, 1627–1639. [CrossRef]
31. Shao, Y.; Lunetta, R.S.; Wheeler, B.; Iiames, J.S.; Campbell, J.B. An evaluation of time-series smoothing algorithms for land-cover classifications using MODIS-NDVI multi-temporal data. *Remote Sens. Environ.* **2016**, *174*, 258–265. [CrossRef]
32. Jiménez-Muñoz, J.C.; Sobrino, J.A. A single-channel algorithm for land-surface temperature retrieval from ASTER data. *IEEE Geosci. Remote Sens. Lett.* **2009**, *7*, 176–179. [CrossRef]
33. Ortez, O.A.; McMechan, A.J.; Hoegemeyer, T.; Ciampitti, I.A.; Nielsen, R.L.; Thomison, P.R.; Abendroth, L.J.; Elmore, R.W. Conditions potentially affecting corn ear formation, yield, and abnormal ears: A review. *Crop Forage Turfgrass Manag.* **2022**, *8*, e20173. [CrossRef]
34. Teasdale, J.R.; Cavigelli, M.A. Meteorological fluctuations define long-term crop yield patterns in conventional and organic production systems. *Sci. Rep.* **2017**, *7*, 688. [CrossRef]
35. Schauberger, B.; Archontoulis, S.; Arneth, A.; Balkovič, J.; Ciais, P.; Deryng, D.; Elliott, J.; Folberth, C.; Khabarov, N.; Müller, C.; et al. Consistent negative response of US crops to high temperatures in observations and crop models. *Nat. Commun.* **2017**, *8*, 13931. [CrossRef] [PubMed]
36. Joshi, V.R.; Kazula, M.J.; Coulter, J.A.; Naeve, S.L.; Garcia y Garcia, A. In-season weather data provide reliable yield estimates of maize and soybean in the US central Corn Belt. *Int. J. Biometeorol.* **2021**, *65*, 489–502. [CrossRef] [PubMed]
37. Atwell, B.; Kriedemann, P.; Turnbull, C. *Plants in Action: Adaption in Nature, Performance in Cultivation*, 1st ed.; Macmillan Education: Melbourne, Australia, 1999.
38. Nleya, T.; Chungu, C.; Kleinjan, J. *Corn Growth and Development*; North Dakota State University: Bismarck, ND, USA, 2019; pp. 5–8.
39. Breiman, L. Random Forests. *Mach. Learn.* **2001**, *45*, 5–32. [CrossRef]
40. Pedregosa, F.; Varoquaux, G.; Gramfort, A.; Michel, V.; Thirion, B.; Grisel, O.; Blondel, M.; Prettenhofer, P.; Weiss, R.; Dubourg, V.; et al. Scikit-learn: Machine learning in Python. *J. Mach. Learn. Res.* **2011**, *12*, 2825–2830.
41. Propastin, P.A.; Kappas, M. Reducing uncertainty in modeling the NDVI-precipitation relationship: A comparative study using global and local regression techniques. *GISci. Remote Sens.* **2008**, *45*, 47–67. [CrossRef]
42. Fernández-Manso, A.; Quintano, C.; Fernández-Manso, O. Forecast of NDVI in coniferous areas using temporal ARIMA analysis and climatic data at a regional scale. *Int. J. Remote Sens.* **2011**, *32*, 1595–1617. [CrossRef]
43. Gao, P.; Du, W.; Lei, Q.; Li, J.; Zhang, S.; Li, N. NDVI Forecasting Model Based on the Combination of Time Series Decomposition and CNN–LSTM. *Water Resour. Manag.* **2023**, *37*, 1481–1497. [CrossRef]
44. Hochreiter, S.; Schmidhuber, J. Long short-term memory. *Neural Comput.* **1997**, *9*, 1735–1780. [CrossRef] [PubMed]
45. Abadi, M.; Barham, P.; Chen, J.; Chen, Z.; Davis, A.; Dean, J.; Devin, M.; Ghemawat, S.; Irving, G.; Isard, M.; et al. Tensorflow: A system for large-scale machine learning. In Proceedings of the 12th {USENIX} Symposium on Operating Systems Design and Implementation ({OSDI} 16), Savannah, GA, USA, 2–4 November 2016; pp. 265–283.
46. Huber, P.J. Robust Estimation of a Location Parameter. *Ann. Stat.* **1964**, *53*, 73–101. [CrossRef]

47. Rudorff, B.; Batista, G. Spectral response of wheat and its relationship to agronomic variables in the tropical region. *Remote Sens. Environ.* **1990**, *31*, 53–63. [CrossRef]
48. Laipelt, L.; Kayser, R.H.B.; Fleischmann, A.S.; Ruhoff, A.; Bastiaanssen, W.; Erickson, T.A.; Melton, F. Long-term monitoring of evapotranspiration using the SEBAL algorithm and Google Earth Engine cloud computing. *ISPRS J. Photogramm. Remote Sens.* **2021**, *178*, 81–96. [CrossRef]

remote sensing MDPI

Review

Monitoring Yield and Quality of Forages and Grassland in the View of Precision Agriculture Applications—A Review

Abid Ali [1,*] and Hans-Peter Kaul [2]

[1] Department of Agricultural and Food Sciences, University of Bologna, Viale Fanin 44, 40127 Bologna, Italy
[2] Department of Crop Sciences, Institute of Agronomy, University of Natural Resources and Life Sciences Vienna (BOKU), 3430 Tulln, Austria; hans-peter.kaul@boku.ac.at
* Correspondence: abid.ali4@unibo.it

Abstract: The potential of precision agriculture (PA) in forage and grassland management should be more extensively exploited to meet the increasing global food demand on a sustainable basis. Monitoring biomass yield and quality traits directly impacts the fertilization and irrigation practises and frequency of utilization (cuts) in grasslands. Therefore, the main goal of the review is to examine the techniques for using PA applications to monitor productivity and quality in forage and grasslands. To achieve this, the authors discuss several monitoring technologies for biomass and plant stand characteristics (including quality) that make it possible to adopt digital farming in forages and grassland management. The review provides an overview about mass flow and impact sensors, moisture sensors, remote sensing-based approaches, near-infrared (NIR) spectroscopy, and mapping field heterogeneity and promotes decision support systems (DSSs) in this field. At a small scale, advanced sensors such as optical, thermal, and radar sensors mountable on drones; LiDAR (Light Detection and Ranging); and hyperspectral imaging techniques can be used for assessing plant and soil characteristics. At a larger scale, we discuss coupling of remote sensing with weather data (synergistic grassland yield modelling), Sentinel-2 data with radiative transfer modelling (RTM), Sentinel-1 backscatter, and Catboost–machine learning methods for digital mapping in terms of precision harvesting and site-specific farming decisions. It is known that the delineation of sward heterogeneity is more difficult in mixed grasslands due to spectral similarity among species. Thanks to Diversity-Interactions models, jointly assessing various species interactions under mixed grasslands is allowed. Further, understanding such complex sward heterogeneity might be feasible by integrating spectral un-mixing techniques such as the super-pixel segmentation technique, multi-level fusion procedure, and combined NIR spectroscopy with neural network models. This review offers a digital option for enhancing yield monitoring systems and implementing PA applications in forages and grassland management. The authors recommend a future research direction for the inclusion of costs and economic returns of digital technologies for precision grasslands and fodder production.

Keywords: moisture sensors; mowing events; impact sensors; near-infrared spectroscopy; remote sensing; grassland; decision support systems; digital agriculture

1. Introduction

Forage and grassland crops provide a wide range of ecosystem services (soil, water, carbon) through producing food and feed, and show impacts on agricultural systems and the environment. Mainly forages and grassland communities involve interest in maximizing the utilization intensity and overall profits for advanced farming at the sustainable dimension. Farm management practises that sustain variable rate nitrogen (N) applications can increase N use efficiency of biomass and fodder crops without reducing economic yield [1,2]. Sustainable livestock production is dependent on animal feed and farm productivity levels, allowing the maximum utilization of daily fresh dry matter (DM) contents per animal [3]. The maximum utilization of fresh DM has become the cheapest feeding source for livestock production [4].

For the sustainable production and maximum utilization of forages and grassland species, it is very important to employ precision technologies in monitoring biomass yield and quality components that directly impact fertilization and irrigation practises [5–7]. However, the application of digital farming in forages and grassland management is highly dependent on monitoring systems and the procedure involved [8–10]. Additionally to this, monitoring technologies should not only be low-cost but also equipped with an easy installation procedure [11,12].

Traditional methods are labour-intensive and high-cost, and with low accuracy [11], grouped into direct and indirect methods [12]. The direct methods consist of harvesting by hand or machine, visual estimation, and a subjective approach as ocular estimates [13,14], recommended for vegetation cover over smaller area [11]. Manual methods are not applicable on large areas [12], and evaluating their performance on a regular basis is not very common [15]. For quality measurements, mostly laboratory-based methods are used that are time-consuming such as grinding, oven drying, and subsequent laboratory analyses [16]. The measurements by these systems could be improved by comparing with an indirect method [15,17]. Such indirect methods are based on calibration procedures like linear or multiple regressions [18,19]. In this system, yield monitors are used on self-propelled forage harvesters (SPFHs) to measure yield and moisture data of alfalfa, grass species, and corn silage [20,21]. However, these machines require precise calibration [22]. Addition to this, calibration must be performed based on different field conditions, and crop type, at least once a day or if changes occurred during harvesting. In this case, Rayburn et al. [18] used a rising plate meter to estimate forage and grass biomass through pre-calibrated regression models.

Currently, applications of digital monitoring technologies are becoming popular for grassland canopy vegetation; however, these systems are highly dependent on the procedure used, ease of implementation, and expertise [9,10]. For example, Seefeldt and Booth [23] used digital photographs for grassland canopy vegetation. However, this digital monitoring system has shown inaccuracy, much like traditional methods, and low spatial resolution and is time-consuming; thus, it is not considered feasible for large-scale canopy vegetation monitoring. Moreover, it does not consider multiple scattering between vegetation and soil, therefore creating uncertainties if applied on large and heterogeneous areas under mixed grassland [24]. Physical modelling approaches based on RTM have potential to change the physical relations between spectral values and grassland canopy cover [25]. These modelling-based measurement approaches have advantages in terms of transferability among different study areas; however, they have limited invert accuracy [26]. Furthermore, a large number of datasets are necessary to run complex RTM, especially over larger area [27]. Regression-based approaches rely on empirical relationships between plant spectral reflectance and canopy cover [28]. Many regression methods have been applied successfully in this field, such as linear/non-linear regression, random forest, back-propagation neural networks, and support vector machines [29,30]. Recently, regression-based approaches are being broadly employed for retrieving remotely sensed vegetation indices because of their high accuracy and transferability [31].

This review went further and analyzed the developments in digital monitoring systems, offering cost-effective solutions for challenges associated with regression-based approaches in mixed grasslands. For example, Un-manned Aerial Vehicle (UAV) or drone technology can estimate grassland canopy cover over on-demand high-throughput imagery (millimetre–centimetre resolution), using a similar approach like digital photograph-based classification [32–34]. Some of the other latest digital studies emphasizing high-resolution biomass monitoring and leaf area index (LAI) estimation for grassland-specific PA applications are the integration of Sentinel-2 and RTM [35]; combined Sentinel-1 (S1) SAR (Synthetic Aperture Radar) and daily weather data, through a machine learning-based model (Catboost) for grassland cut detection [36]; integration of MODIS (Moderate-Resolution Imaging Spectroradiometer) NDVI (Normalized Difference Vegetation Index) and weather information, through a synergistic grassland yield model, for determining the start of the growing season and phenological variables [37]; and integration of drone imagery and satellite data for grassland mapping [38].

Besides the advantages of these digital monitoring systems, precision technologies are comparatively less employed in grassland than forages [9]. Schellberg and Hejcman [39] noticed that the low adoption of digital farming in grassland might be due to high species diversity caused by spectral similarities among species, especially in mixed grassland. Currently, these issues are being resolved through digital models in mixed grasslands as highlighted via Diversity-Interactions mixed models, determining effects of various species interactions on agricultural ecosystems [40,41]. The main aim of the review is to highlight the advanced monitoring systems, available sensors, potential remote sensing technologies, and NIR spectroscopy applications in precision forages and grassland measurements. The review provides a wealth of case studies for precision forages and grassland management, a critical analysis on the applications of these technologies, an in-depth discussion of potential challenges, and a vision of future research directions. This review serves as a valuable source that will help researchers and decision makers to employ precise forages and grassland monitoring systems and to adopt digital farming in this field.

2. Review Search Methodology

The review includes scientific papers published from 2000 to 2024. It focused on the current monitoring technologies, including remote sensing methods, for forages and grassland measurements in terms of precision farming applications. In order to guarantee a thorough and objective evaluation, the articles were screened and chosen by following the individual sections in the planned review. The selection of relevant scientific papers was performed using Scopus, Springer, Wiley journal, ScienceDirect, and Google Scholar search engines. The selected papers within the search domain comprised life sciences, remote sensing, agriculture engineering, PA, agronomy, environmental science, biological sciences, and livestock production. The review focused on in-depth literature reviews, within the scope of updated forages and grassland monitoring systems in terms of digital agriculture applications, published from the years 2010 to 2024. The included articles were filtered by query strings. The keywords to search methodology are "PA", "precision forage management", "sensor used in agricultural machines", "biomass sensors", "grassland monitoring systems", "satellite and proximal sensing", "simulation modelling in precision forages and grassland agriculture", "precision harvest", "delineating sward heterogeneity", and "DSS". First, we analyze the monitoring systems of biomass and quality measurements, providing the advancements in remote sensing technologies, helping researchers to adopt digital options in forage and grassland management. Thereafter, the review discusses the challenges in delineating sward heterogeneity and their possible solutions in relation to the application of DSS in precision forages and grassland agriculture. The conclusion provides digital solutions to forage and grassland communities that will have positive impact on increasing their farm productivity on a sustainable basis while protecting the environment.

3. Current Monitoring Technologies

Among monitoring systems, remote sensing and UAV technology applications have been widely employed in monitoring biomass and quality traits [42–49]. LiDAR, terrestrial laser scanning, and airborne-based LiDAR are equally important in this field [50]. These robust technologies can provide accurate data concerning plant horizontal and vertical structure. Théau et al. [48] used UAV-based multispectral data for estimating fresh and dry biomass of grasslands. Some yield and quality traits of alfalfa and grasslands were estimated by canopy reflectance and biophysical traits [51]. Forage and grassland canopy vegetation were measured by Sentinel-2 data [52]. The DM and N contents of grasses were estimated by hyperspectral sensors [53]. However, aerial photogrammetry can also be used successfully in vegetation monitoring [54]. Some studies have employed the photogrammetry technique for pasture management [55,56].

Besides sensing technologies, many different sensors, load cells, and capacitance-controlled oscillators are employed on harvesting machines for measuring mass flow and impact force of the material [57,58]. In these studies, pull-type forage harvesters were used as compared to the latest SPFHs. Shinners et al. [59] installed sensors successfully on SPFHs. The John Deere moisture sensor, also known as HarvestLabTM (HL), works with the company's GreenStarTM global positioning system (GPS) to measure the mass flow rate. However, the sensors that measure on-the-go biomass, such as a pendulum sensor, have shown more accurate results [60]. A forage-throughput sensor and large square baler can be used to measure the mass flow rate and moisture data of herbage crops [58,61–63].

Some harvesting machines and tools used in measurement systems are as follows: (i) the constantly weighing method using windrower machines [64], (ii) measuring the mass flow rate of grass and alfalfa using curved plates on a mowing machine [65]. Biomass yield of grass can also be measured by quantifying the torque and pressure of windrower devices or mowing machines [66]. New Holland's bale-weighing system, if joined with GPS, has potential to develop forage maps. Waggon-based silage yield maps were produced by Lee et al. [57]. Larger and conditioner mowers can be used for monitoring hay and silage, whereas windrowers could be used in grass species. Demmel et al. [64] used a windrowing device to estimate yield of forage and hay crops. Furthermore, the application of a feed roller on a silage chopper, a harvesting machine, was used for picking up forage material from the ground, cutting them into small pieces and transferring them into a storage container for further processing. In this system, the amount of spring force was measured when crop material passed through the feed roller [59].

The DM contents of a silage crop were measured by sampling and the weighing method [67].

NIR spectroscopy is also considered as quick, non-invasive, accurate, and high-throughput technology for monitoring biomass and forage quality of switchgrass [68], and Napier grass [69]. This technology has some limitations such as the complex calibration, selection of spectra, type of available tissue, and external weather conditions. To solve these issues, understanding about the sample pre-processing and chemometrics techniques is required [70]. Akins et al. [71] measured the DM contents of forages using near-infrared spectroscopy (NIR) technology. Field spectrometers were used to estimate biomass cover and quality traits [72–74]. Currently, advanced monitoring systems such as the integration of satellite remote sensing and UAV technology with machine learning algorithms (such as random forest, convolutional neural network, and neural networks) have been widely used in predictive modelling for biomass crops and grassland management [30,75–77]. In this context, the integration of optical and SAR remote sensing data with machine learning models (such as random forest, and gradient boosting regression tree (GBRT) model) was used successfully to measure grass height with accuracy [78]. The DM yield of temperate grassland species was estimated broadly through the integration of UAV data with machine learning algorithms such as random forest, support vector machine, and partial least squares regression models [79].

3.1. *Mass Flow Measurement Sensors*

In mass flow measurement, weight and flow speed of the material are the key components. Basically, mass flow is measured based on the combination of weight and speed of the crop material during harvesting. Some examples are

- A laboratory-based dielectric sensor [80] in forage crops.
- Capacitance sensors with artificial neural network modelling, an Adaptive Neuro-Fuzzy Inference System, and multiple regression techniques in paddy rice [81].
- A parallel plate capacitance-throughput sensor in forage crops [82].
- Kormann [83] measured the mass flow based on the X-ray absorption technique.
- Huenink [62] used a large square baler that uses feed-roll displacement (flow speed) and bale weight to measure the mass flow rate.
- Shinners et al. [59] measured the bale weight while fitting the load cells with large square balers. They found that the dynamic bale weight method had the highest prediction accuracy (R^2 = 0.99) as compared to SPFH and self-propelled windrower machines.
- Shinners et al. [84] used a self-propelled forage windrower machine on forage cutting equipment, with a prediction accuracy of R^2 = 0.83–0.90.

3.2. *Volume Flow Measurement Sensors*

The volume flow rate can be measured by estimating the flowing speed of the material through sensors. It depends on the thickness of the material and growing conditions. However, mass density is important to be measured in volumetric measurement. Volume flow must be linked with the amount of plant material being harvested based on the collective integration of flowing speed, moisture content, speed, and width of the harvester (forage harvester, Green-star John Deere system, or other harvesters). These parameters influence the resulting maps of a harvested crop.

- The volume and mass flow rate of miscanthus and switchgrass were measured by a commercial auger [85].
- A linear potentiometer (Shinners et al., 2003), a vertical displacement transducer [58], and fixed load cells with springswere installed on forage harvesters, to measure volumetric flow of biomass materials [61].

3.3. *Impact Sensors Installed on Harvesting Machines*

There are different impact sensors available that can be used to measure crop quantity and quality contents. These sensors consist of compact plates, which are located in the windrowing apparatus of the harvesting machine (i.e., forage harvester). The efficiency of such yield measurement systems depends on the type of sensor used [86]. Many forage and grass species have different phonological and morphological characteristics (such as green-chopped, wilted hay or straw; perennial rye grass; etc.), which require a multi-sensor approach to assess their heterogeneity at the field scale. Therefore, future sensors should be developed based on multiple information that can be used over a wide range of plant diversity. Some examples are

- Kumhála et al. [87] used torque sensors and curved impact plates on a mowing machine for the measurements of forage yields. They found very good coefficients of determination (R^2 = 0.95) between the conditioner's power, impact force, and material flow rate.
- Torque sensors and impact-type yield sensors were used to estimate mass flow and spatial maps of grass species [87].

- Kumhála and Prosek [65] measured the mass flow of grass and alfalfa crops through a curved impact plate mounted on a mowing machine.
- Savoie et al. [58] used a curved-type impact plate to determine the mass flow rate of hay material with 5% error, based on the moisture data through prediction modelling.

3.4. Other Available Sensors for Monitoring Biomass and Moisture Data

Several sensors can be employed successfully on forage harvesters to measure biomass traits and moisture contents [88]: a linear potentiometer (measures feed-roll displacement/volume flow), a capacitance sensor (measures moisture and mass flow rate) and NIR spectrometer (measures crop spectral properties and quality traits, as well as DM of hay and corn silage), a torque sensor (determines torque and shaft speed), load cells (measure weight of materials, crop flow, and feed-roll displacement), a strain gauge (assesses weight of crop and crop flow), and curved impact plates (measure crop impact force). These sensors must be calibrated for each plant species separately and at the optimum moisture level.

Recently, Digman and Shinners [89] used NIR sensors (also known as North American moisture calibration sensors) on John Deere SPFHs to measure the moisture and DM yield contents. Vomax stated that microwave sensing was more accurate for moisture measurement as compared to NIR sensing. However, NIR has shown optimal results in monitoring DM contents under field conditions [89] and laboratory studies [71]. Forage quality such as protein, starch, and fibre contents can be assessed by NIR sensors, in view of improving feed quality in field conditions [90].

NIR sensors installed on yield monitors can be used to measure DM contents of silage maize as well as mixed species (alfalfa and grasses) [20].

Digman and Shinners [91] used three sensors (capacitance, NIR, and microwave sensors) for moisture data.

Mechanical and electronic pasture sensors are used to measure pasture biomass [11]. Some examples of sensors that can be used to make moisture measurements are presented in Table 1.

Table 1. Available sensors used for moisture measurements.

Measurement Systems	Conductivity Sensor	Dielectric Sensor	Capacitance Sensor	Microwave Sensor				NIR Sensor		Nuclear resonance Sensors	Nuclear radiation Sensors
				Transmission	Reflection	Resonator	Time area reflection	Transmission	Reflection		
influencing factors	Performance										
material temperature	M	M	M	M	M	M	M	L	L	VH	VH
material density	M	L	L	M	M	M	M	L	VH	H	M
layer thickness	M	M	M	M	VH	M	M	L	VH	H	M
corn size	M	M	M	M	M	M	M	M	H	M	M
sensor pressure	M	M	M	VH	VH	VH	VH	H	H	M	VH
material similarity	L	L	M	H	L	H	H	H	H	H	M

Table 1. *Cont.*

Measurement Systems	Conductivity Sensor	Dielectric Sensor	Capacitance Sensor	Microwave Sensor: Transmission	Microwave Sensor: Reflection	Microwave Sensor: Resonator	Microwave Sensor: Time area reflection	NIR Sensor: Transmission	NIR Sensor: Reflection	Nuclear resonance Sensors	Nuclear radiation Sensors
electrolyte content	L	M	H	H	H	H	H	VH	VH	VH	VH
material colour	VH	VH	VH	VH	VH	VH	VH	M	M	VH	VH
extraneous light	VH	VH	VH	VH	VH	VH	VH	L	L	VH	VH
biomass material	H	M	M	L	L	L	L	M	M	M	M
radiation	M	M	M	L	L	M	M	H	H	M	H
material velocity	H	H	H	M	M	VL	VL	H	H	M	H

Performance based on ranking "low (L), medium (M), high (H) and very high (VH)" on moisture measurement systems [88].

3.5. Remote Sensing Technology Applications for Monitoring Biomass Yield and Quality Traits

Recent advancements in remote sensing technologies have significantly enhanced our ability to estimate yield and quality dynamics in forages and grasslands (Table 2). These cutting-edge technologies, including satellite imagery, UAV-based sensors, and ground-based optical sensors, offer unprecedented efficiency in monitoring and utilizing (cuts of) forages and grasslands [92,93]. Remote sensing provides valuable insights into the spatial and temporal estimation of biomass, which is crucial for optimizing grazing management and ensuring sustainable land use [45,94]. As remote sensing approaches have received great attention in the advancements of PA and sustainable development, therefore, its role is becoming more and more critical in digital agriculture [95,96].

Table 2. Remote sensing technologies used for monitoring biomass and quality traits in terms of precision forages and grassland management.

Methods/Sensors	Applications	Advantages	Limitations and Solutions	References
Optical remote sensing technology	Uses optical and hyperspectral images to calculate vegetation indices like NDVI for biomass estimation	Large coverage area, repeatable data collection	Lower spatial resolution as compared to drones, weather dependency. Solution: combination of multi-sensors (e.g., microwave and optical sensors)	[44,97,98]
Combined Sentinel-1 (S1) SAR and daily weather data, using a machine learning-based model (Catboost)	Detection of precision grassland cuts, for making fertilization and irrigation decisions	Enable cut detection over entire calendar year, eliminates missing and false cuts	Persistent cloud cover may affect results Solution: readings should be taken under clear sky conditions	[36]
Sentinel-1 backscatter time series	To determine the precise cut events such as harvesting frequency or utilization efficiency of grasslands by livestock	Determine the limiting factors that influence the cut detection	Factors like size, shape, and slope of plants may affect the detection of mowing events. Solution: requires dense in situ measurements to understand the behaviour of C-band backscatter of plants	[99]

Table 2. *Cont.*

Methods/Sensors	Applications	Advantages	Limitations and Solutions	References
Sentinel-2 data based on RTM of vegetation	Determination of biophysical variables, such as LAI calculated by advanced algorithm techniques, for biomass and LAI of grasslands	Observed lower root-mean-square error than direct measurements	Results might be affected under adverse climatic conditions. Solution: clear sky readings, coupling of cloud-free remote sensing with field spectrometer data	[35]
Sentinel-2 time series	Grassland cut detection for highest fodder utilization and production	Higher resolution over multispectral bands, high accuracy, frequent revisiting time	Can be affected by cloud cover. Solution: aggregation of Sentinel-2 and ground data with process-based algorithms	[100,101]
RGB (Red, Green, Blue) resolution sensors mountable on drone	Captures high-resolution colour images to estimate vegetation cover	High spatial resolution, flexible data acquisition	Limited battery life. Solution: further improvements in navigation and the miniaturization of these technologies	[102]
Multispectral cameras	Captures data in several spectral bands for biomass acquisition	High image quality, ease of use	Ambient conditions, requires ideal camera settings Solution: requires multiple measurements [78]	[103,104]
LiDAR (UAV/airborne biomass mapping)	Aid in assessing vegetation height, canopy structure, biomass, and LAI estimation	Detailed 3D data (resolution: 532–1064 nm), low cost, high accuracy and speed	Climatic factors such as fog, rain, snow, complex image pre-processing, segmentation, extraction, detection, and image quality assessment. Solution: specialized software and training	[105,106]
Hyperspectral imaging	Captures detailed spectral information (0.4–2.5 μm) for assessing biomass and quality traits	Accurate, time saving, broad spectral bands, produces large datasets, environmentally safe, high signal-to-noise ratio	Difficult for detecting different items within one image [107], ambient conditions. Solution: analyzing hyperspectral data with deep learning approaches [108]	[109–111]
Machine learning/Artificial Intelligence (AI)	Uses algorithms to analyze remote sensing data and predict biomass	Can improve accuracy and predictive capabilities	Requires large datasets and computational resources	[112,113]
Synthetic Aperture Radar (SAR) sensors	Microwave radar to estimate biomass by measuring surface roughness and vegetation structure	Effective under all-weather conditions, works day/night, penetrates canopy	Recommended for high aboveground biomass species [114]	[115,116]
Ground-based laser scanning	Laser pulses to measure vegetation structure and density	Rapid, providing dense spatial data of plant species	High cost, sensitive to extreme weather, low area coverage. Solution: Airborne Laser Scanning [117]	[118]
AVHRR (Advanced Very-High-Resolution Radiometer) and MODIS sensors	Plant health and vegetation assessment	Estimating the spatial variability of vegetation cover and plant health	Limited resolution, complex calibration, difficulties in assessing consistent time series data [119]. Solution: coupling of AVHRR and MODIS data [120]	[121,122]

Table 2. *Cont.*

Methods/Sensors	Applications	Advantages	Limitations and Solutions	References
Integrated remote sensing data (MODIS NDVI) and weather information through synergistic approach for grassland yield modelling	Detection of start of the grassland growing seasons for phenological observations, and biomass estimation of grasslands	Daily revisiting frequency, cover large area, time saving	This method gives high NDVI values that may fall outside of the growing season. Solution: requires future research for further validation of this technique	[37]
Small-sat technology	Equipped with sensors that can determine vegetation canopy cover, soil moisture, growth stages, and weed and pest detection	Cost effectiveness, high resolution, frequent revisit times	Limited sensor capacity, short lifespan, less coverage. Solution: digital smart technology powered by the IoT, AI, and machine learning techniques [123]	[124,125]

Optical remote sensing data are not only influenced by external weather, and diseases and insect pest attacks, but also by crop growth responses, the crop type, and the manner of farming [44]). Therefore, one should quantify the uncertainty in satellite sensing data produced under diversified climatic conditions and other factors [126], in order to avoid biasedness in the spatial data produced [127,128]. In this context, NIR technology is playing a promising role, which could be installed on drones, SPFHs, or combine harvesters [129]. This sensing system enables us to monitor biomass and quality traits [130]. Currently, researchers are developing new sensors using blue, green, and NIR spectra. For example, Fan et al. [131] changed the setting of a visible NIR camera for measuring LAI of Italian ryegrass.

Integrated cloud-free high-resolution data with machine learning modelling equipped us with the precise estimation of several spatial parameters that could be useful for grassland management [10,132,133]. These technologies are not only being used in understanding grassland dynamics but also in decision-making processes for sustainable development [42,134].

In the end, UAVs or LiDAR equipped with high-resolution cameras such as multispectral, hyperspectral, and thermal sensors evaluate crop and soil health and insect pest infestation in real time. The data produced from such systems and analyzed through AI and machine leaning models offer us a programme to make innovative farming decisions for maximizing resource use efficiency and advanced farm management practises. These benefits may include those such as saving labour costs, improved farm productivity, and reduced environmental impacts from irrigation, fertilization, and pesticide applications. There are some successful examples of these technologies (UAVs, LiDAR) that should be adopted in resource-constrained contexts such as integrated UAVs with AI and machine learning used for variable rate fertilization and automated crop health assessment in real time, based on data-driven farming decision making that improves resource use efficiency [135]. UAV data are analyzed by Deep Neural Networks to make decisions through resource-constrained devices [136]. The integration of UAV hyperspectral data with machine learning algorithms enables effectively eliminating noise issues, providing spatio-temporal maps of water bodies with high prediction accuracy (R^2 of 0.91) [137]. This method may improve farm conservation practises for agro-ecosystems. Besides this technology, infrared thermal imaging has also shown promising results in agricultural water monitoring and irrigation scheduling [138].

4. Fodder Quality Aspects Using NIR Spectroscopy

Most conventional methods for the measurement of herbage quality require plant samples from the field and pre-processing for laboratory analyses. The recent advancements in NIR have reduced the pre-processing analysis by means of the in situ NIR technique. This technique has not required the external chemical input as compared to traditional methods such as wet chemistry. The advanced NIR is a rapid method reducing laboratory workload for the chemical analysis of herbage crops [139].

It is well known that an NIR technique requires precise calibration with the individual species. Parrini et al. [140] used 105 samples of pastures, and demonstrated that NIR technology can actually measure the chemical composition of pasture species. However, laboratory-based measurements are required for robust calibration in view of restraining the error in actual data, providing a source of reference for accuracy in final results [141]. It is important to include an adequate number of samples in calibration [142]. During spectra collection, it is recommended that fresh samples should be weighed at regular intervals. In general, fresh samples are not always available; so, spectra are mostly taken from dried samples, and calibration performed is mostly based on already processed samples. Furthermore, special attention is required for drying and grinding the samples, because water is a strong absorbent to NIR wavelengths, and particle size influences the whole spectrum. The NIR spectrum must be obtained in uniform conditions through including appropriate sample quantity [143]. All these premises of NIR calibration are important for a quality analysis [140].

Thereafter, calibration should be performed by introducing the spectro-chemical modelling technique that builds an association between the spectral region and compositional trait. In this context, Decruyenaere et al. [141] used the spectro-chemical model for the prediction of chemical traits such as crude protein (CP) and DM contents. In general, high R^2 and low error parameters in prediction indicate the correct use of the calibration procedure [16,144,145].

4.1. Assessment of Quality Traits

Recently, forage and grass quality analyses by means of NIR have become popular in the agricultural industry [146]. The improvement in the development of the NIR technique is achieved through coupling of NIR with machine learning for detecting quality traits [147]. The main quality parameters are as follows:

4.1.1. Moisture Contents

Harvesting plant material at proper moisture is important for feed conservation and can increase the nutrient quality of the animals' feed. Therefore, accuracy in moisture measurement is very crucial. In this context, moisture contents were measured by NIR spectroscopy in corn stover silage [148] and forage soybean [149]. Jin et al. [150] determined the leaf water contents of Miscanthus grass species by integrating visible and NIR spectrometry with multivariate modelling techniques.

4.1.2. Organic Dry Matter Contents

The DM contents can be analyzed by the NIR technique [151]. Fernandes [152] measured the DM of fresh and processed samples, with high prediction accuracy (R^2) from 0.87 to 0.89. Castro et al. [153] measured the organic matter contents of dried and ground samples with $R^2 = 0.92$. Higher correlations showed the similarity between both spectral treatments and principal component analysis techniques. By comparing the pre-processing of samples for NIR calibration, the overall performance was better when processed with dried and grounded samples than fresh samples [152].

4.1.3. Protein Contents

The CP often calculated as N × 6.25 and total nitrogen are widely measured by the NIR method with very high prediction accuracy (R^2), due to strong absorption bonds among –N-H in the NIR spectrum [143]. This fact is confirmed by considerably good association among plant species (R^2 = 0.60 to 0.99) [145,152,154]. The NIR calibration curves were developed for each chemical trait using a group of samples, and predictive equations were developed. Finally, validation was performed using external samples, and a correlation coefficient over 0.99 was evaluated for the used samples except those that showed the accuracy of the predictive equations [145].

4.1.4. Fibre Contents

The neutral detergent fibre (NDF), acid detergent fibre (ADF), and acid detergent lignin (ADL) are some of the other most expedient quality traits of forages measured by NIR technology. Many past studies developed robust calibration equations using a large number of external samples for NIR calibration, which showed R^2 ranging from 0.62 to 0.99 for NDF, 0.75 to 0.97 for ADF, and 0.86 to 0.94 for ADL [145,152,154,155]. These methods are highly dependent on spectra used, calibration equations, and environmental conditions [155]. Hence, the robustness in their prediction accuracy was achieved based on a large number of datasets. Additionally to this, Molano et al. [145] recommended further research to improve the accuracy of analytical methods for ADF determination.

4.1.5. Fat/Lipid Contents

The determination of lipids in plants is not very common with NIR technology, due to their low concentration in forage species. Additionally, fats do not show quite specific reflection or absorption bands in NIR. However, the experiments conducted so far have been heterogeneous, as Ullmann et al. [154] analyzed 245 samples of forage species and obtained prediction accuracy with R^2 = 0.94 and standard error of calibration (SEC = 2.17), whereas Fernandes [152] used 115 samples of forages and achieved medium-level prediction accuracy (R^2 = 0.51 and SEC = 0.60), which may be due to high similarities between databases used that should be treated as a single model for each species.

4.2. Limitations, Potential Problems, and Room for Improvement of NIR Spectroscopy

Besides technological advantages, NIR has some limitations and potential problems such as that standard error in the laboratory samples can cause error in final prediction accuracy, and other biological related measurements [156]. Other challenges and potential problems might be related to spectral measurements, regression models, and machine learning approaches, which can pose problems such as overfitting [157], pattern recognition problems [158], or a lack of understanding of choosing the right samples and processing steps. To solve these problems, sample pre-processing and prediction techniques need to be upgraded that can adapt the pattern; otherwise, the robust model must be taught each year again from a scratch version. Machine leaning-based neural network issues could be resolved through using 'back propagation neural network', 'particle swarm optimization', and 'generalized regression neural network' methods [158].

The necessity of measuring moisture content of samples poses challenges in a quality analysis, especially while assessing protein and sugar content. For solid samples, procedures such as lyophilization or drying are recommended for solving this issue. For liquid-based samples, the transflection spectrum method is advised [159].

Temperature plays an important role in prediction models with NIR spectrometry. It modifies the location as well as intensity of the NIR spectral bands, affecting the predictive accuracy of calibration models. This prevailing problem could be solved through the global temperature compensation technique. Global models use spectra over a full temperature range that gives better predictive results than local models [160].

Hong and Chia [161] published a review on prevailing challenges and their possible solutions of NIR measurements.

Future studies are recommended to discuss the impacts of samples on light scattering that can predict NIR measurements with high accuracy [162]. Besides this, NIR spectroscopy concerning temperature equilibrium should be conducted based on robust calibrated models or recently developed techniques like a 'light emitting diode' for the ease of a rapid NIR analysis. Machine learning-based neural network models can become a next-level solution in this field.

5. Precision Forages and Grassland Management

Precision and digital technologies are becoming popular for biomass and quality estimation and management in the site-specific dimension, aiming to increase the resource use efficiency for sustainable development [9,16]. Currently, this technology has been employed over a wide range of agri-food industries [163,164]. In contrast to PA technology, traditional estimation methods are often time-consuming and equipped with low accuracy [16]. For maximum benefits, low inputs and accurate biomass-monitoring methods like the estiGrass3D+ model could be a great opportunity in this field [165].

Most grasslands are semi-natural, which evolved under livestock grazing and regular mowing events. These semi-natural grasslands are highly influenced by diverse plant species, frequency of harvest intensity, and fertilization practises. Concerning this, Ineichen et al. [7] used forage samples, such as yield, botanical composition, and concentrations of net energy, utilizable crude protein, and different phenolic fractions, from three long-term fertilization experiments over a 2-year duration. This study concluded that combining the fertilization of phosphorus and potassium can offer an increment in forage productivity and enhance forage quality under mountain grasslands. Further, most pastures are located on hilly areas [166], resulting in variability in the excrement of grazed livestock and finally soil spatial nutrient availability. This can cause inaccuracy in biomass measurement, as demonstrated by Schori [167]. Borra-Serrano et al. [168] suggested to include weather data like Growth Degree Days (GDDs), as the GDD difference between harvesting and measurement dates can improve herbage biomass estimations. Additionally, the random forest algorithm is the most promising approach that could be used for monitoring grass and forage biomass [169]. Generally, the random forest method is considered more precise for biomass prediction as compared to linear models [170].

Furthermore, sensor technology, crop-modelling-based DSS, and robotics are recommended precision technologies for forage and grassland monitoring and management [39,59,171–173].

Most farmers are busy, having varied levels of skills and knowledge of PA technologies [9]. However, precision farming in forages and grassland is still low due to its complexity and high cost [174–176]. To solve these issues, the training and education of farmers by expert researchers may increase the awareness of positive impacts of digital farming on enhancing farm productivity, overall feed quality, and societal and environmental benefits [10].

5.1. Delineating Sward Heterogeneity

Delineating agricultural farm heterogeneity is crucial in precision farming [10]. It is well known that spatio-temporal heterogeneity always occurs in field conditions, due to internal and external factors that influence crop behaviours throughout the growing season [10]. These influencing factors can be related to soil entities, human intervention, biotic agents, external weather conditions, species interaction effects, previous cropping history at a particular field, and the manner of farming practises [10,40]. For understanding such complex heterogeneity influenced by various factors as mentioned above, it is recommended to cultivate crop species over a single-type or defined crop mixture, whose intrinsic behaviour could be assumed as a reference source in terms of influencing exertion occurring by these factors.

Thus, it is important to determine each influencing factor that affects sward heterogeneity over forages and grassland cultivation. In general, heterogeneity is considered higher under intensive grazing, which increases uncertainty in biomass measurement and management [177]. Previous research suggested that the integration of plant species helped in maintaining field biodiversity [178]. However, delineating sward heterogeneity has been quite difficult under mixed grasslands due to spectral similarities among species, or the limitation of remote indices like NDVI. These challenges can be resolved through integrating spectral un-mixing techniques or using hyperspectral imaging. Examples are the super-pixel segmentation technique used by Sun et al. [179]; hyperspectral un-mixing [180]; and un-mixing a Sentinel-2 median composite using a multi-level fusion procedure [181]. Authors also suggested to employ the "diversity-interactions mixed model" for evaluating growth and productivity of different species in mixed grasslands. These interaction models use species proportions and their interactions as predictors using a regression equation to determine biodiversity and ecosystem function relationships. However, it can be difficult to model numerous interactions over various species; also, interactions can be temporally variable, and may be dependent on spatial planting or other farming information. McDonnell et al. [40] developed a new Diversity-Interactions model through including weed invasion data from 16 species as random effects and within a plot species planting pattern over three years. The random effects were determined based on pairwise species within plots, rather than the more common plot level in mixed models. Brophy et al. [41] developed the Diversity-Interactions model using a data fusion from grassland and microbial experiment based on fixed coefficients and random effects to determine the interaction effects of species diversity on an agricultural ecosystem. This model offers improved standard errors for testing fixed coefficients and includes lack-of-fit tests for diversity effects. Therefore, these Diversity-Interactions models limit the complexities concerning how species interactions contribute in species-rich grassland ecosystems.

Sward height of herbage crops was measured by a rising plate meter for pasture management [16]. Barthram et al. [182] determined the sward height (variability over 30–70% of total 100%) in perennial ryegrass (Lolium perenne), because of selective grazing by sheep. Rook et al. [183] stated that continuous grazing may raise problems in managing field heterogeneity. To overcome this problem, mowing can be used to homogenize swards. Such events can be detected using remote sensing technologies in grassland farming such as integrating SAR and Sentinel-1 data [184]. This NIR technology has been widely used for monitoring biomass, quality traits, and species identification [132,185]. Xue and Su [186] produced high-quality vegetation indices during different growth stages of crop plants, but image quality might be affected by the presence of clouds and fog. Such issues could be resolved through coupling of remote sensing data with simulation modelling for developing DSS in forages and grassland agriculture [187,188]. Further, determining sward height with UAVs is highly recommended, using digital terrain model and machine learning approaches [165].

Global Navigation Satellite System (known as GNSS) technology allows for monitoring structural health of plant species [189]. Mistele and Schmidhalter [190] developed a spectral sensor with specific configuration to assess the N status of various plant species. Precision yield trends (genetic and non-genetic traits) were determined in forage maize and perennial ryegrass [191]. Li et al. [192] developed variable rate irrigation zones based on apparent soil electrical conductivity data for improving forage yields of Sudangrass (Sorghum sudanense) and alfalfa (Medicago sativa). Bailey et al. [193] collected plant samples manually for producing spatial maps of DM and nutrient contents of grass species. Spatial maps of phosphorus and potassium could be considered as a reference for evaluating nutrient status of grass species [194,195]. These sensors detect the reflectance properties of plants, which can help in managing mineral nutrients in a site-specific way. Schellberg and Hejcman [39] used the NIR technique to determine the application rates of slurry as an organic fertilizer.

5.2. Decision Support Systems

Currently, DSSs are becoming popular for PA applications in forage and grassland management. A broad spectrum of DSS has depicted many advantages within the domain of nutrient, insect pest, and livestock management [196]. Many successful studies are available that can be employed to promote digital farming in grasslands. For example, Herrmann et al. [197] integrated various factors such as N fertilizers, the defoliation rate, the type of grass species, and meteorological data to estimate biomass and CP of herbage species. AgroMANAGER is a web-based application that is supportive in keeping farm records, and helps in monitoring farms online [198]. Different European countries are offering access to open-source DSS tools such as MesParcelles in France, AgrarCommander in Austria, NMP Online in Ireland, MarkOnline in Denmark [199], and Web Module Düngung in Germany. The European project GrassQ was developed for precision grassland measurement and management, using proximal and satellite data [200].

Some recent studies demonstrated that remote sensing data integrated into DSS workflows could be used for tasks like grazing rotation or fertilization scheduling in grassland farming. Concerning this, integrating Sentinel-1 (S1), SAR, and optical remote sensing (Landsat-8 and Sentinel-2) data through multiple linear regression, support vector machine, and random forest was used to determine LAI and aboveground biomass. The results demonstrated that combining S1, Landsat-8, and S2 has a potential to monitor a grazing tallgrass prairie to offer timely and accurate data for precision grassland management [201]. Thereafter, Klingler et al. [35] developed an optimal grassland growth model as a support system based on the integration of Sentinel-2 data with RTM. The study concluded that seasonal changes showed a significant effect on canopy LAI and morphology; therefore, attention should be given during grassland management, i.e., fertilization or other decision making. Watzig et al. [100] developed a support system for improving Austrian fodder management and production systems, in terms of detecting accurate timing of grassland cuts (post-season, in-season) using Copernicus Sentinel-2 imagery. Cuts were detected based on threshold differences between a fitted NDVI curve (using Whittaker smoother) and actual NDVI observations. Moreover, false cuts were eliminated, using a gradient boosting algorithm that may be introduced due to missed cloud cover. Thereafter, Dujakovic et al. [36] went further and integrated Sentinel-1 (S1) SAR and daily weather data using a machine learning-based model (Catboost) for determining grassland cut events over an entire year and it eliminates the need for fixed growing season start/end dates. This method provides cost-effective grassland monitoring that serves as a DSS to determine mowing events influenced by fertilization inputs.

Bochtis et al. [202] developed a DSS for soil-sensitive fields to support farmers for minimizing damages caused by large agricultural machines. Delaby et al. [203] developed a DSS for precision pasture management. Navarro-Hellin et al. [204] recommended the 'smart irrigation decision support system'. This support system is considered more accurate and efficient for developing smart irrigation plans with the minimum amount of water usage compared to traditional irrigation planning. Thereafter, Lopez-Riquelme et al. [205] developed a decision support PA application based on the FIWARE cloud, demonstrating a beneficial service in smart irrigation. Some studies on delineating field productivity use online crop and weather databases for site-specific farming in grassland management [206,207].

Besides these systems, Asher and Brosh [208] introduced a new monitoring technology, called the herd management system, which can be used to measure energy balance and forage quality for livestock production. This monitoring system is efficient for improving land management decisions on animal grazing. Subhashree et al. [209] developed a web tool including a forage economics calculator that supports forage management (alfalfa, corn). Their economic analysis includes the net return, break-even ratio, pay-back period, and final economic returns. This multi-device-based web tool can be installed on an automatic bale picker, recommended for large field areas.

Recently, advanced machine learning algorithms (such as random forest and convolutional neural networks) are being employed in predictive modelling, which help with specific decisions for grassland management [8,10,188]. Examples are the Ecological Succession Model [210], Agricultural Land Management Alternative with Numerical Assessment Criteria model [8], Simulation of Production and Utilization of Rangelands model [211], and Pasture Growth Simulation using grass measurements and weather data [212]. These models have capacity to combine various factors over complex grassland species for canopy biomass and aligned robust mappings. Xie et al. [213] employed Landsat remote sensing data to develop prediction models like artificial neural network and multiple linear regression models. Enough research has been performed on grasslands; however, the limited number of available remote indices and the support from long-term experimental data remain as potential challenges [30].

Current Challenges in the Adoption of DSS and Their Possible Solutions

The price of an agricultural commodity can be affected by factors such as gross productivity, logistics and inventory costs, demand and supply, etc. Some changes in single or more factors can cause uncertainty. Therefore, it is recommended that DSS must pay attention to prevailing uncertainty because of dynamic factors. Further to this, re-planning is a critical challenge in the development of DSS. This is because unexpected issues may arise after some time, like mechanical failures of a harvesting machine and abrupt weather changes. These issues may affect decisions on agricultural tasks such as precision harvesting and spraying, or precision nutrient or smart irrigation operations [214]. Therefore, DSS should be adjusted based on smart strategies or a new solution shall be developed for further decisions for farmers and stakeholders to complete the assigned tasks.

The application of digital DSS by smallholder farmers depends on three factors: (i) people (knowledge and skills), (ii) developed technology (ease of use and affordable price), (iii) rate of efficiency (high performance, high profits). High costs of digital DSS are the main challenge in the adoption of such high-tech methods for smallholder farmers due to the current financial crisis. To solve this issue, Oliveira-Jr et al. [215] developed an Internet of Things (IoT)-based digital system for small-scale farms. This support system is easy to use and has a low price, working on the principle 'do-it-yourself'. The DSS can be employed on a smartphone.

Some researchers are trying to develop DSS that can solve agricultural problems autonomously. However, currently available DSSs are not yet developed with such an intelligent level because of complex agriculture and limited computation time. Further, these systems may provide inaccurate, even wrong, suggestions to farmers. So, experienced researchers are needed for validating the feasibility and correcting the mistakes of developed decision supports [216].

Further, implementation costs and potential economic returns of DSS have not been fully investigated yet in the published literature. Authors recommended the future research to further investigate the digital farming applications in terms of implementation costs and economic returns that will open new opportunities in precision grasslands and fodder research [9].

6. Conclusions and Future Research Directions

This review offers practical advice for enhancing biomass yield, and implementing quality monitoring systems and digital farming techniques in precision forages and grassland management. In this context, the authors provide current monitoring technologies and methods, including mainly mass flow sensors, volume flow sensors, impact sensors, moisture sensors, and remote sensing and NIR technology, field heterogeneity, and digital DSS, for assessing biomass, fodder, and grass quality traits, providing some clues for utilization intensity of fodder crops for livestock production.

The authors suggest that integrating multiple sensors' data, produced by hyperspectral and thermal sensors, on-the-go sensors, torque sensors, radar sensors, NIR sensors, and optical sensors, could be a potential solution in evaluating biomass and quality traits. These types of data could be used in the decision-making process during the growing season, especially on the time of precision harvesting (cuts). It is required to search sensors for all types of plant species and mixed stands. In this context, the available digital sensors must be upgraded, and equipped with multiple data information (crop species, soil characteristics, weather information), especially for mixed grassland species.

Among plant sensing technologies, hyperspectral–UAV imaging and field spectroscopy can contribute accurately in monitoring yield and quality dynamics; however, substantial expertise is required in image processing and analyses to detect relevant crop traits. Most advanced remote sensing technologies should be employed in monitoring biomass and quality traits: the integration of remote sensing and weather data through a synergistic approach such as grassland yield modelling by combining Sentinel-2 with RTM, Sentinel-1 backscatter, and Catboost–machine learning techniques.

The NIR high-throughput technology offers advanced solutions in monitoring forage quality (such as moisture, DM contents, protein, fibre, lipids, and metabolizable energy contents); however, care should be taken during spectra collection and calibration for tissue type or temperature conditions. To solve these NIR calibration and validation issues, machine learning-based neural network models or pre-calibrated regression models are suggested.

It is quite challenging to delineate sward heterogeneity in mixed grasslands, which might occur due to spectral similarities among species or the limitation of remote indices like NDVI. To solve these problems in mixed grasslands, specific solutions are recommended like the application of mentioned Diversity-Interactions models, and integration of un-mixing spectral techniques, or hyperspectral imaging, such as the super-pixel segmentation technique, and multi-level fusion procedure.

For sustainable grassland management and production, integrating a multifunctional approach will be supportive in managing trade-offs for regulating socioeconomics and policy measures across grassland communities around the world, including implementation costs, impacts, and overall benefits of managing ecosystem services and agricultural systems. Further, there is need to include digital technologies for data acquisition or augmentation in diverse environments, such as Electronic Identification Systems, RFID, Soil and Plant Sensors, Smart Phones, and High-Throughput Proximal/Remote Sensing platforms, Diversity-Interactions models, machine learning and AI-based algorithms, and Automatic Application Controllers.

The authors recommend future studies on the advancement of the development of sensor technology by increasing their adaptability, dependability, and ease of implementation for improving biomass yield and quality in mixed grasslands. In addition, economic analyses in digital farming are also missing in the current literature; research on this subject may help farmers to adopt advanced farming in grasslands and fodder production.

Author Contributions: A.A.: conceptualization, investigation, writing—original draft, review and editing, data curation, validation. H.-P.K.: review and editing, visualization, resources. All authors have read and agreed to the published version of the manuscript.

Funding: This research received no external funding.

Data Availability Statement: No new data were created or analyzed in this study. Data sharing is not applicable to this article.

Acknowledgments: We thank Andreas Schaumberger for his useful suggestions to improve the quality of this paper.

Conflicts of Interest: The authors declare no conflicts of interest.

References

1. McCarthy, J.; Delaby, L.; Hennessy, D.; McCarthy, B.; Ryan, W.; Pierce, K.M.; Brennan, A.; Horan, B. The effect of stocking rate on soil solution nitrate concentrations beneath a free-draining dairy production system in Ireland. *J. Dairy Sci.* **2015**, *98*, 4211–4224. [CrossRef] [PubMed]
2. Dentler, J.; Kiefer, L.; Hummler, T.; Bahrs, E.; Elsaesser, M. The impact of low-input grass-based and high-input confinement-based dairy systems on food production, environmental protection and resource use. *Agroecol. Sustain. Food Syst.* **2020**, *44*, 1089–1110. [CrossRef]
3. Wilkinson, J.M.; Lee, M.R.; Rivero, M.J.; Chamberlain, A.T. Some challenges and opportunities for grazing dairy cows on temperate pastures. *Grass Forage Sci.* **2020**, *75*, 1–17. [CrossRef] [PubMed]
4. Finneran, E.; Crosson, P.; O'kiely, P.; Shalloo, L.; Forristal, D.; Wallace, M. Stochastic simulation of the cost of home-produced feeds for ruminant livestock systems. *J. Agric. Sci.* **2012**, *150*, 123–139. [CrossRef]
5. Hanrahan, L.; Geoghegan, A.; O'Donovan, M.; Griffith, V.; Ruelle, E.; Wallace, M.; Shalloo, L. PastureBase Ireland: A grassland decision support system and national database. *Comput. Electron. Agric.* **2017**, *136*, 193–201. [CrossRef]
6. Beukes, P.C.; McCarthy, S.; Wims, C.M.; Gregorini, P.; Romera, A.J. Regular estimates of herbage mass can improve profitability of pasture-based dairy systems. *Anim. Prod. Sci.* **2019**, *59*, 359–367. [CrossRef]
7. Ineichen, S.; Marquardt, S.; Kreuzer, M.; Reidy, B. Forage quality of species-rich mountain grasslands subjected to zero, PK and NPK mineral fertilization for decades. *Grass Forage Sci.* **2020**, *75*, 385–397. [CrossRef]
8. Kiniry, J.R.; Kim, S.; Meki, M.N.; Johnson, M.V.V. Forage Yield Estimation with a Process-Based Simulation Model. In *Forage Groups*; IntechOpen: London, UK, 2018.
9. Shalloo, L.; O'Donovan, M.; Leso, L.; Werner, J.; Ruelle, E.; Geoghegan, A.; Delaby, L.; O'leary, N. Grass-based dairy systems, data and precision technologies. *Animal* **2018**, *12*, s262–s271. [CrossRef]
10. Ali, A.; Hassan, M.U.; Kaul, H.P. Broad Scope of Site-Specific Crop Management and Specific Role of Remote Sensing Technologies Within It—A Review. *J. Agron. Crop Sci.* **2024**, *210*, e12732. [CrossRef]
11. Sanderson, M.A.; Rotz, C.A.; Fultz, S.W.; Rayburn, E.B. Estimating forage mass with a commercial capacitance meter, rising plate meter, and pasture ruler. *Agron. J.* **2001**, *93*, 1281–1286. [CrossRef]
12. de Morais, L.F.; Cavalcante, A.C.R.; Aquino, D.D.N.; Candido, M.J.D. Remote sensing applied to grassland ecosystems in regions with climatic vulnerability. *Ed. Cient. Digit.* **2022**, *7*, 151–164.
13. Mannetje, L. Measuring biomass of grassland vegetation in Field and Laboratory Methods for Grassland and Animal Production. *L. Tmannetje RM Jones. Cap.* **2000**, *7*, 24–31.
14. Godínez-Alvarez, H.; Herrick, J.E.; Mattocks, M.; Toledo, D.; Van Zee, J. Comparison of three vegetation monitoring methods: Their relative utility for ecological assessment and monitoring. *Ecol. Indic.* **2009**, *9*, 1001–1008. [CrossRef]
15. Rayburn, E.B.; Lozier, J.D.; Sanderson, M.A.; Smith, B.D.; Shockey, W.L.; Seymore, D.A.; Fultz, S.W. Alternative methods of estimating forage height and sward capacitance in pastures can be cross calibrated. *Forage Grazinglands* **2007**, *5*, 1–6. [CrossRef]
16. Murphy, D.J.; O'Brien, B.; Hennessy, D.; Hurley, M.; Murphy, M.D. Evaluation of the precision of the rising plate meter for measuring compressed sward height on heterogeneous grassland swards. *Precis. Agric.* **2021**, *22*, 922–946. [CrossRef]

17. López-Guerrero, I.; Fontenot, J.P.; García-Peniche, T.B. Comparisons between four methods of biomass estimation in tall fescue grasslands. *Mex. J. Livest. Sci.* **2011**, *2*, 209–220.
18. Rayburn, E.B.; Shockey, W.L.; Seymour, D.A.; Smith, B.D. Calibration of pasture forage mass to plate meter compressed height is a second-order response with a zero intercept. *Crop Forage Turfgrass Manag.* **2017**, *3*, 3192. [CrossRef]
19. Cho, W.; Brorsen, B.W.; Biermacher, J.T.; Rogers, J.K. Rising plate meter calibrations for forage mass of wheat and rye. *Agric. Environ. Lett.* **2019**, *4*, 180057. [CrossRef]
20. Long, E.A.; Ketterings, Q.M.; Russell, D.; Vermeylen, F.; DeGloria, S. Assessment of yield monitoring equipment for dry matter and yield of corn silage and alfalfa/grass. *Precis. Agric.* **2016**, *17*, 546–563. [CrossRef]
21. Worek, F.; Thurner, S. Yield measurement of wilted forage and silage maize with forage harvesters. In *Precision agriculture'21*; Wageningen Academic: Budapest, Hungary, 2021; Chapter 11; pp. 103–110.
22. Anderson, E.; Arundale, R.; Maughan, M.; Oladeinde, A.; Wycislo, A.; Voigt, T. Growth and agronomy of *Miscanthus* × *giganteus* for biomass production. *Biofuels* **2011**, *2*, 167–183. [CrossRef]
23. Seefeldt, S.S.; Booth, D.T. Measuring plant cover in sagebrush steppe rangelands: A comparison of methods. *Environ. Manag.* **2006**, *37*, 703–711. [CrossRef] [PubMed]
24. Yan, K.; Gao, S.; Chi, H.; Qi, J.; Song, W.; Tong, Y.; Mu, X.; Yan, G. Evaluation of the vegetation-index-based dimidiate pixel model for fractional vegetation cover estimation. *IEEE Trans. Geosci. Remote Sens.* **2021**, *60*, 1–14. [CrossRef]
25. Quan, X.; He, B.; Yebra, M.; Yin, C.; Liao, Z.; Zhang, X.; Li, X. A radiative transfer model-based method for the estimation of grassland aboveground biomass. *Int. J. Appl. Earth Obs. Geoinf.* **2017**, *54*, 159–168. [CrossRef]
26. Tu, Y.; Jia, K.; Liang, S.; Wei, X.; Yao, Y.; Zhang, X. Fractional vegetation cover estimation in heterogeneous areas by combining a radiative transfer model and a dynamic vegetation model. *Int. J. Digit. Earth* **2020**, *13*, 487–503. [CrossRef]
27. Jacquemoud, S.; Verhoef, W.; Baret, F.; Bacour, C.; Zarco-Tejada, P.J.; Asner, G.P.; François, C.; Ustin, S.L. PROSPECT+ SAIL models: A review of use for vegetation characterization. *Remote Sens. Environ.* **2009**, *113*, S56–S66. [CrossRef]
28. Li, J.; Xu, B.; Yang, X.; Jin, Y.; Zhao, L.; Zhao, F.; Chen, S.; Guo, J.; Qin, Z.; Ma, H. Characterizing changes in grassland desertification based on Landsat images of the Ongniud and Naiman Banners, Inner Mongolia. *Int. J. Remote Sens.* **2015**, *36*, 5137–5149. [CrossRef]
29. Lehnert, L.W.; Meyer, H.; Wang, Y.; Miehe, G.; Thies, B.; Reudenbach, C.; Bendix, J. Retrieval of grassland plant coverage on the Tibetan Plateau based on a multi-scale, multi-sensor and multi-method approach. *Remote Sens. Environ.* **2015**, *164*, 197–207. [CrossRef]
30. Morais, T.G.; Teixeira, R.F.; Figueiredo, M.; Domingos, T. The use of machine learning methods to estimate aboveground biomass of grasslands: A review. *Ecol. Indic.* **2021**, *130*, 108081. [CrossRef]
31. Ge, J.; Meng, B.; Liang, T.; Feng, Q.; Gao, J.; Yang, S.; Huang, X.; Xie, H. Modeling alpine grassland cover based on MODIS data and support vector machine regression in the headwater region of the Huanghe River, China. *Remote Sens. Environ.* **2018**, *218*, 162–173. [CrossRef]
32. Geng, X.; Wang, X.; Fang, H.; Ye, J.; Han, L.; Gong, Y.; Cai, D. Vegetation coverage of desert ecosystems in the Qinghai-Tibet Plateau is underestimated. *Ecol. Indic.* **2022**, *137*, 108780. [CrossRef]
33. Rueda-Ayala, V.P.; Peña, J.M.; Höglind, M.; Bengochea-Guevara, J.M.; Andújar, D. Comparing UAV-based technologies and RGB-D reconstruction methods for plant height and biomass monitoring on grass ley. *Sensors* **2019**, *19*, 535. [CrossRef] [PubMed]
34. Yu, T.; Ni, W.; Zhang, Z.; Liu, Q.; Sun, G. Regional sampling of forest canopy covers using UAV visible stereoscopic imagery for assessment of satellite-based products in Northeast China. *J. Remote Sens.* **2022**, *2022*, 9806802. [CrossRef]
35. Klingler, A.; Schaumberger, A.; Vuolo, F.; Kalmár, L.B.; Pötsch, E.M. Comparison of direct and indirect determination of leaf area index in permanent grassland. PFG–Journal of Photogrammetry. *Remote Sens. Geoinf. Sci.* **2020**, *88*, 369–378.
36. Dujakovic, A.; Schaumberger, A.; Klingler, A.; Mayer, K.; Atzberger, C.; Klisch, A.; Vuolo, F. Growth unveiled: Decoding the start of grassland seasons in Austria. *Eur. J. Remote Sens.* **2024**, *57*, 2323633. [CrossRef]
37. Dujakovic, A.; Watzig, C.; Schaumberger, A.; Klingler, A.; Atzberger, C.; Vuolo, F. Enhancing Grassland Cut Detection Using Sentinel-2 Time Series Through Integration of Sentinel-1 Sar and Weather Data. *SRRN* **2024**. [CrossRef]
38. Hu, T.; Cao, M.; Zhao, X.; Liu, X.; Liu, Z.; Liu, L.; Huang, Z.; Tao, S.; Tang, Z.; Guo, Y.; et al. High-resolution mapping of grassland canopy cover in China through the integration of extensive drone imagery and satellite data. *ISPRS J. Photogramm. Remote Sens.* **2024**, *218*, 69–83. [CrossRef]
39. Schellberg, J.; Hejcman, M. Precision fertilizer management on grassland. In *Fertilizers: Properties, Applications and Effects*; Nova Science Publisher: New York, NY, USA, 2008; p. 107.
40. McDonnell, J.; McKenna, T.; Yurkonis, K.A.; Hennessy, D.; de Andrade Moral, R.; Brophy, C. A mixed model for assessing the effect of numerous plant species interactions on grassland biodiversity and ecosystem function relationships. *J. Agric. Biol. Environ. Stat.* **2023**, *28*, 1–19. [CrossRef]
41. Brophy, C.; Dooley, Á.; Kirwan, L.; Finn, J.A.; McDonnell, J.; Bell, T.; Cadotte, M.W.; Connolly, J. Biodiversity and ecosystem function: Making sense of numerous species interactions in multi-species communities. *Ecology* **2017**, *98*, 1771–1778. [CrossRef]

42. Ali, I.; Cawkwell, F.; Dwyer, E.; Barrett, B.; Green, S. Satellite remote sensing of grasslands: From observation to management. *J. Plant Ecol.* **2016**, *9*, 649–671. [CrossRef]
43. Cevallos, L.N.M.; García, J.L.R.; Suárez, B.I.A.; González, C.A.L.; González, I.S.; Campoverde, J.A.Y.; Guzmàn, J.A.M.; Toulkeridis, T. A NDVI analysis contrasting different spectrum data methodologies applied in pasture crops previous grazing–A case study from Ecuador. In Proceedings of the 2018 International Conference on eDemocracy & eGovernment (ICEDEG), Ambato, Ecuador, 4–6 April 2018; pp. 126–135.
44. Ali, A.; Martelli, R.; Lupia, F.; Barbanti, L. Assessing multiple years' spatial variability of crop yields using satellite vegetation indices. *Remote Sens.* **2019**, *11*, 2384. [CrossRef]
45. Reinermann, S.; Asam, S.; Kuenzer, C. Remote sensing of grassland production and management—A review. *Remote Sens.* **2020**, *12*, 1949. [CrossRef]
46. Geipel, J.; Bakken, A.K.; Jørgensen, M.; Korsaeth, A. Forage yield and quality estimation by means of UAV and hyperspectral imaging. *Precis. Agric.* **2021**, *22*, 1437–1463. [CrossRef]
47. Michez, A.; Philippe, L.; David, K.; Sébastien, C.; Christian, D.; Bindelle, J. Can low-cost unmanned aerial systems describe the forage quality heterogeneity? Insight from a timothy pasture case study in southern Belgium. *Remote Sens.* **2020**, *12*, 1650. [CrossRef]
48. Théau, J.; Lauzier-Hudon, É.; Aube, L.; Devillers, N. Estimation of forage biomass and vegetation cover in grasslands using UAV imagery. *PLoS ONE* **2021**, *16*, e0245784. [CrossRef] [PubMed]
49. Peng, J.; Zeiner, N.; Parsons, D.; Féret, J.B.; Söderström, M.; Morel, J. Forage Biomass Estimation Using Sentinel-2 Imagery at High Latitudes. *Remote Sens.* **2023**, *15*, 2350. [CrossRef]
50. Eitel, J.U.; Magney, T.S.; Vierling, L.A.; Brown, T.T.; Huggins, D.R. LiDAR based biomass and crop nitrogen estimates for rapid, non-destructive assessment of wheat nitrogen status. *Field Crop. Res.* **2014**, *159*, 21–32. [CrossRef]
51. Post, C.J.; DeGloria, S.D.; Cherney, J.H.; Mikhailova, E.A. Spectral measurements of alfalfa/grass fields related to forage properties and species composition. *J. Plant Nutr.* **2007**, *30*, 1779–1789. [CrossRef]
52. Dusseux, P.; Guyet, T.; Pattier, P.; Barbier, V.; Nicolas, H. Monitoring of grassland productivity using Sentinel-2 remote sensing data. *Int. J. Appl. Earth Obs. Geoinf.* **2022**, *111*, 102843. [CrossRef]
53. Nishikawa, H.; Oenema, J.; Sijbrandij, F.; Jindo, K.; Noij, G.J.; Hollewand, F.; Meurs, B.; Hoving, I.; van der Vlugt, P.; Bouten, M.; et al. Dry Matter Yield and Nitrogen Content Estimation in Grassland Using Hyperspectral Sensor. *Remote Sens.* **2023**, *15*, 419. [CrossRef]
54. Mukherjee, A.; Misra, S.; Raghuwanshi, N.S. A survey of unmanned aerial sensing solutions in precision agriculture. *J. Netw. Comput. Appl.* **2019**, *148*, 102461. [CrossRef]
55. Michez, A.; Lejeune, P.; Bauwens, S.; Herinaina, A.A.L.; Blaise, Y.; Castro Muñoz, E.; Lebeau, F.; Bindelle, J. Mapping and monitoring of biomass and grazing in pasture with an unmanned aerial system. *Remote Sens.* **2019**, *11*, 473. [CrossRef]
56. Lussem, U.; Schellberg, J.; Bareth, G. Monitoring forage mass with low-cost UAV data: Case study at the Rengen grassland experiment. PFG–Journal of Photogrammetry. *Remote Sens. Geoinf. Sci.* **2020**, *88*, 407–422.
57. Lee, W.S.; Schueller, J.K.; Burks, T.F. Wagon-based silage yield mapping system. *Agric. Eng. Int. CIGR J.* **2005**, *7*, 1–14.
58. Savoie, P.; Lemire, P.; Thériault, R. Evaluation of five sensors to estimate mass–flow rate and moisture of grass in a forage harvester. *Appl. Eng. Agric.* **2002**, *18*, 389–397. [CrossRef]
59. Shinners, K.J.; Huenink, B.M.; Behringer, C.B. Precision agriculture as applied to North American hay and forage production. In Proceedings of the International Conference on Crop Harvesting and Processing, Louisville, KY, USA, 9–11 February 2003; American Society of Agricultural and Biological Engineers: St. Joseph, MI, USA, 2003. 20p.
60. Ehlert, D.; Hammen, V.; Adamek, R. On-line sensor pendulum-meter for determination of plant mass. *Precis. Agric.* **2003**, *4*, 139–148. [CrossRef]
61. Forristal, P.D.; Keppel, D. *The Application of Harvester-Mounted Forage Yield Sensing Devices*; Teagasc: Carlow, Ireland, 2001.
62. Huenink, B.M. *Developing Yield Maps of Hay and Forage Productions*; University of Wisconsin—Madison: Madison, WI, USA, 2003.
63. Kumhala, F.; Prosek, V.; Kroulik, M. Capacitive sensor for chopped maize throughput measurement. *Comput. Electron. Agric.* **2010**, *70*, 234–238. [CrossRef]
64. Demmel, M.; Schwenke, T.; Heuwinkel, H.; Locher, F.; Rottmeier, J. Development and field test of a yield measurement system in a mower conditioner. In *Proc. AgEng*; MKIFK Hungarian Gazette Publishing and Legal Translation Centre Ltd.: Budapest, Hungary, 2002; pp. 159–160.
65. Kumhála, F.; Prošek, V. Laboratory measurement of mowing machine material feed rate. *Precis. Agric.* **2003**, *4*, 413–419. [CrossRef]
66. Wild, K.; Ruhland, S.; Haedicke, S. Local yield detection of grass in a mower conditioner. In Proceedings of the 2005 ASAE Annual Meeting, Providence, RI, USA, 24–27 July 2005; American Society of Agricultural and Biological Engineers: St. Joseph, MI, USA, 2005.
67. Miller, C.M.; Fadel, J.G.; Heguy, J.M.; Karle, B.M.; Price, P.L.; Meyer, D. Optimizing accuracy of protocols for measuring dry matter and nutrient yield of forage crops. *Sci. Total Environ.* **2018**, *624*, 180–188. [CrossRef]

68. Razar, R.M.; Makaju, S.; Missaoui, A.M. QTL mapping of biomass and forage quality traits measured using near-infrared reflectance spectroscopy (NIRS) in switchgrass. *Euphytica* **2021**, *217*, 51. [CrossRef]
69. Anderson, W.F.; Dien, B.S.; Masterson, S.D.; Mitchell, R.B. Development of near-infrared reflectance spectroscopy (NIRS) calibrations for traits related to ethanol conversion from genetically variable napier grass (*Pennisetum purpureum* Schum.). *BioEnergy Res.* **2019**, *12*, 34–42. [CrossRef]
70. Cozzolino, D. Advantages and limitations of using near infrared spectroscopy in plant phenomics applications. *Comput. Electron. Agric.* **2023**, *212*, 108078. [CrossRef]
71. Akins, M.S.; Dobberstain, M.; Shaver, R.D. Evaluation of on-farm forage DM determined by near infrared spectroscopy. In Proceedings of the American Dairy Science Association®-American Society of Animal Science-Association Mexicana de Producción Animal-Canadian Society of Animal Science-Western Section ASAS Joint Annual Meeting, Champaign, IL, USA, 15–19 July 2012; pp. 279–290.
72. Safari, H.; Fricke, T.; Wachendorf, M. Determination of fibre and protein content in heterogeneous pastures using field spectroscopy and ultrasonic sward height measurements. *Comput. Electron. Agric.* **2016**, *123*, 256–263. [CrossRef]
73. Zhou, Z.; Morel, J.; Parsons, D.; Kucheryavskiy, S.V.; Gustavsson, A.-M. Estimation of yield and quality of legume and grass mixtures using partial least squares and support vector machine analysis of spectral data. *Comput. Electron. Agric.* **2019**, *162*, 246–253. [CrossRef]
74. Fernández-Habas, J.; Cañada, M.C.; Moreno, A.M.G.; Leal-Murillo, J.R.; González-Dugo, M.P.; Oar, B.A.; Gómez-Giráldez, P.J.; Fernández-Rebollo, P. Estimating pasture quality of Mediterranean grasslands using hyperspectral narrow bands from field spectroscopy by Random Forest and PLS regressions. *Comput. Electron. Agric.* **2022**, *192*, 106614. [CrossRef]
75. Castro, W.; Marcato Junior, J.; Polidoro, C.; Osco, L.P.; Gonçalves, W.; Rodrigues, L.; Santos, M.; Jank, L.; Barrios, S.; Valle, C.; et al. Deep learning applied to phenotyping of biomass in forages with UAV-based RGB imagery. *Sensors* **2020**, *20*, 4802. [CrossRef]
76. Grüner, E.; Wachendorf, M.; Astor, T. The potential of UAV-borne spectral and textural information for predicting aboveground biomass and N fixation in legume-grass mixtures. *PLoS ONE* **2020**, *15*, e0234703. [CrossRef]
77. Muro, J.; Linstädter, A.; Magdon, P.; Wöllauer, S.; Männer, F.A.; Schwarz, L.M.; Ghazaryan, G.; Schultz, J.; Malenovský, Z.; Dubovyk, O. Predicting plant biomass and species richness in temperate grasslands across regions, time, and land management with remote sensing and deep learning. *Remote Sens. Environ.* **2022**, *282*, 113262. [CrossRef]
78. Zhang, L.; Ren, H. Estimation of grassland height using optical and SAR remote sensing data. *Adv. Space Res.* **2023**, *72*, 4298–4310. [CrossRef]
79. Lussem, U.; Bolten, A.; Kleppert, I.; Jasper, J.; Gnyp, M.L.; Schellberg, J.; Bareth, G. Herbage Mass, N Concentration, and N Uptake of Temperate Grasslands Can Adequately Be Estimated from UAV-Based Image Data Using Machine Learning. *Remote Sens.* **2022**, *14*, 3066. [CrossRef]
80. Kiviharju, K.; Salonen, K.; Moilanen, U.; Meskanen, E.; Leisola, M.; Eerikäinen, T. On-line biomass measurements in bioreactor cultivations: Comparison study of two on-line probes. *J. Ind. Microbiol. Biotechnol.* **2007**, *34*, 561–566. [CrossRef]
81. Tahmasebi, M.; Tabatabaei-kolor, R. Measuring of Paddy mass flow using capacitive sensor and modeling with using multiple regression, ANN, and ANFIS models. *Iran. J. Biosyst. Eng.* **2017**, *48*, 221–227.
82. Kumhala, F.; Prosek, V.; Kroulik, M.; Kviz, Z. Parallel plate mass flow sensor for forage crops and sugar beet. In Proceedings of the 2008, Providence, RI, USA, 29 June–2 July 2008; 1p; American Society of Agricultural and Biological Engineers: St. Joseph, MI, USA, 2008.
83. Kormann, G. Mass flow measurement based on X-ray absorption. *Agric. Eng. Conf.* **2004**, *2004*, 34. [CrossRef]
84. Shinners, K.J.; Barnett, N.G.; Schlesser, W.M. Measuring Mass-Flow-Rate on Forage Cutting Equipment. 2000, 1–20. Available online: https://www.researchgate.net/publication/288661644_Measuring_mass-flow-rate_on_forage_cutting_equipment (accessed on 29 December 2024).
85. Miao, Z.; Grift, T.E.; Hansen, A.C.; Ting, K.C. Flow performance of ground biomass in a commercial auger. *Powder Technol.* **2014**, *267*, 354–361. [CrossRef]
86. Bailey, J.S.; Higgins, A.; Jordan, C. Empirical models for predicting the dry matter yield of grass silage swards using plant tissue analyses. *Precis. Agric.* **2000**, *2*, 131–145. [CrossRef]
87. Kumhála, F.; Kroulík, M.; Prošek, V. Development and evaluation of forage yield measure sensors in a mowing-conditioning machine. *Comput. Electron. Agric.* **2007**, *58*, 154–163. [CrossRef]
88. Kormann, G.; Auernhammer, H. Continuous moisture measurements in self-propelled forage harvesters. *Landtechnik* **2002**, *57*, 264–265.
89. Digman, M.F.; Shinners, K.J. Real-time moisture measurement on a forage harvester using near-infrared reflectance spectroscopy. *Trans. ASABE* **2008**, *51*, 1801–1810. [CrossRef]
90. Schroeder, J.W. *Silage Fermentation and Preservation*; NDSU Extension Service: Fargo, ND, USA, 2004; 8p.

91. Digman, M.F.; Shinners, K.J. Technology background and best practices: Yield mapping in hay and forage. In Proceedings of the University of California Cooperative Extension Proceeding, California Alfalfa and Grain Symposium, Sacramento, CA, USA, 10–12 December 2012.
92. Sishodia, R.P.; Ray, R.L.; Singh, S.K. Applications of remote sensing in precision agriculture: A review. *Remote Sens.* **2020**, *12*, 3136. [CrossRef]
93. Jung, J.; Maeda, M.; Chang, A.; Bhandari, M.; Ashapure, A.; Landivar-Bowles, J. The potential of remote sensing and artificial intelligence as tools to improve the resilience of agriculture production systems. *Curr. Opin. Biotechnol.* **2021**, *70*, 15–22. [CrossRef]
94. Sharma, P.; Leigh, L.; Chang, J.; Maimaitijiang, M.; Caffé, M. Above-ground biomass estimation in oats using UAV remote sensing and machine learning. *Sensors* **2022**, *22*, 601. [CrossRef]
95. Badreldin, N.; Prieto, B.; Fisher, R. Mapping grasslands in mixed grassland ecoregion of Saskatchewan using big remote sensing data and machine learning. *Remote Sens.* **2021**, *13*, 4972. [CrossRef]
96. Gevaert, C.M.; Tang, J.; García-Haro, F.J.; Suomalainen, J.; Kooistra, L. Combining hyperspectral UAV and multispectral Formosat-2 imagery for precision agriculture applications. In Proceedings of the 2014 6th Workshop on Hyperspectral Image and Signal Processing: Evolution in Remote Sensing (WHISPERS), Lausanne, Switzerland, 24-27 June 2014; pp. 1–4.
97. Pettorelli, N.; Laurance, W.F.; O'Brien, T.G.; Wegmann, M.; Nagendra, H.; Turner, W. Satellite remote sensing for applied ecologists: Opportunities and challenges. *J. Appl. Ecol.* **2014**, *51*, 839–848. [CrossRef]
98. Askari, M.S.; McCarthy, T.; Magee, A.; Murphy, D.J. Evaluation of grass quality under different soil management scenarios using remote sensing techniques. *Remote Sens.* **2019**, *11*, 1835. [CrossRef]
99. Reuß, F.; Navacchi, C.; Greimeister-Pfeil, I.; Vreugdenhil, M.; Schaumberger, A.; Klingler, A.; Mayer, K.; Wagner, W. Evaluation of limiting factors for SAR backscatter based cut detection of alpine grasslands. *Sci. Remote Sens.* **2024**, *9*, 100117. [CrossRef]
100. Watzig, C.; Schaumberger, A.; Klingler, A.; Dujakovic, A.; Atzberger, C.; Vuolo, F. Grassland cut detection based on Sentinel-2 time series to respond to the environmental and technical challenges of the Austrian fodder production for livestock feeding. *Remote Sens. Environ.* **2023**, *292*, 113577. [CrossRef]
101. Spoto, F.; Martimort, P.; Drusch, M. Sentinel-2: ESA's optical high-resolution mission for GMES Operational Services. In Proceedings of the First Sentinel-2 Preparatory Symposium, Frascati, Italy, 23–27 April 2012; Volume 707, 2p.
102. Anderson, K.; Gaston, K.J. Lightweight unmanned aerial vehicles will revolutionize spatial ecology. *Front. Ecol. Environ.* **2013**, *11*, 138–146. [CrossRef]
103. Mulla, D.J. Twenty five years of remote sensing in precision agriculture: Key advances and remaining knowledge gaps. *Biosyst. Eng.* **2013**, *114*, 358–371. [CrossRef]
104. Gholizadeh, H.; Gamon, J.A.; Townsend, P.A.; Zygielbaum, A.I.; Helzer, C.J.; Hmimina, G.Y.; Yu, R.; Moore, R.M.; Schweiger, A.K.; Cavender-Bares, J. Detecting prairie biodiversity with airborne remote sensing. *Remote Sens. Environ.* **2019**, *221*, 38–49. [CrossRef]
105. Obanawa, H.; Yoshitoshi, R.; Watanabe, N.; Sakanoue, S. Portable LiDAR-based method for improvement of grass height measurement accuracy: Comparison with SfM methods. *Sensors* **2020**, *20*, 4809. [CrossRef]
106. Rivera, G.; Porras, R.; Florencia, R.; Sánchez-Solís, J.P. LiDAR applications in precision agriculture for cultivating crops: A review of recent advances. *Comput. Electron. Agric.* **2023**, *207*, 107737. [CrossRef]
107. Bhargava, A.; Sachdeva, A.; Sharma, K.; Alsharif, M.H.; Uthansakul, P.; Uthansakul, M. Hyperspectral Imaging and Its Applications: A Review. *Heliyon* **2024**, *10*, e33208. [CrossRef] [PubMed]
108. Tejasree, G.; Agilandeeswari, L. An extensive review of hyperspectral image classification and prediction: Techniques and challenges. *Multimed. Tools Appl.* **2024**, *83*, 80941–81038. [CrossRef]
109. Goel, P.K.; Prasher, S.O.; Landry, J.A.; Patel, R.M.; Viau, A.A.; Miller, J.R. Estimation of crop biophysical parameters through airborne and field hyperspectral remote sensing. *Trans. ASAE* **2003**, *46*, 1235.
110. Wijesingha, J.; Astor, T.; Schulze-Brüninghof, D.; Wengert, M.; Wachendorf, M. Predicting forage quality of Grasslands using UAV-borne imaging spectroscopy. *Remote Sens.* **2020**, *12*, 126. [CrossRef]
111. Wengert, M.; Wijesingha, J.; Schulze-Brüninghoff, D.; Wachendorf, M.; Astor, T. Multisite and Multitemporal Grassland Yield Estimation Using UAV-Borne Hyperspectral Data. *Remote Sens.* **2022**, *14*, 2068. [CrossRef]
112. Kumar, L.; Mutanga, O. Remote sensing of above-ground biomass. *Remote Sens.* **2017**, *9*, 935. [CrossRef]
113. Han, L.; Yang, G.; Dai, H.; Xu, B.; Yang, H.; Feng, H.; Li, Z.; Yang, X. Modeling maize above-ground biomass based on machine learning approaches using UAV remote-sensing data. *Plant Methods* **2019**, *15*, 10. [CrossRef]
114. Sainuddin, F.V.; Chirakkal, S.; Asok, S.V.; Das, A.K.; Putrevu, D. Evaluation of multifrequency SAR data for estimating tropical above-ground biomass by employing radiative transfer modeling. *Environ. Monit. Assess.* **2023**, *195*, 1102. [CrossRef]
115. Barrett, B.; Nitze, I.; Green, S.; Cawkwell, F. Assessment of multi-temporal, multi-sensor radar and ancillary spatial data for grasslands monitoring in Ireland using machine learning approaches. *Remote Sens. Environ.* **2014**, *152*, 109–124. [CrossRef]
116. Szigarski, C.; Jagdhuber, T.; Baur, M.; Thiel, C.; Parrens, M.; Wigneron, J.P.; Piles, M.; Entekhabi, D. Analysis of the radar vegetation index and potential improvements. *Remote Sens.* **2018**, *10*, 1776. [CrossRef]

117. McRoberts, R.E.; Andersen, H.E.; Næsset, E. Using airborne laser scanning data to support forest sample surveys. In *Forestry Applications of Airborne Laser Scanning: Concepts Case Study*; Springer: Berlin/Heidelberg, Germany, 2014; pp. 269–292.
118. Xu, K.; Su, Y.; Liu, J.; Hu, T.; Jin, S.; Ma, Q.; Zhai, Q.; Wang, R.; Zhang, J.; Li, Y.; et al. Estimation of degraded grassland aboveground biomass using machine learning methods from terrestrial laser scanning data. *Ecol. Indic.* **2020**, *108*, 105747. [CrossRef]
119. Dech, S.; Holzwarth, S.; Asam, S.; Andresen, T.; Bachmann, M.; Boettcher, M.; Dietz, A.; Eisfelder, C.; Frey, C.; Gesell, G.; et al. Potential and challenges of harmonizing 40 years of AVHRR data: The TIMELINE experience. *Remote Sens.* **2021**, *13*, 3618. [CrossRef]
120. Mao, D.; Wang, Z.; Luo, L.; Ren, C. Integrating AVHRR and MODIS data to monitor NDVI changes and their relationships with climatic parameters in Northeast China. *Int. J. Appl. Earth Obs. Geoinf.* **2012**, *18*, 528–536. [CrossRef]
121. Assmann, J.J.; Kerby, J.T.; Cunliffe, A.M.; Myers-Smith, I.H. Vegetation monitoring using multispectral sensors—Best practices and lessons learned from high latitudes. *J. Unmanned Veh. Syst.* **2018**, *7*, 54–75. [CrossRef]
122. Kawamura, K.; Akiyama, T.; Yokota, H.O.; Tsutsumi, M.; Yasuda, T.; Watanabe, O.; Wang, S. Comparing MODIS vegetation indices with AVHRR NDVI for monitoring the forage quantity and quality in Inner Mongolia grassland, China. *Grassl. Sci.* **2005**, *51*, 33–40. [CrossRef]
123. Shahab, H.; Iqbal, M.; Sohaib, A.; Khan, F.U.; Waqas, M. IoT-based agriculture management techniques for sustainable farming: A comprehensive review. *Comput. Electron. Agric.* **2024**, *220*, 108851. [CrossRef]
124. Curnick, D.J.; Davies, A.J.; Duncan, C.; Freeman, R.; Jacoby, D.M.; Shelley, H.T.; Shelley, H.T.; Rossi, C.; Wearn, O.R.; Williamson, M.J.; et al. A new technological frontier in ecology and conservation? *Remote Sens. Ecol. Conserv.* **2022**, *8*, 139–150. [CrossRef]
125. Chopping, M.; Laliberte, A.; Rango, A. Multi-angle data from CHRIS/Proba for determination of canopy structure in desert rangelands. In Proceedings of the 2004 IEEE International Geoscience and Remote Sensing Symposium, Anchorage, AK, USA, 20–24 September 2004; Volume 7, pp. 4742–4745.
126. Cofta, P.; Karatzas, K.; Orłowski, C. A conceptual model of measurement uncertainty in IoTiot sensor networks. *Sensors* **2001**, *21*, 1827. [CrossRef]
127. Muangprathub, J.; Boonnam, N.; Kajornkasirat, S.; Lekbangpong, N.; Wanichsombat, A.; Nillaor, P. IoT and agriculture data analysis for smart farms. *Comput. Electron. Agric.* **2019**, *156*, 467–474. [CrossRef]
128. Teh, H.Y.; Kempa-Liehr, A.W.; Wang, K.I.K. Sensor data quality: A systematic review. *J. Big Data* **2020**, *7*, 11. [CrossRef]
129. Saari, H.; Akujärvi, A.; Holmlund, C.; Ojanen, H.; Kaivosoja, J.; Nissinen, A.; Niemeläinen, O. Visible, very near IR and short wave IR hyperspectral drone imaging system for agriculture and natural water applications. *Int. Arch. Photogramm. Remote Sens. Spat. Inf. Sci.* **2017**, *42*, 165–170. [CrossRef]
130. Hunt, E.R., Jr.; Hively, W.D.; Fujikawa, S.J.; Linden, D.S.; Daughtry, C.S.; McCarty, G.W. Acquisition of NIR-green-blue digital photographs from unmanned aircraft for crop monitoring. *Remote Sens.* **2010**, *2*, 290–305. [CrossRef]
131. Fan, X.; Kawamura, K.; Guo, W.; Xuan, T.D.; Lim, J.; Yuba, N.; Kurokawa, Y.; Obitsu, T.; Lv, R.; Tsumiyama, Y.; et al. A simple visible and near-infrared (V-NIR) camera system for monitoring the leaf area index and growth stage of Italian ryegrass. *Comput. Electron. Agric.* **2018**, *144*, 314–323. [CrossRef]
132. Liu, H.; Dahlgren, R.A.; Larsen, R.E.; Devine, S.M.; Roche, L.M.; O'Geen, A.T.; Wong, A.J.; Covello, S.; Jin, Y. Estimating rangeland forage production using remote sensing data from a small unmanned aerial system (sUAS) and planetscope satellite. *Remote Sens.* **2019**, *11*, 595. [CrossRef]
133. Pande, C.B.; Moharir, K.N. Application of hyperspectral remote sensing role in precision farming and sustainable agriculture under climate change: A review. In *Climate Change Impacts on Natural Resources, Ecosystems and Agricultural Systems*; Springer: Berlin/Heidelberg, Germany, 2023; pp. 503–520.
134. Lyu, X.; Li, X.; Dang, D.; Dou, H.; Wang, K.; Lou, A. Unmanned aerial vehicle (UAV) remote sensing in grassland ecosystem monitoring: A systematic review. *Remote Sens.* **2022**, *14*, 1096. [CrossRef]
135. Agrawal, J.; Arafat, M.Y. Transforming Farming: A Review of AI-Powered UAV Technologies in Precision Agriculture. *Drones* **2024**, *8*, 664. [CrossRef]
136. Jouhari, M.; Al-Ali, A.K.; Baccour, E.; Mohamed, A.; Erbad, A.; Guizani, M.; Hamdi, M. Distributed CNN inference on resource-constrained UAVs for surveillance systems: Design and optimization. *IEEE Internet Things J.* **2021**, *9*, 1227–1242. [CrossRef]
137. Ge, X.; Ding, J.; Jin, X.; Wang, J.; Chen, X.; Li, X.; Liu, J.; Xie, B. Estimating agricultural soil moisture content through UAV-based hyperspectral images in the arid region. *Remote Sens.* **2021**, *13*, 1562. [CrossRef]
138. García-Tejero, I.F.; Rubio, A.E.; Viñuela, I.; Hernández, A.; Gutiérrez-Gordillo, S.; Rodríguez-Pleguezuelo, C.R.; Durán-Zuazo, V.H. Thermal imaging at plant level to assess the crop-water status in almond trees (cv. Guara) under deficit irrigation strategies. *Agric. Water Manag.* **2018**, *208*, 176–186. [CrossRef]
139. Boschma, S.P.; Murphy, S.R.; Harden, S. Growth rate and nutritive value of sown tropical perennial grasses in a variable summer-dominant rainfall environment, Australia. *Grass Forage Sci.* **2017**, *72*, 234–247. [CrossRef]

140. Parrini, S.; Acciaioli, A.; Crovetti, A.; Bozzi, R. Use of FT-NIRS for determination of chemical components and nutritional value of natural pasture. *Ital. J. Anim. Sci.* **2018**, *17*, 87–91. [CrossRef]
141. Decruyenaere, V.; Planchon, V.; Dardenne, P.; Stilmant, D. Prediction error and repeatability of near infrared reflectance spectroscopy applied to faeces samples in order to predict voluntary intake and digestibility of forages by ruminants. *Anim. Feed. Sci. Technol.* **2015**, *205*, 49–59. [CrossRef]
142. Pujol, S.; Pérez-Vendrell, A.M.; Torrallardona, D. Evaluation of prediction of barley digestible nutrient content with near-infrared reflectance spectroscopy (NIRS). *Livest. Sci.* **2007**, *109*, 189–192. [CrossRef]
143. Stuth, J.; Jama, A.; Tolleson, D. Direct and indirect means of predicting forage quality through near infrared reflectance spectroscopy. *Field Crops Res.* **2003**, *84*, 45–56. [CrossRef]
144. Landau, S.; Glasser, T.; Dvash, L. Monitoring nutrition in small ruminants with the aid of near infrared reflectance spectroscopy (NIRS) technology: A review. *Small Rumin. Res.* **2006**, *61*, 1–11. [CrossRef]
145. Molano, M.L.; Cortés, M.L.; Ávila, P.; Martens, S.D.; Muñoz, L.S. Ecuaciones de calibración en espectroscopía de reflectancia en el infrarrojo cercano (NIRS) para predicción de parámetros nutritivos en forrajes tropicales. *Trop. Grassl.-Forrajes Trop.* **2016**, *4*, 139–145. [CrossRef]
146. Lahart, B.; McParland, S.; Kennedy, E.; Boland, T.M.; Condon, T.; Williams, M.; Galvin, N.; Buckley, F. Predicting the dry matter intake of grazing dairy cows using infrared reflectance spectroscopy analysis. *J. Dairy Sci.* **2019**, *102*, 8907–8918. [CrossRef]
147. Baath, G.S.; Baath, H.K.; Gowda, P.H.; Thomas, J.P.; Northup, B.K.; Rao, S.C.; Singh, H. Predicting forage quality of warm-season legumes by near infrared spectroscopy coupled with machine learning techniques. *Sensors* **2020**, *20*, 867. [CrossRef]
148. Zhang, M.; Zhao, C.; Shao, Q.; Yang, Z.; Zhang, X.; Xu, X.; Hassan, M. Determination of water content in corn stover silage using near-infrared spectroscopy. *Int. J. Agric. Biol. Eng.* **2019**, *12*, 143–148. [CrossRef]
149. Asekova, S.; Han, S.I.; Choi, H.J.; Park, S.J.; Shin, D.H.; Kwon, C.H.; Shannon, J.G.; Lee, J.D. Determination of forage quality by near-infrared reflectance spectroscopy in soybean. *Turk. J. Agric. For.* **2016**, *40*, 45–52. [CrossRef]
150. Jin, X.; Shi, C.; Yu, C.Y.; Yamada, T.; Sacks, E.J. Determination of leaf water content by visible and near-infrared spectrometry and multivariate calibration in Miscanthus. *Front. Plant Sci.* **2017**, *8*, 721. [CrossRef] [PubMed]
151. Cozzolino, D. Use of infrared spectroscopy for in-field measurement and phenotyping of plant properties: Instrumentation, data analysis, and examples. *Appl. Spectrosc. Rev.* **2014**, *49*, 564–584. [CrossRef]
152. Fernandes, A.M.F. Uso da espectroscopia de reflectância do infravermelho próximo (NIRS) para previsão da composição bromatológica de vagens de algaroba e palma forrageira. Master's Thesis, Universidade Estadual Vale do Acaraú, Sobral, Brazil, 2015.
153. Castro, P.; Fernández-Lorenzo, B.; Valladares, J. Pasture Analysis Using NIRS. SERIDA: Villaviciosa, Spain, 2005; Volume 1, pp. 73–80. Available online: https://agris.fao.org/search/en/records/6472461e53aa8c896304638b (accessed on 29 December 2024).
154. Ullmann, I.; Herrmann, A.; Hasler, M.; Taube, F. Influence of the critical phase of stem elongation on yield and forage quality of perennial ryegrass genotypes in the first reproductive growth. *Field Crops Res.* **2017**, *205*, 23–33. [CrossRef]
155. Simeone, M.; Gontijo Neto, M.M.; Guimaraes, C.D.C.; Medeiros, E.; Barrocas, G.; Pasquini, C. Use of NIR and PLS to predict chemical composition of Brachiaria. In Proceedings of the 17th International Conference On Near Infrared Spectroscopy, Foz do Iguassu, Brazil, 18–23 October 2015.
156. Xia, Y.; Huang, W.; Fan, S.; Li, J.; Chen, L. Effect of spectral measurement orientation on online prediction of soluble solids content of apple using Vis/NIR diffuse reflectance. *Infrared Phys. Technol.* **2019**, *97*, 467–477. [CrossRef]
157. Li, X.; Huang, J.; Xiong, Y.; Zhou, J.; Tan, X.; Zhang, B. Determination of soluble solid content in multi-origin 'Fuji'apples by using FT-NIR spectroscopy and an origin discriminant strategy. *Comput. Electron. Agric.* **2018**, *155*, 23–31. [CrossRef]
158. Wu, Y.; Li, L.; Liu, L.; Liu, Y. Nondestructive measurement of internal quality attributes of apple fruit by using NIR spectroscopy. *Multimed. Tools Appl.* **2019**, *78*, 4179–4195. [CrossRef]
159. Weishaupt, I.; Zimmer, M.; Neubauer, P.; Schneider, J. Model based optimization of transflection near infrared spectroscopy as a process analytical tool in a continuous flash pasteurizer. *J. Food Sci.* **2020**, *85*, 2020–2031. [CrossRef]
160. Campos, M.I.; Antolin, G.; Debán, L.; Pardo, R. Assessing the influence of temperature on NIRS prediction models for the determination of sodium content in dry-cured ham slices. *Food Chem.* **2018**, *257*, 237–242. [CrossRef]
161. Hong, F.W.; Chia, K.S. A review on recent near infrared spectroscopic measurement setups and their challenges. *Measurement* **2021**, *171*, 108732. [CrossRef]
162. Dahm, D.J. Explaining some light scattering properties of milk using representative layer theory. *J. Near Infrared Spectrosc.* **2013**, *21*, 323–339. [CrossRef]
163. McNunn, G.; Heaton, E.; Archontoulis, S.; Licht, M.; VanLoocke, A. Using a crop modeling framework for precision cost-benefit analysis of variable seeding and nitrogen application rates. *Front. Sustain. Food Syst.* **2019**, *3*, 108. [CrossRef]
164. Akhter, R.; Sofi, S.A. Precision agriculture using IoT data analytics and machine learning. *J. King Saud Univ.-Comput. Inf. Sci.* **2022**, *34*, 5602–5618. [CrossRef]

165. Aebischer, P.; Sutter, M.; Birkinshaw, A.; Nussbaum, M.; Reidy, B. Herbage biomass predictions from UAV data using a derived digital terrain model and machine learning. *Grass Forage Sci.* **2024**, *79*, 1–13. [CrossRef]
166. Lüscher, A.; Grieder, C.; Huguenin-Elie, O.; Klaus, V.; Reidy, B.; Schneider, M.K.; Schubiger, F.; Suter, D.; Suter, M.; Kölliker, R. Grassland systems in Switzerland with a main focus on sown grasslands. *Improv. Sown Grassl. Through Breed. Manag.* **2019**, *24*, 3–16.
167. Schori, F. Mit Herbometer und Pasturemeter die Wuchshöhe von Weiden messen und die Grasmasse schätzen. *Agrar. Schweiz* **2020**, *11*, 46–52.
168. Borra-Serrano, I.; De Swaef, T.; Muylle, H.; Nuyttens, D.; Vangeyte, J.; Mertens, K.; Saeys, W.; Somers, B.; Roldàn-Ruiz, I.; Lootens, P. Canopy height measurements and non-destructive biomass estimation of Lolium perenne swards using UAV imagery. *Grass Forage Sci.* **2019**, *74*, 356–369. [CrossRef]
169. Subhashree, S.N.; Igathinathane, C.; Akyuz, A.; Borhan, M.; Hendrickson, J.; Archer, D.; Halvorson, J. Tools for predicting forage growth in rangelands and economic analyses—A systematic review. *Agriculture* **2023**, *13*, 455. [CrossRef]
170. Schonlau, M.; Zou, R.Y. The random forest algorithm for statistical learning. *Stata J.* **2020**, *20*, 3–29. [CrossRef]
171. Schellberg, J.; Verbruggen, E. Frontiers and perspectives on research strategies in grassland technology. *Crop Pasture Sci.* **2014**, *65*, 508–523. [CrossRef]
172. Shaw, R.; Lark, R.M.; Williams, A.P.; Chadwick, D.R.; Jones, D.L. Characterising the within-field scale spatial variation of nitrogen in a grassland soil to inform the efficient design of in-situ nitrogen sensor networks for precision agriculture. *Agric. Ecosyst. Environ.* **2016**, *230*, 294–306. [CrossRef]
173. Higgins, S.; Schellberg, J.; Bailey, J.S. Improving productivity and increasing the efficiency of soil nutrient management on grassland farms in the UK and Ireland using precision agriculture technology. *Eur. J. Agron.* **2019**, *106*, 67–74. [CrossRef]
174. Gandorfer, M.; Meyer-Aurich, A. Economic potential of site-specific fertiliser application and harvest management. In *Precision Agriculture: Technology and Economic Perspectives*; Springer: Cham, Switzerland, 2017; pp. 79–92.
175. Bao-wei, S.; Geng-xing, Z.; Chao, D. Spatio-temporal variability of soil nutrients and the responses of growth during growth stages of winter wheat in the north of China. *bioRxiv* **2018**, 398701. [CrossRef]
176. Pathak, H.S.; Brown, P.; Best, T. A systematic literature review of the factors affecting the precision agriculture adoption process. *Precis. Agric.* **2019**, *20*, 1292–1316. [CrossRef]
177. Klootwijk, C.W.; Holshof, G.; Van den Pol-van Dasselaar, A.; van Helvoort, K.L.; Engel, B.; de Boer, I.J.; van Middelaar, C.E. The effect of intensive grazing systems on the rising plate meter calibration for perennial ryegrass pastures. *J. Dairy Sci.* **2019**, *102*, 10439–10450. [CrossRef] [PubMed]
178. Houseman, G.R. Aggregated seed arrival alters plant diversity in grassland communities. *J. Plant Ecol.* **2014**, *7*, 51–58. [CrossRef]
179. Sun, X.; Zhang, F.; Yang, L.; Zhang, B.; Gao, L. A hyperspectral image spectral unmixing method integrating slic superpixel segmentation. In Proceedings of the 2015 7th Workshop on Hyperspectral Image and Signal Processing: Evolution in Remote Sensing (WHISPERS), Tokyo, Japan, 2–5 June 2015; pp. 1–4.
180. Su, Y.; Zhang, S.; Xu, X.; Li, J. Spectral unmixing for geospatial image analysis. In *Advances in Machine Learning and Image Analysis for GeoAI*; Elsevier: Amsterdam, The Netherlands, 2024; pp. 255–279.
181. Fernández-Guisuraga, J.M.; González-Pérez, I.; Reguero-Vaquero, A.; Marcos, E. Estimating Grassland Biophysical Parameters in the Cantabrian Mountains Using Radiative Transfer Models in Combination with Multiple Endmember Spectral Mixture Analysis. *Remote Sens.* **2024**, *16*, 4547. [CrossRef]
182. Barthram, G.T.; Duff, E.I.; Elston, D.A.; Griffiths, J.H.; Common, T.G.; Marriott, C.A. Frequency distributions of sward height under sheep grazing. *Grass Forage Sci.* **2005**, *60*, 4–16. [CrossRef]
183. Rook, A.J.; Dumont, B.; Isselstein, J.; Osoro, K.; WallisDeVries, M.F.; Parente, G.; Mills, J. Matching type of livestock to desired biodiversity outcomes in pastures–a review. *Biol. Conserv.* **2004**, *119*, 137–150. [CrossRef]
184. De Vroey, M.; Radoux, J.; Defourny, P. Grassland mowing detection using sentinel-1 time series: Potential and limitations. *Remote Sens.* **2021**, *13*, 348. [CrossRef]
185. Wachendorf, M. Advances in remote sensing for monitoring grassland and forage production. Burleigh Dodds Science Publishing: Cambridge, UK,, 2018; pp. 353–362.
186. Xue, J.; Su, B. Significant remote sensing vegetation indices: A review of developments and applications. *J. Sens.* **2017**, *2017*, 1353691. [CrossRef]
187. Della Nave, F.N.; Ojeda, J.J.; Irisarri, J.G.N.; Pembleton, K.; Oyarzabal, M.; Oesterheld, M. Calibrating APSIM for forage sorghum using remote sensing and field data under sub-optimal growth conditions. *Agric. Syst.* **2022**, *201*, 103459. [CrossRef]
188. Wang, Y.; Qin, R.; Cheng, H.; Liang, T.; Zhang, K.; Chai, N.; Gao, J.; Feng, Q.; Hou, M.; Liu, J.; et al. Can machine learning algorithms successfully predict grassland aboveground biomass? *Remote Sens.* **2022**, *14*, 3843. [CrossRef]
189. Shen, N.; Chen, L.; Liu, J.; Wang, L.; Tao, T.; Wu, D.; Chen, R. A review of global navigation satellite system (GNSS)-based dynamic monitoring technologies for structural health monitoring. *Remote Sens.* **2019**, *11*, 1001. [CrossRef]

190. Mistele, B.; Schmidhalter, U. Spectral measurements of the total aerial N and biomass dry weight in maize using a quadrilateral-view optic. *Field Crop. Res.* **2008**, *106*, 94–103. [CrossRef]
191. Schils, R.L.M.; Van den Berg, W.; Van der Schoot, J.R.; Groten, J.A.M.; Rijk, B.; Van de Ven, G.W.J.; Groten, J.A.M.; Rijk, B.; Van de Ven, G.W.J.; Van Middelkoop, J.C.; et al. Disentangling genetic and non-genetic components of yield trends of Dutch forage crops in the Netherlands. *Field Crop. Res.* **2020**, *249*, 107755. [CrossRef]
192. Li, M.; Wang, Y.; Guo, H.; Ding, F.; Yan, H. Evaluation of variable rate irrigation management in forage crops: Saving water and increasing water productivity. *Agric. Water Manag.* **2023**, *275*, 108020. [CrossRef]
193. Bailey, J.S.; Wang, K.; Jordan, C.; Higgins, A. Use of precision agriculture technology to investigate spatial variability in nitrogen yields in cut grassland. *Chemosphere* **2001**, *42*, 131–140. [CrossRef]
194. Shi, Z.; Wang, K.; Bailey, J.S.; Jordan, C.; Higgins, A.H. Temporal changes in the spatial distributions of some soil properties on a temperate grassland site. *Soil Use Manag.* **2002**, *18*, 353–362. [CrossRef]
195. Feng, Q.; Liu, Y.; Mikami, M. Geostatistical analysis of soil moisture variability in grassland. *J. Arid. Environ.* **2004**, *58*, 357–372. [CrossRef]
196. Jones, J.W.; Antle, J.M.; Basso, B.; Boote, K.J.; Conant, R.T.; Foster, I.; Godfray, H.C.J.; Herrero, M.; Howitt, R.E.; Janssen, S.; et al. Toward a new generation of agricultural system data, models, and knowledge products: State of agricultural systems science. *Agric. Syst.* **2017**, *155*, 269–288. [CrossRef]
197. Herrmann, A.; Kelm, M.; Kornher, A.; Taube, F. Performance of grassland under different cutting regimes as affected by sward composition, nitrogen input, soil conditions and weather—A simulation study. *Eur. J. Agron.* **2005**, *22*, 141–158. [CrossRef]
198. Bournaris, T.; Manos, B.; Vlachopoulou, M.; Manthou, V. AgroMANAGER, a web application for farm management. *Int. J. Bus. Inf. Syst.* **2011**, *8*, 440–455. [CrossRef]
199. Bligaard, J. Mark online, a full scale GIS-based Danish farm management information system. *Int. J. Food Syst. Dyn.* **2014**, *5*, 190–195.
200. O'Brien, B.; Murphy, D.; Askari, M.S.; Burke, R.; Magee, A.; Umstätter, C.; McCarthy, T. Modelling precision grass measurements for a web-based decision platform to aid grassland management. *Precis. Livest. Farming* **2019**, *9*, 858–863.
201. Wang, J.; Xiao, X.; Bajgain, R.; Starks, P.; Steiner, J.; Doughty, R.B.; Chang, Q. Estimating leaf area index and aboveground biomass of grazing pastures using Sentinel-1, Sentinel-2 and Landsat images. *ISPRS J. Photogramm. Remote Sens.* **2019**, *154*, 189–201. [CrossRef]
202. Bochtis, D.D.; Sørensen, C.G.; Green, O. A DSS for planning of soil-sensitive field operations. *Decis. Support Syst.* **2012**, *53*, 66–75. [CrossRef]
203. Delaby, L.; Duboc, G.; Cloet, E.; Martinot, Y. Pastur'Plan: A dynamic tool to support grazing management decision making in a rotational grazing system. In Proceedings of the 18th Symposium of the European Grassland Federation, Wageningen, The Netherlands, 14–17 June 2015; Wageningen Academic Publishers: Wageningen, The Netherlands, 2015. Volume 20.
204. Navarro-Hellín, H.; Martinez-del-Rincon, J.; Domingo-Miguel, R.; Soto-Valles, F.; Torres-Sánchez, R. A decision support system for managing irrigation in agriculture. *Comput. Electron. Agric.* **2016**, *124*, 121–131. [CrossRef]
205. López-Riquelme, J.A.; Pavón-Pulido, N.; Navarro-Hellín, H.; Soto-Valles, F.; Torres-Sánchez, R. A software architecture based on FIWARE cloud for Precision Agriculture. *Agric. Water Manag.* **2017**, *183*, 123–135. [CrossRef]
206. Ruelle, E.; Hennessy, D.; Delaby, L. Development of the Moorepark St Gilles grass growth model (MoSt GG model): A predictive model for grass growth for pasture based systems. *Eur. J. Agron.* **2018**, *99*, 80–91. [CrossRef]
207. McDonnell, J.; Brophy, C.; Ruelle, E.; Shalloo, L.; Lambkin, K.; Hennessy, D. Weather forecasts to enhance an Irish grass growth model. *Eur. J. Agron.* **2019**, *105*, 168–175. [CrossRef]
208. Asher, A.; Brosh, A. Decision support system (DSS) for managing a beef herd and its grazing habitat's sustainability: Biological/agricultural basis of the technology and its validation. *Agronomy* **2022**, *12*, 288. [CrossRef]
209. Subhashree, S.N.; Igathinathane, C.; Hendrickson, J.; Archer, D.; Liebig, M.; Halvorson, J.; Kronberg, S.; Toledo, D.; Sedivec, K.; Peck, D. Forage economics calculator web tool: A decision support system for forage management. *Comput. Electron. Agric.* **2023**, *208*, 107775. [CrossRef]
210. Siehoff, S.; Lennartz, G.; Heilburg, I.C.; Roß-Nickoll, M.; Ratte, H.T.; Preuss, T.G. Process-based modeling of grassland dynamics built on ecological indicator values for land use. *Ecol. Model.* **2011**, *222*, 3854–3868. [CrossRef]
211. Corson, M.S.; Skinner, R.H.; Rotz, C.A. Modification of the SPUR rangeland model to simulate species composition and pasture productivity in humid temperate regions. *Agric. Syst.* **2006**, *87*, 169–191. [CrossRef]
212. Romera, A.J.; Beukes, P.; Clark, C.; Clark, D.; Levy, H.; Tait, A. Use of a pasture growth model to estimate herbage mass at a paddock scale and assist management on dairy farms. *Comput. Electron. Agric.* **2010**, *74*, 66–72. [CrossRef]
213. Xie, Y.; Sha, Z.; Yu, M.; Bai, Y.; Zhang, L. A comparison of two models with Landsat data for estimating above ground grassland biomass in Inner Mongolia, China. *Ecol. Model.* **2009**, *220*, 1810–1818. [CrossRef]
214. Evers, L.; Barros, A.I.; Monsuur, H.; Wagelmans, A. Online stochastic UAV mission planning with time windows and time-sensitive targets. *Eur. J. Oper. Res.* **2014**, *238*, 348–362. [CrossRef]

215. Oliveira-Jr, A.; Resende, C.; Pereira, A.; Madureira, P.; Gonçalves, J.; Moutinho, R.; Soares, F.; Moreira, F. Iot sensing platform as a driver for digital farming in rural Africa. *Sensors* **2020**, *20*, 3511. [CrossRef]
216. Kamali, F.P.; Borges, J.A.; Meuwissen, M.P.; de Boer, I.J.; Lansink, A.G.O. Sustainability assessment of agricultural systems: The validity of expert opinion and robustness of a multi-criteria analysis. *Agric. Syst.* **2017**, *157*, 118–128. [CrossRef]

remote sensing

MDPI

Article

Open and Free Sentinel-2 Mowing Event Data for Austria

Petra Miletich [1], Marco Kirchmair [2], Janik Gregory Deutscher [1], Alexander Schippl [1,2] and Manuela Hirschmugl [1,2,*]

1 Joanneum Research, DIGITAL, 8010 Graz, Austria; petra.miletich@joanneum.at (P.M.); janik.deutscher@joanneum.at (J.G.D.); alexander.schippl@edu.uni-graz.at (A.S.)

2 Department of Geography and Regional Science, University of Graz, 8010 Graz, Austria; marco.kirchmair@uni-graz.at

* Correspondence: manuela.hirschmugl@joanneum.at or manuela.hirschmugl@uni-graz.at

Abstract: The accurate detection of mowing events is important in many applications, including in agricultural contexts such as yield and fodder production, as well as biodiversity assessments, habitat modeling, and protected area monitoring. This work presents the first free and open dataset of mowing events covering the entire Austrian territory for the year 2023 at a spatial resolution of 10 × 10 m. We use the Sentinel-2 time series of the Normalized Difference Vegetation Index (NDVI) to detect mowing events, and additionally, we use the mean of the two ShortWave InfraRed (SWIR) bands to exclude misclassification caused by remaining cloud artifacts and shadows. The validation procedure builds on a visual interpretation of the Panomax webcam archive complemented by a selection of field observations. The final validation dataset consists of 211 mowing events recorded in 85 different locations across Austria. In total, 77.73% of these mowing events were detected with a mean time delay of 4 days. The detection delay in summer was smaller than the values recorded in spring and fall. The pixel-based approach exhibited superior efficacy, especially for meadows with three or more mowing events, compared to the polygon-based approach. The results of our study are consistent with those of previous works demonstrating the capacity to produce high-quality mowing event data for various grassland areas in a fully automated manner, independent from training datasets. The results could be used in research on biodiversity or in practical applications such as agricultural policy support and control, fodder supply evaluation, or impact assessment in nature restoration efforts.

Keywords: Sentinel-2 time series; mowing events; grassland; monitoring; agriculture; land use intensity

1. Introduction

Grasslands play a significant role in land use in Austria, comprising 1.2 million hectares of land in 2020, which accounts for approximately half of Austria's agricultural lands [1]. These grasslands serve multiple essential functions, including erosion prevention, carbon storage, the maintenance of soil quality and water regulation, the provision of fodder for livestock, and supporting diverse habitats for various species. Therefore, understanding mowing events and intensities is crucial in applications such as fodder yield and quality assessment, biodiversity evaluation, habitat modeling, and protected area monitoring.

Several studies in recent years have demonstrated the potential of remote sensing in detecting mowing events [2–6]. A comprehensive review by Reinermann et al. [7] revealed that around 40% of grassland-related studies focused on mowing and grazing management options. However, previous studies on mowing event detection were either limited to small areas [3–5,7] or conducted on a polygon level rather than on an individual pixel level [4,6]. In most cases, these polygons were derived from the European Union's common agricultural policy (CAP) integrated administration and control system (IACS). Some studies relied on extensive training data, which can be challenging and costly to obtain [3,6]. In terms of spectral features, 46% of the studies investigated mowing events using remote sensing data by employing the Normalized Difference Vegetation Index (NDVI) [7]. Furthermore, studies that provided pixel-level results for larger areas using automated methods were unable to offer independent validation data and relied on the time series itself to interpret the mowing events [2]. Such a validation approach makes it impossible to calculate the detection delay for the mowing events. Further, not all datasets are open and freely available.

The research questions of this study are as follows:

(1) What detection rate for mowing events can be expected from a fully automated, wall-to-wall, nationally applied approach independent from training data?
(2) Is a pixel-based or a polygon-based application more suitable?
(3) What is the detection delay overall and for the individual cut dates?

To address these research questions, we produced and evaluated the first dataset of mowing events for the entire Austrian territory in the year 2023, at a 10 × 10 m resolution, which is freely and openly accessible. Our methodology builds upon the approach introduced by Kolecka et al. [5], with some adaptations and extensions. We also compare the results derived per pixel and per IACS polygon for a smaller test site. To validate the national product, we use the completely independent Panomax webcam archive to digitize reference meadows and record the corresponding mowing events. Additionally, selected field observations are incorporated into the validation dataset. One main step forward is to show which accuracy is achievable without training data for a whole country and using independent validation data. The scientific sub-objectives are threefold: first, to show the benefit of including the SWIRMean into the analysis, second, to compare pixel-based and polygon-based approaches, and third, to analyze the detection delay for different seasons. The practical significance of our study output is the national wall-to-wall coverage, the provision of the results freely and openly, and the easy repeatability for other years, as no costly training data are needed.

2. Materials

2.1. Study Area and Input Data

This study is conducted in Austria, employing a wall-to-wall approach covering the entire territory, which spans approximately 84,000 square kilometers, encompassing flat lowlands in the east to the Central Alps in the western provinces. The diverse topography and different climatic conditions result in a wide range of grassland types and utilization intensities across Austria. The climatic conditions range from the Continental-Pannonian climate in the east and Illyrian climate in the south-east to the harsher Alpine climate in the main parts of central Austria. The Central European transitional climate dominates the north and we find some areas with Atlantic influences in the west. Grasslands are found in about 90% of the Austrian farms, but the main focus of dairy farming is in the Alpine and northern regions [8]. In this topographically, pedologically, and climatically diverse setting, favorable production sites and disadvantaged sites are often closely intertwined [6]. In combination with small ownership structures, where two thirds of the farmed units of grasslands are smaller than 0.5 ha [8], a large variety of grassland use intensity can be expected. Unlike other studies that focused solely on cultivated grasslands with at least one cut per year [6], our approach processes the entire country in a wall-to-wall manner, thus allowing a broader range of applications to benefit from our results. This includes biodiversity assessments in areas such as Natura 2000 sites or alpine grasslands, which are seldom cut. For this study, we utilized all Sentinel-2 datasets from the year 2023 available through the Copernicus Data Space Ecosystem (CDSE), accessible at https://dataspace.copernicus.eu/ (accessed on 5 August 2024).

2.2. Validation Data

Validation data were derived from the freely accessible Panomax webcam archive (www.panomax.com, accessed on 18 February 2025), which currently comprises 830 active cameras worldwide, with 721 of them being accessible to the public (as of 6 December 2024). In Austria, there are approximately 200 cameras; however, not all of them were operational in 2023 or do not display archived imagery from that year. Furthermore, many of the webcams are situated in urban areas or high-altitude regions, limiting the number of usable cameras for mowing event detection.

Using the available webcam data, the precise date and even time of the cutting event can be determined through visual interpretation, as most Panomax webcams store images in a 15 or 30 min interval. The meadows visible in the webcams were digitized within a Geographic Information System (GIS) environment, and the corresponding mowing events were recorded as related attributes. Additionally, field observations were included in the dataset. In total, 211 mowing events were documented across 85 mapped areas throughout Austria, with the majority located in Carinthia (29%), followed by Salzburg (24%) and Styria (17%). For three provinces (Burgenland, Vorarlberg, and Lower Austria), only one area per province is currently available (see Figure 1).

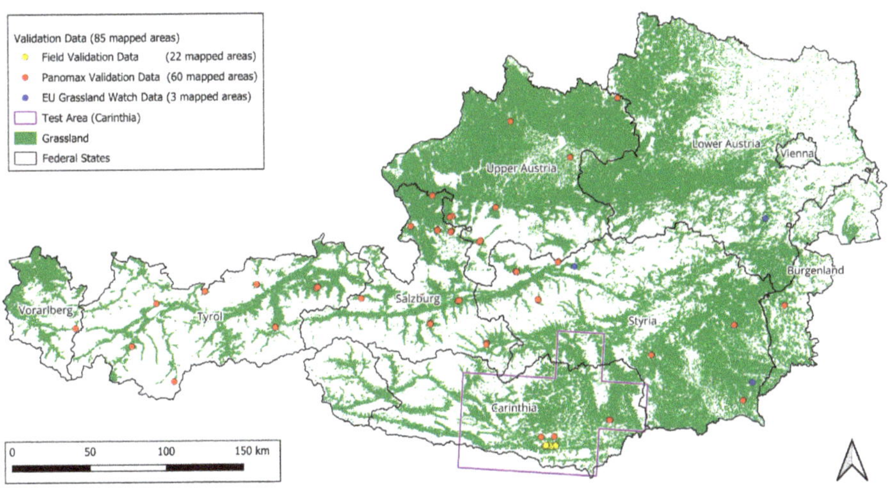

Figure 1. Map of Austria with the distribution of grassland areas (from the Integrated Administration and Control System for 2023) with at least one cut per season (in green) and the validation data-categorized into field-based and Panomax validation datasets as well as data from the EU Grassland Watch Project.

2.3. IACS Data for Comparison

The IACS dataset typically includes the LPIS (Land Parcel Identification System) and GSAA (GeoSpatial Aid Application) data, both of which constitute spatial information that supports IACS as a spatial reference for aid applications under the Common Agricultural Policy (CAP). The LPIS data comprises georeferenced blocks of agricultural areas that have been identified and digitized primarily from ortho-imagery. These areas are potentially eligible for EU aid application. On the other hand, GSAA consists of georeferenced parcels that are in agricultural use and for which aid has been applied. Therefore, GSAA is usually a component of LPIS. In the context of Austria and this study, we exclusively consider LPIS as the most comprehensive dataset within the IACS 2023 dataset used. The IACS categories used in this study as well as their definitions are given in Table 1. Please note the ambiguity between the IACS category definition and the actual number of mowing/use events.

Table 1. IACS categories and their definitions [9].

IACS Category	Definition
pasture (*Hutweide*)	The pasture is a low-yielding, grazed permanent grassland (usually without maintenance cutting) on which mechanical fodder production or maintenance is not possible or is not carried out due to the nature of the soil. These areas must be fully grazed **at least once per year**.
mountain meadow (*Bergmähder*)	Mountain meadows are extensive mowing areas above the local permanent settlement boundary, whereby these areas must be above the altitude of the home farm and generally not directly adjacent to home farm areas of the same farm. The majority of the area must be above 1200 m.a.s.l. Mountain meadows must be fully mown **at least once every two years** and the mown material must be removed.
single-mown meadow (*Mähwiese 1*)	Single-mown meadows are areas on which the entire surface is mown once a crop year and the mown material is removed from the area.
mown meadow/pasture with two uses (*Mähwiese 2*)	Mown meadows/pastures with two uses are areas on which full-surface **mowing is carried out either twice** (including the removal of the mown material) **or only once but combined with a full-surface grazing** in the same year. A selective maintenance cut does not count.
mown meadow/pasture with three or more uses (*Mähwiese 3*)	Mown meadows/pastures with three or more uses are areas on which full-surface mowing is carried out **either three times** (including the removal of the mown material) **or a combination of one or two mowing events** (including the removal of the mown material) **with full-surface grazing** in the same year to reach at least three full-surface uses.

3. Methods

3.1. Mowing Event Detection Method

Similar to prior studies [5,10], the approach relies on time series data from Sentinel-2 to identify declines in the time series curves, with stringent pre-processing being crucial for successful mowing detection. This pre-processing includes atmospheric correction, cloud masking, and topographic correction. For atmospheric correction, we used Sen2Cor [11,12]. The Sen2Cor processor has enhanced the ability to convert Level-1C data (top-of-atmosphere reflectance) to Level-2A (bottom-of-atmosphere reflectance), including maps of aerosol optical thickness, water vapor, and scene classification for clouds and snow [13]. Sentinel-2 data undergo rigorous geometric correction using a Global Reference Image (GRI) database and automatic ground control point matching, improving geolocation accuracy to sub-pixel levels [14]. Cloud masking was done using Function of Mask (FMASK) [15], an automated algorithm that detects and masks clouds, cloud shadows, and snow in optical satellite imagery. Finally, topographic normalization is used to correct terrain-induced brightness variations in satellite imagery. The approach applies iterative classification to spectrally distinct clusters and derives specific Minnaert constants for each. This allows fully automatic, reference-free correction of topographic effects, which is essential for large-scale processing in mountainous areas [16,17]. In addition to previous methods [5], the ShortWave InfraRed (SWIR) bands are utilized alongside the Normalized Difference Vegetation Index (NDVI). The NDVI is calculated as the ratio between the difference and sum of the NIR and the red band (NDVI = (Band 8/NIR − Band 4/red)/(Band 8/NIR + Band 4/red). The two SWIR bands at a spatial resolution of 20 × 20 m are band 11 with a central wavelength at 1610 nm and band 12 with a central wavelength at 2190 nm. The SWIRMean is defined as the mean pixel value of the two SWIR bands.

The method initially examines the NDVI time series to identify significant drops, with a drop greater than 0.2 in NDVI flagged as a potential mowing event [5]. To account for noise or minor vegetation changes, two consecutive smaller NDVI drops, each less than 0.04, are also considered as potential mowing events. However, NDVI drops do not only occur due to mowing events, but also due to unmasked shadows, haze, or thin cirrus clouds, which reduce visible and near-infrared reflectance, leading to a spurious decline in NDVI. When NDVI drops due to thin cirrus clouds, cloud shadows, or haze in the data, the SWIRMean typically remains unchanged or drops.

In contrast, a genuine mowing event results in a drop in NDVI and a simultaneous increase in the SWIRMean due to changes in vegetation structure and moisture content. This complementary behavior is used to differentiate between atmospheric artifacts and actual mowing events, enhancing the robustness of the detection approach. The behavior differences in a 'look-alike' and a real mowing event are shown in Figure 2. For the NDVI drop on 30 May (marked with a black vertical line), there is no corresponding SWIRMean increase, leading to the filtering out of this drop as a 'look-alike' mowing event. In contrast, the detected mowing event on 19 June (marked with a green vertical line in Figure 2) shows the typical behavior of both indices and was also identified in the Panomax reference images on 11 June (red line). In addition to this visual check, we also tracked the amount of 'look-alike' mowing events for the small test site to see how much the SWIRMean test actually sorted out. We found 226.896 valid mowing events and 236.274 'look-alike' events. Based on these numbers, the importance of the SWIRMean as an additional criterion becomes evident. While the SWIRMean effectively helps mitigate misclassifications caused by atmospheric artifacts such as haze, shadows, and thin cirrus clouds, its performance may be affected by complex terrain. In mountainous areas, topographic effects—such as varying slopes and aspects—can influence reflectance even after correction. Although the applied Minnaert correction reduces terrain-induced illumination variability, residual effects may remain.

Upon detecting a mowing event, a regeneration phase of two weeks is assumed [5], during which no further mowing events are identified. This temporal buffer allows for the natural recovery of vegetation and prevents the misclassification of subsequent vegetation changes within this timeframe as separate mowing events. The outputs of the mowing detection process include the total count of mowing events, the dates of (a maximum of six) detected mowing events, and the number of valid observations available in the time series for each detection, providing a comprehensive overview of mowing activity during the analyzed period.

The method is designed to accommodate both vector and raster data. When applied to vector data, mowing events can be detected within predefined field boundaries, such as those provided by the IACS system. For pixel-level analysis, the method identifies mowing events on a per-pixel basis. Nonetheless, additional post-processing is recommended to smooth the results and minimize noise, ensuring spatial consistency and accuracy in the detection outcomes. We applied a simple 3 × 3 majority filter. The dual capability allows the method to be flexibly applied across different spatial scales while maintaining robustness in mowing event detection. Although the method is capable of working both retrospectively and in near-real time, it was not processed in near-real time for this study; thus, differences in accuracy from real-time processing cannot be assessed.

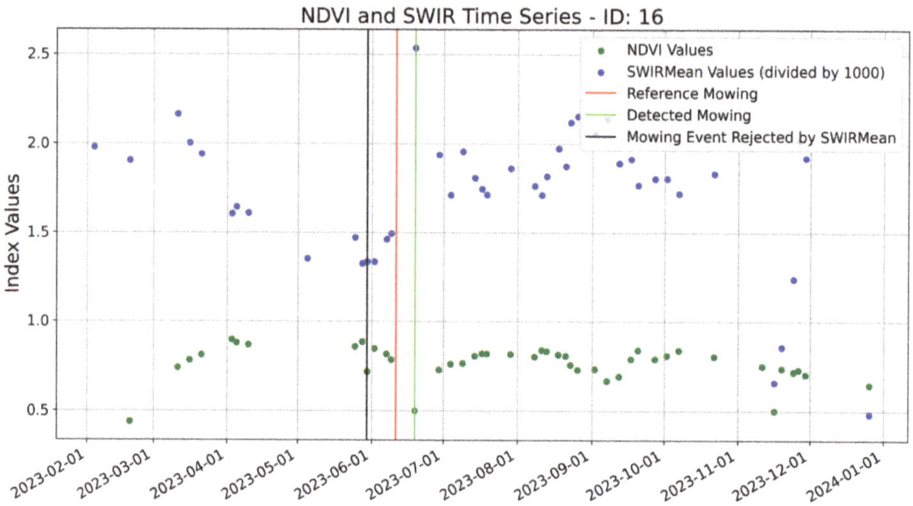

Figure 2. Visualization of the SWIRMean inclusion in an example time series.

3.2. Validation Method

Ensuring the accuracy of the analysis, the validation of mowing detection results is crucial. This process involved leveraging completely independent and high-resolution geotagged imagery from Panomax webcams (www.panomax.com, accessed on 18 February 2025) to identify and delineate mowing areas. Figure 2 illustrates the validation workflow, demonstrating the integration of webcam observations with geographic data. As depicted in Figure 3A, the georeferenced location of the Panomax webcam and its field of view were analyzed to outline specific mowing areas. These visible areas in the imagery were digitized in QGIS as polygons to represent the observed extent of mowing activities. The alignment of the polygons with the actual mowing fields was facilitated using Panomax's satellite map and live-view imagery (Figure 3B).

To enhance the precision of the analysis, a 5 m inverse buffer was applied to each polygon. This adjustment aimed to minimize the influence of potential inaccuracies from digitization and the effect of mixed pixels by focusing on the core regions of the validation polygons.

To evaluate the detection rate and delay, the most frequently detected mowing date in each mapped meadow was compared to the reference mowing date. The reference mowing event was confirmed when the grassland in the Panomax image visibly changed from tall, uniform vegetation to a cut appearance—often accompanied by visible swath rows, drying hay, or mowing machinery. The meadow areas visible in the webcams were digitized as polygons based on perspective-corrected aerial views from the webcams static map layer. We then linked each observed mowing date to the corresponding polygon. For the 24 field datasets and the three EU grassland watch observations, a window of −3 days was allowed to still be considered correct, as during the field survey, it was not always possible to ascertain if the mowing occurred exactly on the day of observation or a few days earlier. For logistical reasons, no field surveys were carried out over the weekends, which means mowing events from, e.g., a Friday evening, may only have been surveyed and reported on the following Monday. This procedure to allow a −3 day window during the analysis aligns with the methodology employed in previous works [6,18]. Any detections that occurred more than three days before the v date were classified as 'not detected'. Moreover, any detections with a delay exceeding one of the specified thresholds (30 days, 15 days, and 7 days) were also deemed 'not detected' in the respective statistics. For the remaining detections, the delay [days] in comparison to the validation date was calculated.

Figure 3. Validation workflow using Panomax webcam data (Panomax, St. Gilgen am Wolfgangsee). (**A**) Panomax camera location and its field of view (green box), with the red polygon indicating the digitized mowing area derived from the imagery. (**B**) Live-view imagery from the Panomax archive on 21 May 2023 at 1:30 p.m., showing the mowing activities and the camera's archive settings in the blue box.

To assess commission errors, a full screening of the entire Panomax archive inspecting each individual day is necessary. Due to the time-consuming nature of this full screening, we randomly selected 20 out of the 85 areas for the full screening. For these 20 areas, we screened each individual day and evaluated grazing events alongside the mowing events.

3.3. Method for Comparison with IACS

We utilized five categories from the IACS dataset [9], and the definitions for these categories are provided in Section 2.3. However, these definitions do not allow a clear assessment of the number of mowing events per category in the year 2023. Therefore, we had to make assumptions regarding the typical number of mowing events for the purpose of comparison:

- No mowing events were assumed for pastures (*Hutweide*).
- One annual mowing event was assumed for mountain meadows (*Bergmähder*).
- It was assumed that there was one mowing event per year for single-mown meadows (*Mähwiese 1*).
- Two annual mowing events were assumed for mown meadows/pastures with two uses (*Mähwiese 2*).
- It was assumed that mown meadows/pastures with three or more uses (*Mähwiese 3*) experienced three or more mowing events per year.

It is important to note that this estimation of mowing events per category may not be entirely reliable, as there are cases where mountain meadows, for example, are mown twice a year, while others are mown only every second year. Therefore, we consider this analysis as a plausibility check rather than an accuracy assessment.

4. Results

4.1. Mapping Result

We applied the pixel-based approach to process Austria wall-to-wall, and the resulting map clipped to IACS categories (as listed in Table 1) is presented in Figure 4. The map effectively illustrates the more intensively used grassland areas in the northern foreland, predominantly in Upper Austria, and in the inner-Alpine valleys compared to the less intensively used areas in Carinthia, for example.

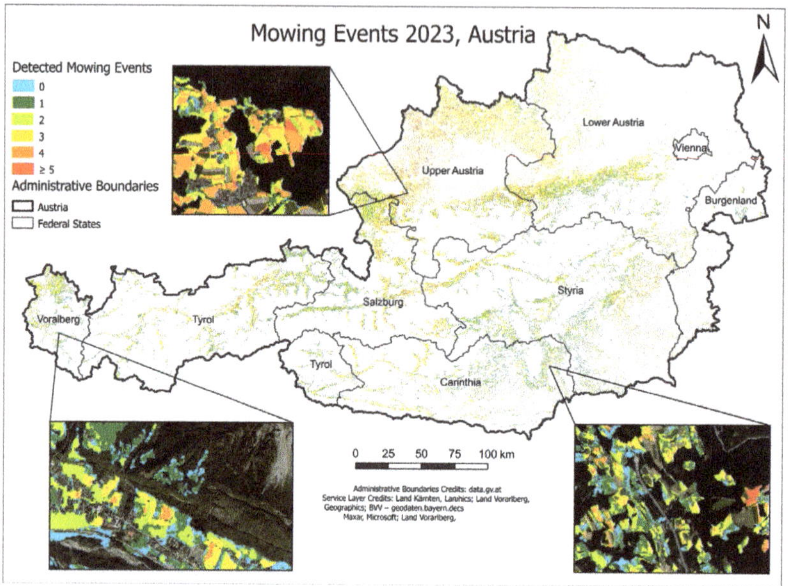

Figure 4. Map of Austria with detected mowing events using the pixel-based approach.

4.2. Accuracy Assessment Based on Validation Dataset

We conducted a comparison between the results of our pixel-based mowing event detection for Austria and all the validation data obtained using the methods described earlier. An example for the successful detection is shown in Figure 5 with all reference events detected with a maximum delay, in this case, of 10 days.

The findings revealed a detection rate of above 80% (80.57% for a maximum delay of 15 days and 81.99% for a maximum delay of 30 days). The minimal difference between the 15- and 30-day delays indicates that most mowing events are typically detected within 15 days. Regarding detection delay, the average delay (mean absolute error, MAE) for all detections was approximately 4 days (4.07 for 15 days and 4.30 for 30 days, as shown in Table 2). Within 7 days, 63.98% of the mowing events were detected with an MAE of 2.8 days. This result is noteworthy considering the global 5-day revisit time of Sentinel-2, although it is important to note that the revisit time is not globally consistent. Due to overlapping orbits at higher latitudes, certain areas in Austria receive a 3-day revisit time [19].

Further, we assessed the delay for individual cuts separately. Table 2 shows a slightly higher delay for the first and the last cut compared to the summer cuts (for both 15- and 30-day maximum allowed delay). For the 7-day maximum allowed delay, the third cut shows the smallest delay. The timeliness of mowing event detection strongly depends on the available number and temporal density of cloud-free observations. Areas and/or periods of high cloud cover show longer time lags between the mowing event and the next available interpretable optical satellite image. Previous studies have shown significant spatial (e.g., from single or double Sentinel-2 orbit coverage) and temporal (inter-annual and seasonal) differences in the availability of cloud-free Sentinel-2 data. A recent study on mowing event mapping with Sentinel-2 and Landsat-8 at national scale in Germany reported differences in mowing detection accuracies related to the inhomogeneous availability of cloud-free datasets [18].

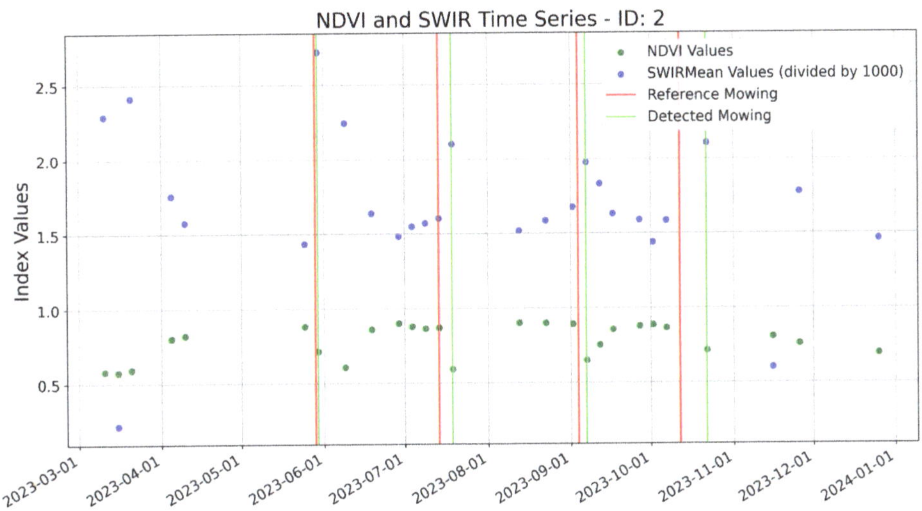

Figure 5. Time series of a successful detection of mowing events using both NDVI and SWIRMean.

Table 2. Mean absolute error (detection delays) for individual cuts and all cuts [days] and overall detection rate.

Maximum Allowed Delay	Delay in Cut Detection (Mean Absolute Error-MAE) [Days]						Detection Rate [%]
	1	2	3	4	5	All Cuts	
30 days	5.3	3.3	4.1	3.9	5.7	4.30	78.67
15 days	5.3	3.3	3.5	3.2	5.7	4.07	77.73
7 days	3.28	2.6	2.5	2.9	3	2.84	63.98

For the randomly selected 20 areas of full screening, we evaluated grazing events alongside the mowing events. We found an average commission error of 0.6 events when considering grazing as a land use event and an average commission error of 0.9 events when grazing is not considered an event (see Table A1 in Appendix A).

4.3. Plausibility Check in Comparison with IACS Dataset

Due to the unclear number of mowing events per category (see Section 2.3), we can consider this comparison as a plausibility check rather than an accuracy assessment.

(a) Pixel-based versus polygon-based for the test site only

To evaluate the suitability of a polygon-based versus a pixel-based approach, we conducted a comparative analysis using a test area in Carinthia (as indicated by the purple outline in Figure 1). Within this area, we separately processed the region using the same settings and methods for polygons and individual pixels. Due to the small number of Panomax webcams and field data, a full-fledged accuracy assessment was not representative. Instead, we compared the results only to the IACS data. Figure 6 displays the percentage deviation of the summed mowing event areas from the IACS (referred to as the reference values) for both polygon-based and raster-based results. In this visual representation, each bar represents the deviation from the IACS, with positive values indicating overestimation (areas with more mowing events than assumed using the IACS categories) and negative values reflecting underestimation (areas with fewer mowing events than assumed using the IACS categories).

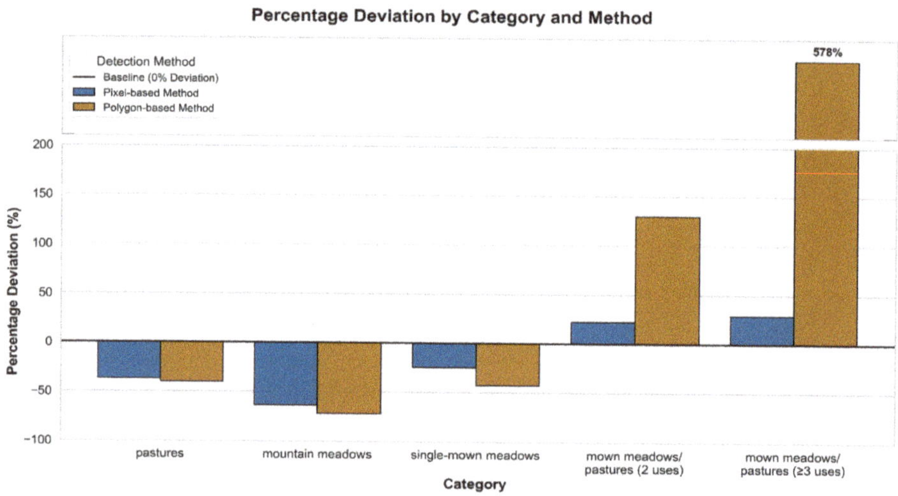

Figure 6. Percentage deviation of summed mowing event areas from IACS values by category and method for the test site located in Carinthia, Austria. Deviations are calculated relative to IACS areas, with the baseline (0% deviation) shown as a black horizontal line.

The results revealed substantial overestimation in the category "mown meadows/pastures with three or more uses" by the polygon method, reaching nearly 600 percent. This indicates that the polygon method classifies approximately six times more area in this category than the IACS. Such a significant overestimation serves as an indicator of substantial commission errors in the polygon-based approach. In contrast, other categories exhibited smaller deviations, with both methods often underestimating areas, particularly in cases like pastures and mountain meadows, where we assumed "no mowing or other use event" in accordance with the IACS.

Across all categories, the pixel-based method demonstrated smaller deviations compared to the polygon-based approach. These findings align with the results of the accuracy assessment, further supporting the conclusion that the pixel-based approach is superior to the polygon-based method.

(b) Pixel-based comparison with IACS for whole of Austria

Building upon the earlier findings that favored the pixel-based approach for its superior performance, we expanded the comparison conducted in the small test site to encompass the entirety of Austria using the pixel-based approach. The deviations per category are displayed in Figure 7, following the same format as Figure 6.

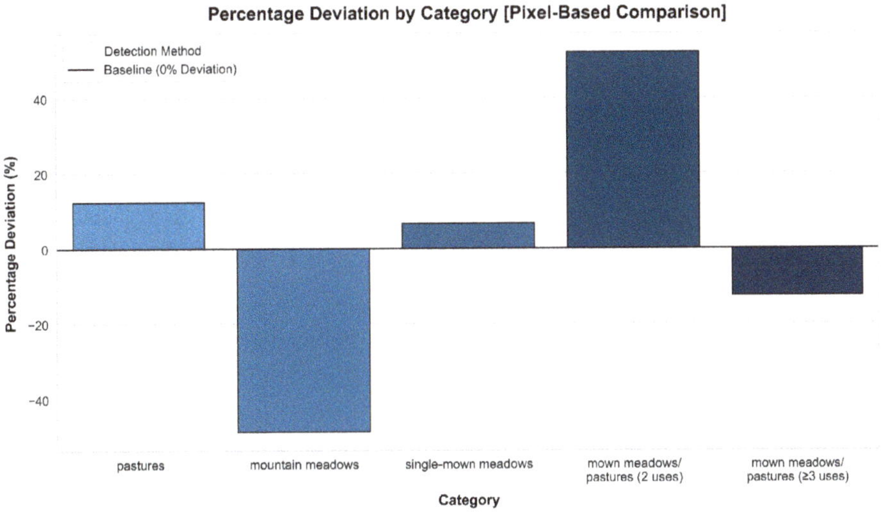

Figure 7. Percentage deviation of summed mowing event areas from IACS values by category for Austria. Deviations are calculated relative to IACS areas, with the baseline (0% deviation) shown as a black horizontal line.

Notably, similar deviations are observed for category "mountain meadows" (underestimation) and for category "mown meadows with two uses" (overestimation) compared to the small test site. However, distinctive behaviors are observed for other categories. In the small test site, pastures were underestimated, yet for the full Austrian coverage, there is a slight overestimation in area. This same pattern is also observed for the categories "single-mown meadows" and "mown meadows with three or more uses". Notably, the latter two categories exhibit excellent results with minimal deviations evident in both the test site and the full coverage.

5. Discussion

This study has explored the feasibility of achieving nationwide and fully automated mowing event detection using Sentinel-2 time series data. Our findings exhibit similar detection delay and overall accuracy to other studies, both with (f-score of 0.79 for [6], 60% for [10]) and without training data (85% for [3], 77% for [5]). Since the overall detection rates are very similar, it seems that our relatively simple, physically determined approach is able to generate the same quality as high-end deep learning tools. Notably, previous studies [4,10] highlighted challenges related to data availability and cloud interference in mowing event detection. We found a moderate relationship between mean monthly cloud cover and detection success. Monthly cloud fraction data from the Sentinel-5P satellite (TROPOMI instrument) were obtained as Level-3 gridded product in NetCDF-format from the Copernicus S5P-PAL portal (https://maps.s5p-pal.com/cloud-fraction/ accessed 28 April 2025). The original grid was interpolated to a regular 0.05° × 0.05° (~5 km) resolution and average cloud cover values were calculated for all reference data locations. The two months with the highest detection rates (August and September) show low (September) and moderate (August) mean cloud cover values (see Table 3). It must be noted that such an analysis has some problems: First, in the case that the reference event occurs towards the end of one month, the satellite data acquisition may already fall into the subsequent month. In that case, the mean cloud cover of reference month would actually not affect detection. Second, in a mountainous country like Austria, cloud cover can vary significantly according to region. Mean monthly cloud cover is not able to depict these regional differences. Figure 8 depicts an example to better understand the effects of missing data. The first mowing event on 25 May (Figure 8, first red vertical line) was preceded by a long period of clouds and, therefore, missing Sentinel-2 data (Figure 8). The event could not be detected by the subsequent observation, as the NDVI drop was too small and there was no second observation to compare to. Similar situations were demonstrated in prior research [5].

Table 3. Comparison of mean monthly cloud cover over reference sites and detection success per month.

Month	No. of Reference Mowing Events	Detected Mowing Events [%]	Monthly Mean Cloud Cover over Reference Sites [%]
April	1	0	80
May	40	50	65
June	46	85	59
July	41	73	53
August	34	97	50
September	30	100	30
October	19	68	48

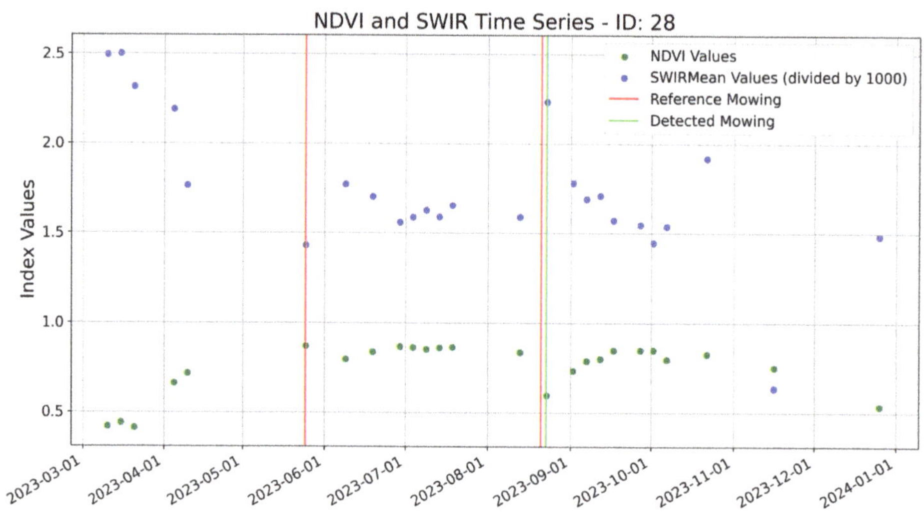

Figure 8. Example time series plot of NDVI and SWIRMean showing a missed mowing event due to cloud induced data gaps in May and early June. The reference area is located in Griffen, Carinthia with one missed reference mowing event (25 May) and one detected mowing event (21 August, detected 23 August).

Other studies integrated SAR data to mitigate the cloud cover problem [20–23]. One study showed slight improvements (MAE 4.1 from 4.6 and f-score 0.81 from 0.77) based on SAR and weather data integration for meadows larger than 2 ha [21]. Other authors found improvements of 10% by including SAR data [20] with a deep learning system for selected test sites in Germany. Our main reason for relying solely on optical data is the wall-to-wall requirement and the Alpine topography of Austria, which poses issues in SAR signals [22]. In our observations using the polygon-based approach, we also encountered excessive detections, as depicted in Figure 6. However, our meticulous preprocessing in conjunction with the pixel-based assessment notably mitigated these false detections (see Figure 6, pixel-based approach). Similar findings were previously noted by Kolecka et al. (2018, [5]). Nevertheless, we recognize that data availability varies annually due to weather patterns, which may still impact detection accuracy, as shown in Figure 8.

The resulting map from our study (see Figure 4) effectively illustrates the distribution of intensively and extensively used grasslands across Austria, with the most heavily used areas situated in the northern provinces of Upper and Lower Austria, followed by the south-eastern foreland of Styria. This distribution closely aligns not only with previous studies [6], but also with the official statistics from Statistics Austria [24], which indicate that these three provinces have the highest shares of intensively used grasslands (see Figure 9).

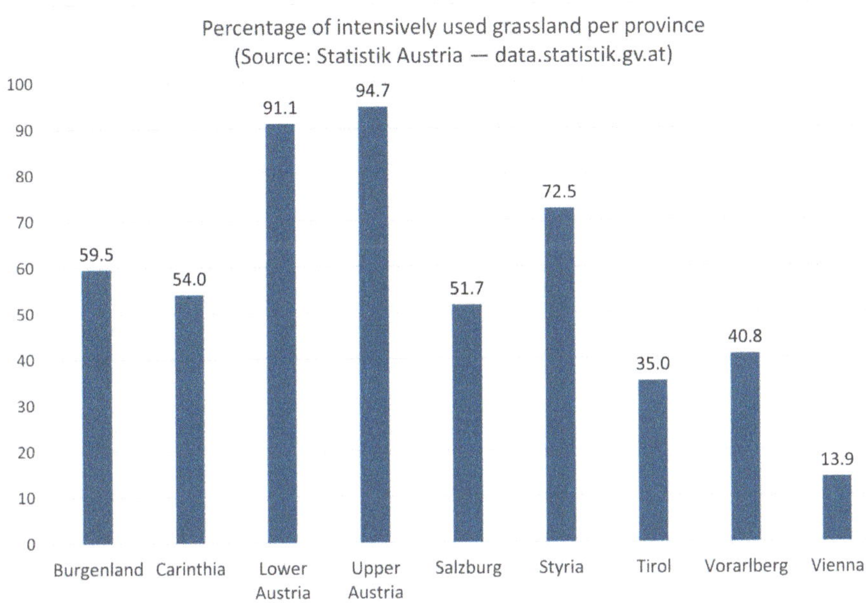

Figure 9. Intensively used grasslands in percent of all grasslands per province [24].

In our accuracy assessment, we deliberately employed much stricter criteria compared to previous authors, who allowed a three- [18] or even a six-day [10] window before an event to be considered correct, whereas we only allowed three days in case of field visits and none for the Panomax data. Our results robustly reinforce their assertion of the superiority of the pixel-based approach over the polygon-based approach (Figure 6 as well as Figure 10), aligning with the findings of other studies [5]. This preference may stem from factors such as (i) the presence of mixed and non-grass pixels within the polygons, including trees, hedges, or artificial surfaces [5] or (ii) different mowing patterns within a single polygon.

Regarding detection delay (MAE in days), Reinermann et al. (2023, [7]) demonstrated the capability to detect 60.3% of all mowing events within 7 days, with an average delay of 2.5 days, using solely Sentinel-2 data. Their reference data were also independent webcam data but building on a small set of reference areas for parametrization. Our findings closely mirror this, with 68.2% of events detected within 7 days and an average delay of 2.6 days. Moreover, with a permissible delay of 15 days, the detection rate exceeds 80%. In a related study, Kolecka et al. (2018, [5]) reported an overall accuracy of slightly less than

80% based on visual interpretation of the utilized Sentinel-2 data. Our results fall within the same range, even when utilizing a wholly independent validation dataset from the Panomax network.

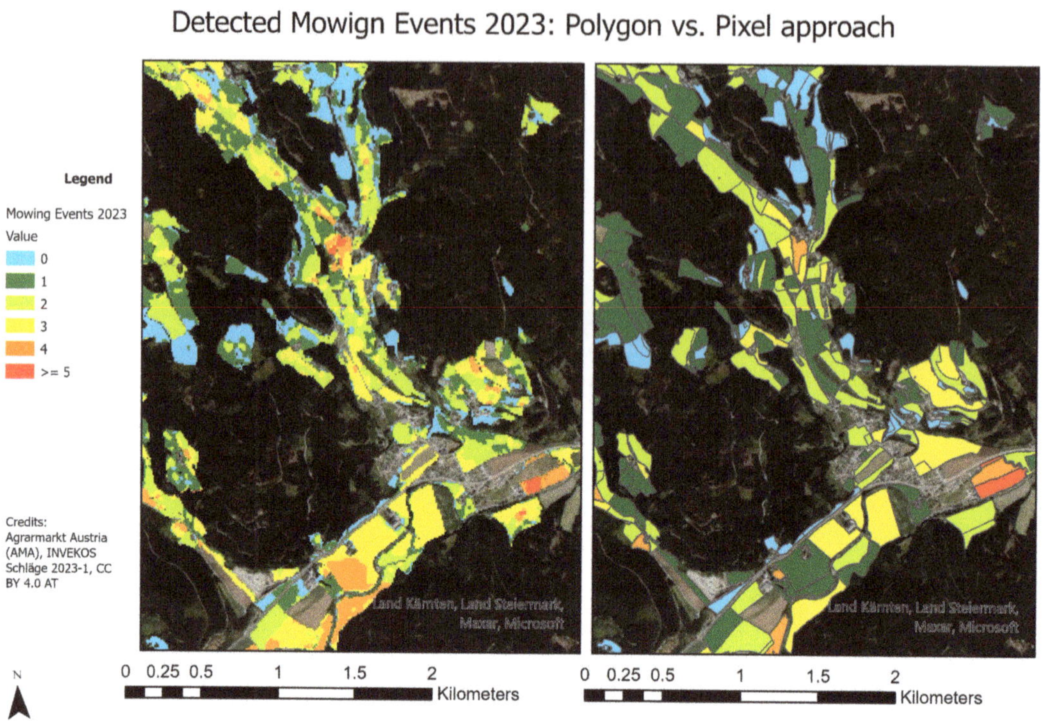

Figure 10. Example area comparing the results of the pixel-based vs. polygon-based approach.

Our study is limited in several ways. First, we recognize that the available Panomax data are not distributed evenly and there might be a regional bias to the results, as the provinces of Lower Austria, Burgenland, and Vorarlberg with significant grassland extent are not well covered. In the case of Vorarlberg, we do not expect significant impacts on the overall validity, as we have good coverage of Tyrol, a similar region. However, for the relatively dry eastern zones of Austria (Lower Austria and Burgenland), there might be more impact on the results once including more reference data from this area. The second study limitation concerns the transferability to other countries and regions. Although covering the whole of Austria means to cover a large variety of grassland types and conditions, our study results are not automatically transferable to other areas. A limitation might be found in the thresholds used for both NDVI drop and SWIRMean increase for much drier or wetter regions. Third, the processing of only one year is a further limitation to the general validity of our findings. This is specifically true for years with extreme weather conditions such as, for example, 2024, when heavy rainfall caused flooding, which could potentially jeopardize the accuracy of mowing event detection. This is, of course, not only true for our approach, but also for other methods.

Therefore, there is scope for future work and improvements. One future work direction would be to expand the reference data to gain a better understanding of the remaining errors. It is important to analyze how accuracy varies with elevation and/or regions. A second future expansion would be to address challenges such as cloud cover and related missing measurements over time by combining Sentinel-2 data with other sensors, such as Planet Scope data, although output quality is expected to be much lower due to the missing SWIR bands. The results for the integration of SAR data, as shown in previous works, are encouraging and should also be further explored. Furthermore, carrying out the processing for additional years would improve the assessment of the method's robustness and its sensitivity to changing weather patterns and evolving management practices. Lastly, conducting a comprehensive comparison of methods (in line with the concept of the mowing detection intercomparison exercise, MODCiX [25]), for a consistent year and the same region would be an important step forward. For this comparison, it is essential to employ independent data for accuracy assessment (which were not utilized for training), as this would represent a significant advancement in effectively comparing and evaluating the applicability of various methods.

6. Conclusions

The main aim of this study was to show, based on independent validation data, how well a fully automated approach for mowing event detection from Sentinel-2 time series works in the diverse Austrian landscape. The method is based on a combination of NDVI drops and SWIRMean increase and works without any kind of training data. In conclusion, this study is an important step forward in mowing event detection, as it proves that detection accuracies comparable to those in the reviewed literature without costly and difficult-to-come-by training data are possible. The detection rate for mowing events reached almost 78% with 15 days of maximum allowable delay. Thanks to rigorous preprocessing schemes and the inclusion of the SWIRMean in the processing, the commission error is less than one (0.9 on average) event. A comparison confirmed previous statements that a pixel-based approach performs better than a polygon-based application. Additionally, the pixel-based approach enables the assessment of areas without polygon boundaries (e.g., outside IACS, see Figure 10) and can differentiate between mowing events on different days within a single IACS polygon (Figure 10). The detection delay throughout the year averages approximately 4 days, with slightly lower delays for cut dates in summer than for those occurring early or late in the season. This fact could mainly be attributed to cloud cover and related missing observations in spring and fall. Our study is limited by the application to one year only, the focus on Austria only, and the available amount of validation data, which could bias the result. Further work should, therefore, focus on processing multiple years, larger areas, and assessing the accuracy based on more validation data for individual regions related to varying weather and growing conditions.

Our results are shared freely and openly to allow for numerous follow-up applications. They can be used for habitat mapping, biodiversity assessments, evaluating usage intensity, or control adherence to specific regulations, such as late mowing (e.g., after 15 June to support biodiversity). We hope this comprehensive study will contribute to advancing the understanding of grassland dynamics and aid in informed decision making for sustainable land-management practices.

Author Contributions: Conceptualization, M.H.; methodology, P.M.; software, A.S.; validation, M.K. and M.H.; formal analysis, P.M. and A.S.; investigation, P.M. and M.H.; writing—original draft preparation, M.H. and P.M.; writing—review and editing, P.M., J.G.D. and M.H.; visualization, M.K. and P.M.; supervision, M.H.; project administration, M.H.; funding acquisition, M.H. All authors have read and agreed to the published version of the manuscript.

Funding: This research was funded by the Austrian Research Promotion Agency FFG under the Austrian Space Application Programme ASAP, project "RestorEO", grant agreement number FO999892628.

Data Availability Statement: Our results are available fully open and free through Zenodo (https://zenodo.org/records/15350942, published on 7 May 2025).

Acknowledgments: During the preparation of this manuscript, the authors used ChatGPT, version 3.5 for the purposes of language polishing. The authors have reviewed and edited the output and take full responsibility for the content of this publication. Further, the authors thank the five anonymous reviewers for their valuable comments and suggestions to improve clarity and quality of this manuscript.

Conflicts of Interest: The authors declare no conflicts of interest.

Appendix A

Table A1. Full evaluation of 20 Panomax areas also assessing grazing activities and commission error.

Location	Panomax Mowing Events	Panomax Grazing Events	Detected Events	Commission (Only Mowing)	Commission (Incl. Grazing)
Griffen	4	0	4	0	0
Pyhrn	5	0	5	0	0
Pertisau	3	0	5	2	2
Liebenau	4	0	3	0	0
Nassereith	2	1	3	1	0
Ramsau im Zillertal	3	1	4	1	0
Pertisau	3	0	4	1	0
Scharnitz	2	1	4	2	1
Donnersbachwald	2	1	3	1	1
Altenmarkt 1	4	0	5	1	0
Altenmarkt 2	4	1	5	1	1
Mondsee	5	0	4	0	0
Bad Mitterndorf	4	0	3	0	0
Großarl 1	4	1	5	1	0
Großarl 2	4	0	6	2	2
Großarl 3	4	0	6	2	2
Großarl 4	3	0	5	2	2
Westendorf 1	4	1	4	0	0
Westendorf 2	4	0	4	0	0
Westendorf 3	3	1	3	0	0
Average				0.9	0.6

References

1. Statistik Austria. *Agrarstrukturdatenerhebung 2020*; Statistik im Fokus; Statistik Austria: Vienna, Austria, 2022; p. 138.
2. Griffiths, P.; Nendel, C.; Pickert, J.; Hostert, P. Towards National-Scale Characterization of Grassland Use Intensity from Integrated Sentinel-2 and Landsat Time Series. *Remote Sens. Environ.* **2020**, *238*, 111124. [CrossRef]
3. Halabuk, A.; Mojses, M.; Halabuk, M.; David, S. Towards Detection of Cutting in Hay Meadows by Using of NDVI and EVI Time Series. *Remote Sens.* **2015**, *7*, 6107–6132. [CrossRef]
4. Hartmann, A.; Sudmanns, M.; Augustin, H.; Baraldi, A.; Tiede, D. Estimating the Temporal Heterogeneity of Mowing Events on Grassland for Haymilk-Production Using Sentinel-2 and Greenness-Index. *Smart Agric. Technol.* **2023**, *4*, 100157. [CrossRef]
5. Kolecka, N.; Ginzler, C.; Pazur, R.; Price, B.; Verburg, P.H. Regional Scale Mapping of Grassland Mowing Frequency with Sentinel-2 Time Series. *Remote Sens.* **2018**, *10*, 1221. [CrossRef]
6. Watzig, C.; Schaumberger, A.; Klingler, A.; Dujakovic, A.; Atzberger, C.; Vuolo, F. Grassland Cut Detection Based on Sentinel-2 Time Series to Respond to the Environmental and Technical Challenges of the Austrian Fodder Production for Livestock Feeding. *Remote Sens. Environ.* **2023**, *292*, 113577. [CrossRef]
7. Reinermann, S.; Asam, S.; Kuenzer, C. Remote Sensing of Grassland Production and Management—A Review. *Remote Sens.* **2020**, *12*, 1949. [CrossRef]

8. Buchgraber, K.; Schaumberger, A.; Pötsch, E.M.; Krautzer, B.; Hopkins, A. Grassland Farming in Austria-Status Quo and Future Prospective. *Grassl. Sci. Eur.* **2011**, *16*, 95–156.
9. AgrarMarkt Austria (AMA). *Österreichisches Programm zur Förderung Einer Umweltgerechten, Extensiven und den Natürlichen Lebensraum Schützenden Landwirtschaft*; AgrarMarkt Austria: Vienna, Austria, 2021; p. 110.
10. Reinermann, S.; Gessner, U.; Asam, S.; Ullmann, T.; Schucknecht, A.; Kuenzer, C. Detection of Grassland Mowing Events for Germany by Combining Sentinel-1 and Sentinel-2 Time Series. *Remote Sens.* **2022**, *14*, 1647. [CrossRef]
11. Roy, D.P.; Li, J.; Zhang, H.K.; Yan, L.; Huang, H.; Li, Z. Examination of Sentinel-2A Multi-Spectral Instrument (MSI) Reflectance Anisotropy and the Suitability of a General Method to Normalize MSI Reflectance to Nadir BRDF Adjusted Reflectance. *Remote Sens. Environ.* **2017**, *199*, 25–38. [CrossRef]
12. European Space Agency (ESA). *S2 MPC: Level-2A Algorithm Theoretical Basis Document*; ESA: Paris, France, 2021.
13. Main-Knorn, M.; Pflug, B.; Louis, J.; Debaecker, V.; Müller-Wilm, U.; Gascon, F. Sen2Cor for Sentinel-2. In Proceedings of the Image and Signal Processing for Remote Sensing XXIII, Warsaw, Poland, 4 October 2017; Bruzzone, L., Ed.; SPIE: San Diego, CA, USA, 2017; Volume 10427, p. 1042704. [CrossRef]
14. Dechoz, C.; Poulain, V.; Massera, S.; Languille, F.; Greslou, D.; de Lussy, F.; Gaudel, A.; L'Helguen, C.; Picard, C.; Trémas, T. Sentinel 2 Global Reference Image. In Proceedings of the Image and Signal Processing for Remote Sensing XXI, Toulouse, France, 15 October 2015; Bruzzone, L., Ed.; SPIE: San Diego, CA, USA, 2015; Volume 9643, p. 96430A. [CrossRef]
15. Frantz, D.; Roder, A.; Udelhoven, T.; Schmidt, M. Enhancing the Detectability of Clouds and Their Shadows in Multitemporal Dryland Landsat Imagery: Extending Fmask. *Geosci. Remote Sens. Lett.* **2015**, *12*, 1242–1246. [CrossRef]
16. Minnaert, M. The Reflection of Light in Rippled Surfaces. *Physica* **1942**, *9*, 925–935. [CrossRef]
17. Gallaun, H.; Schardt, M.; Linser, S. Remote Sensing Based Forest Map of Austria and Derived Environmental Indicators. In Proceedings of the ForestSat Conference, Montpellier, France, 5–7 November 2007.
18. Schwieder, M.; Wesemeyer, M.; Frantz, D.; Pfoch, K.; Erasmi, S.; Pickert, J.; Nendel, C.; Hostert, P. Mapping Grassland Mowing Events across Germany Based on Combined Sentinel-2 and Landsat 8 Time Series. *Remote Sens. Environ.* **2022**, *269*, 112795. [CrossRef]
19. Sudmanns, M.; Tiede, D.; Augustin, H.; Lang, S. Assessing Global Sentinel-2 Coverage Dynamics and Data Availability for Operational Earth Observation (EO) Applications Using the EO-Compass. *Int. J. Digit. Earth* **2020**, *13*, 768–784. [CrossRef] [PubMed]
20. Lobert, F.; Holtgrave, A.-K.; Schwieder, M.; Pause, M.; Vogt, J.; Gocht, A.; Erasmi, S. Mowing Event Detection in Permanent Grasslands: Systematic Evaluation of Input Features from Sentinel-1, Sentinel-2, and Landsat 8 Time Series. *Remote Sens. Environ.* **2021**, *267*, 112751. [CrossRef]
21. Dujakovic, A.; Watzig, C.; Schaumberger, A.; Klingler, A.; Atzberger, C.; Vuolo, F. Enhancing Grassland Cut Detection Using Sentinel-2 Time Series through Integration of Sentinel-1 SAR and Weather Data. *Remote Sens. Appl. Soc. Environ.* **2025**, *37*, 101453. [CrossRef]
22. De Vroey, M.; Radoux, J.; Defourny, P. Grassland Mowing Detection Using Sentinel-1 Time Series: Potential and Limitations. *Remote Sens.* **2021**, *13*, 348. [CrossRef]
23. Tamm, T.; Zalite, K.; Voormansik, K.; Talgre, L. Relating Sentinel-1 Interferometric Coherence to Mowing Events on Grasslands. *Remote Sens.* **2016**, *8*, 802. [CrossRef]
24. Statistik Austria. *Bodennutzung 2020*; Statistik Austria: Vienna, Austria, 2020.
25. Schwieder, M.; Lobert, F.; Muhuri, A.; Oppelt, N.; Asam, S.; Reinermann, S.; Morel, J.; Rossi, M.; Weber, D.; Sarvia, F.; et al. Mowing Detection Intercomparison Exercise—MODCiX. In Proceedings of the Earsel Symposium, Bucharest, Romania, 3–6 July 2023.

remote sensing

MDPI

Article

Extraction of Agricultural Parcels Using Vector Contour Segmentation Network with Hybrid Backbone and Multiscale Edge Feature Extraction

Feiyu Teng, Ling Wu * and Shukuan Liu

School of Information Engineering, China University of Geosciences, Beijing 100083, China; 2004230029@email.cugb.edu.cn (F.T.); 2104230083@email.cugb.edu.cn (S.L.)
* Correspondence: wuling@cugb.edu.cn

Abstract: The accurate acquisition of agricultural parcels from remote sensing images is crucial for agricultural management and crop production monitoring. Most of the existing agricultural parcel extraction methods comprise semantic segmentation through remote sensing images, pixel-level classification, and then vectorized raster data. However, this approach faces challenges such as internal cavities, unclosed boundaries, and fuzzy edges, which hinder the accurate extraction of complete agricultural parcels. Therefore, this paper proposes a vector contour segmentation network based on the hybrid backbone and multi-scale edge feature extraction module (HEVNet). We use the extraction of vector polygons of agricultural parcels by predicting the location of contour points, which avoids the above problems that may occur when raster data is converted to vector data. Simultaneously, this paper proposes a hybrid backbone for feature extraction. A hybrid backbone combines the respective advantages of the Resnet and Transformer backbone networks to balance local features and global features in feature extraction. In addition, we propose a multi-scale edge feature extraction module, which can extract and enhance the edge features of different scales to prevent the possible loss of edge details in down sampling. This paper uses the datasets of Denmark, the Netherlands, iFLYTEK, and Hengyang in China to evaluate our model. The obtained IOU indexes were 67.92%, 81.35%, 78.02%, and 66.35%, which are higher than previous IOU indexes based on the optimal model (DBBANet). The results demonstrate that the proposed model significantly enhances the integrity and edge accuracy of agricultural parcel extraction.

Keywords: remote sensing; contour segmentation; agricultural parcels extraction; edge enhancement

1. Introduction

Farmland is a necessary resource to ensure agricultural development and food production. The quantity and quality of agricultural parcels are decisive factors in ensuring the sustainable development of agriculture and have a direct impact on the development of the national economy [1]. Against the background of accelerated global urbanization and population growth, the expansion of urban construction land has led to the reduction in agricultural land. It is particularly important for agricultural managers to obtain accurate agricultural information [2]. In the field of agricultural remote sensing, agricultural information extraction is a key research topic [3,4]. Carrying out information extraction of agricultural parcels is not only a long-term plan to ensure agricultural development, but also an urgent task to maintain the stability of cultivated land reserve resources [5].

Agricultural parcels generally refer to agricultural land with clear boundaries for the continuous planting of a crop. It is the basic unit for agricultural management, supervision, and statistics, and plays an important role in agricultural management and land use planning [6]. The extraction of agricultural parcels can be applied in various fields such as agricultural yield forecasting [7], farmland area investigation [8], land resource evaluation [9], land use management [10], and the sustainable development of resources and the environment [11,12].

Previous agricultural parcel division mainly depended on site measurements to obtain spatial distribution data and draw plot distribution maps [13]. However, this method often consumes a lot of manpower and time and is inefficient, so it is difficult to extract large-scale agricultural parcels [14]. With the continuous development of remote sensing technology, high-spatial resolution, spectral resolution, and multi temporal remote sensing images have been obtained by us. The extraction of agricultural parcels using remote sensing images has been widely used by researchers [15,16]. Remote sensing has the characteristics of real-time acquisition, rapid monitoring, and wide coverage. Using remote sensing images to automatically delimit agricultural parcels is a hot research topic [17,18].

The traditional remote sensing image extraction method of agricultural parcels is to use the relevant algorithm to classify the edge or region of agricultural parcels at the pixel level. For example, the Sobel or Canny filter operator is used to extract the edges of agricultural parcels in the image, and the extracted edge is used as the field boundary to divide farmland and non-farmland [19]. Conversely, the clustering threshold segmentation algorithm can be used to classify the pixels with similar characteristics into the same category to extract the agricultural parcels in the remote sensing image [20]. Although the above method is simple and easy to implement, it only considers the shallow features and cannot use the high-level semantic features of remote sensing images [21]. In the face of complex environmental backgrounds, the extraction accuracy is poor and it is difficult to extract complete agricultural parcels.

In recent years, with the development of deep learning, the method of extracting land cover categories using deep learning has been widely used, especially for the extraction of single-category features [22–25]. Deep learning semantic segmentation to extract agricultural parcels from remote sensing images has been used by many scholars. Compared with traditional machine learning and threshold segmentation, deep learning networks can extract high-dimensional semantic information and has stronger adaptability to complex and changeable terrain [26,27]. At first, some scholars used convolutional neural network to extract agricultural parcels from remote sensing images for semantic segmentation, such as u-net, FCN, segnet, and other networks [28–30]. Convolutional neural networks extract agricultural parcels by convolutionally checking the color, texture, and other features of the agricultural parcel. However, this is affected by the size of the convolutional kernel receptive field. For convolutional neural networks, it is difficult to effectively integrate the context information of the image, and its extraction ability is weak over a large area. In this regard, some scholars have proposed the Deeplabv3 convolution model [30], which uses hole convolution to expand the receptive field and obtain a wider range of influence characteristics. Some scholars use the Swin Transformer as the backbone network and the Transformer's self-attention mechanism to integrate image contexts [31]. However, in terms of the integration of global features and local details, it is difficult to extract the global features of the image using Deeplabv3, and the Swin transformer trunk has a poor ability to extract local information features. Therefore, integrating global features and local details remains a challenge in agricultural parcel extraction from remote sensing images.

In the process of semantic segmentation for extracting agricultural parcels, the effect of parcel edge segmentation is often poor. In the process of image feature extraction, with the sampling under the feature layer, the edge detail feature is often lost [32]. In order to solve this problem, some scholars have proposed a special edge detection algorithm [24], such as the RCF and HED [33,34] network models, which extract pixels at the edge of the parcel by capturing features at multiple scales. Demarcating agricultural parcels by extracting their edges is essentially the semantic segmentation of edge pixels. Therefore, some scholars have proposed multitask learning network models to extract both the region and edge of agricultural parcels, improving model accuracy by integrating multiple related tasks. For example, François et al. have proposed Resunet-a, a multitask network model that uses three tasks: extracting parcel boundary, extracting parcel area, and extracting distance to boundary. The addition of related tasks improves the shape and boundary of the extraction results [35]. In addition et al. have proposed the SEANet multitask model, which takes into account the uncertainty between different tasks on the basis of designing regional, edge, and distance tasks, and sets the multitask adaptive loss weight to balance the loss between different tasks [36]. Jiang et al. have proposed the BsiNet lightweight multitask network model. On the basis of completing the same task, the parameters of the model are less than those of Resunet-a and SEANet, so the calculation speed is faster and the extraction efficiency is higher [37]. Although the above methods can improve the accuracy of agricultural parcel edge extraction, they are all at the pixel level. This introduced problems such as the fact that the edges of the parcels will not be closed, the extracted adjacent parcel will be bonded, and there will be holes in the middle of the parcel, as well as fragmentation and salt and pepper-like phenomena. The pixel-based extraction method cannot reflect the topological relationship between agricultural parcels, so the agricultural parcels extracted by the model are difficult to be vectorized directly, and need to undergo post processing before mapping.

In the remote sensing community, contour segmentation is often used to extract targets. Jiao et al. have proposed a PolyR-CNN contour segmentation network model to extract polygonal buildings in remote sensing images [38]. Xu et al. proposed to integrate Mask RCNN and DeepSnake to improve the recognition ability of cultivated land [39]. Semantic segmentation extracts agricultural parcels from remote sensing images by first generating raster masks and then using a vectorization scheme to convert these masks into vector polygons of agricultural parcels, as shown in Figure 1. In contrast, contour segmentation generates a set of contour points from remote sensing image features, which are used to represent the vector polygon of the target. This is an object-based extraction method, as shown in Figure 1. For agricultural parcels, the boundary of agricultural parcels extracted by contour segmentation is closed, and can provide topological information between parcels. Therefore, the vector contour extraction is more in line with the actual situation than the raster pixel extraction. But, the existing contour extraction models, such as PolarMask [40] and Point Set [41], are unable to extract complete parcels for large areas. It cannot achieve good results in contour boundary fitting, therefore, contour segmentation is rarely used to extract agricultural parcels from remote sensing images.

To address the issues of voids, fragmentation, and unclosed boundaries in raster masks extracted from agricultural parcels using semantic segmentation, this paper proposes a vector contour prediction network based on a hybrid backbone and multiscale edge feature extraction module (HEVNet) for agricultural parcel extraction in remote sensing images. This solved the problems of poor extraction performance and fine edge refinement for agricultural parcels of different scales, and improved the geographic measurement accuracy of agricultural parcel delineation. The main contributions of this study are summarized as follows:

1. The hybrid backbone combining Resnet 50 and Swin Transformer is used to extract the texture features of agricultural parcels while taking into account the context in-

formation. It breaks through the limitation of the receptive field of the traditional convolutional network and enhances the recognition ability of the network for agricultural parcels of different sizes.

2. The multiscale edge feature extraction module is added to predict and calculate the loss of the parcel edge from the multiscale features, so as to supervise and restrict the prediction of the parcel edge and improve the accuracy of contour prediction.
3. We use the vector contour prediction module based on contour instance segmentation to extract the agricultural parcels in an object-based manner, and directly generate the vector polygon of the agricultural parcels to solve the problems of internal cavity and unclosed edge of the agricultural parcels.

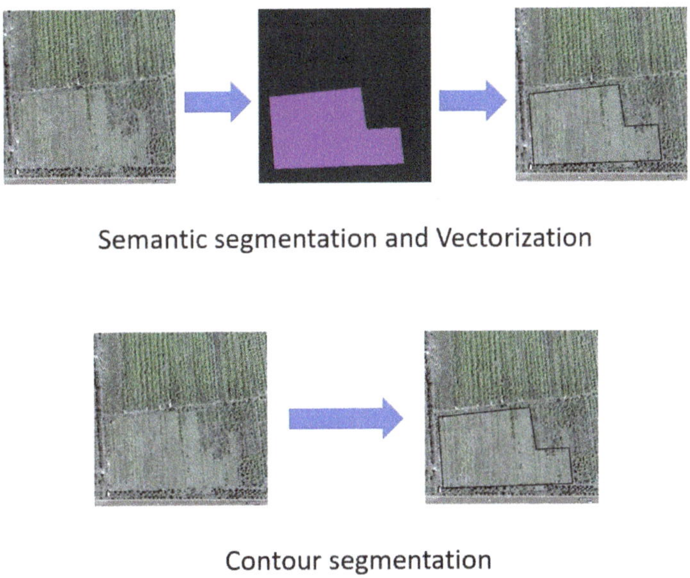

Figure 1. Semantic segmentation vectorization and direct vector extraction.

2. Materials and Methods

2.1. Research Region and Data

In this paper, we selected three public datasets: the cultivated land dataset in the Netherlands, the cultivated land dataset in Denmark, and a cultivated land dataset from the iFLYTEK challenge in 2021. In addition, we also selected Hengyang in Hunan Province as the research area, and outlined the cultivated land in the remote sensing images of Hengyang as the dataset. Data mapping and processing using ARCGIS 10.2. The overview of the four study areas is shown in Figure 2.

Netherlands data: The remote sensing image of the Netherlands is from a Sentinel-2 multispectral 10-m resolution remote sensing image downloaded from GEE. The image was collected in May 2020, including RGB and NIR bands, and the cloud content in the image accounted for less than 10%. Agricultural parcel vector labels are obtained from the key registered crop parcel dataset provided by Publieke Dienstverlening Op de Kaart (PDOK). The registered crop parcel dataset includes farmland, grassland, wasteland, natural areas, and other parcels. Here, we only extract farmland as the label of our dataset. The whole area is located in the northeast of the Netherlands, covering an area of about 15000 square kilometers. The mechanization mode is used in a large area of Dutch agriculture. The parcel is densely distributed, the shape is relatively regular, and there are obvious boundaries or obvious planting patterns.

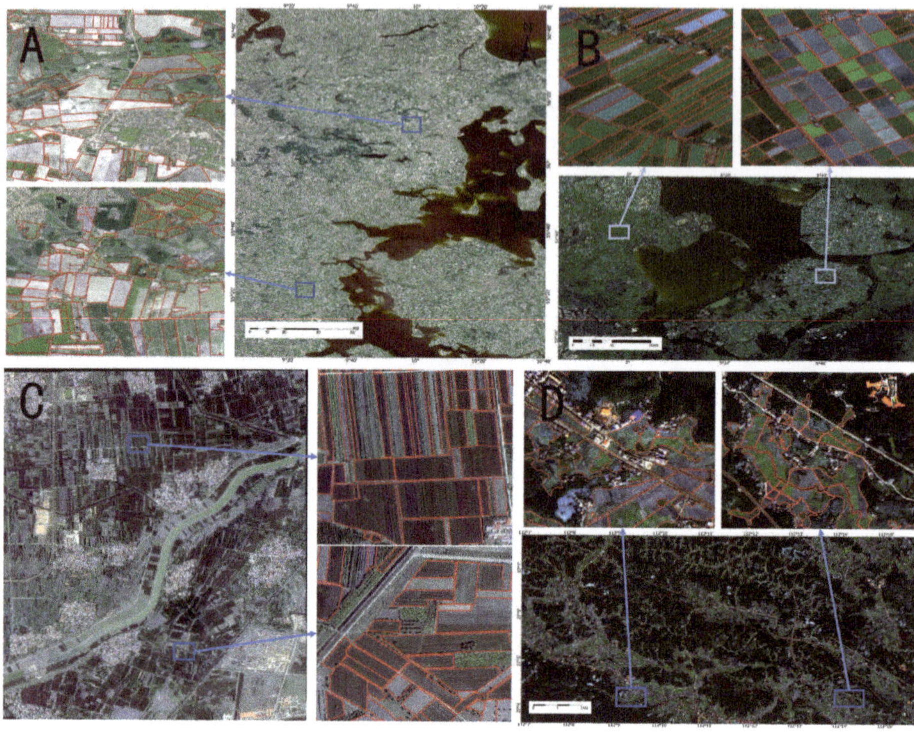

Figure 2. Overview and labeling diagram of the four study areas. (**A**) is Denmark; (**B**) is the Netherlands; (**C**) is iFLYTEK; and (**D**) is Hengyang.

Denmark data: The dataset of Denmark is from the European Union Land Parcel Identification System (LPIS), which includes two Sentinel-2 true-color synthetic images obtained on May 8th, 2016 and the polygon vector files of agricultural parcels manually created in the region. The remote sensing image has a resolution of 10 m and contains three bands of RGB. The whole area is located in the east of Denmark, covering an area of more than 23000 square kilometers. Agricultural parcels in Denmark are relatively densely distributed, but the area is different, and the shapes of parcels are diverse, with many irregular parcels.

iFLYTEK data: iFLYTEK dataset is from the iFLYTEK Challenge 2021 cultivated land extraction from high-resolution remote sensing image. The image data is composed of high-spatial resolution images obtained by the Jilin-1 satellite, with a total of 31 images, of which the spatial resolution is between 0.75 m and 1.1 m. Each image contains RGB and NIR bands. The parcels in the whole dataset are unevenly distributed in the image, but the shape is regular and the structure is simple, without obvious fragmentation, and the boundary between adjacent parcels is not obvious.

Hengyang data: We collected the remote sensing images obtained by the GF-2 satellite during the receiving season of Hengyang in 2023, highlighting the diversified farmland landscape in the region. The spatial resolution of the panchromatic band and the multi-spectral band (RGB and NIR) of the remote sensing images provided by the GF-2 satellite is 0.8m and 3.2m, respectively. The downloaded remote sensing image is calibrated by radiation and atmospheric correction, and the panchromatic sharpening method is used to fuse the multispectral band and panchromatic band in the image, and its spatial resolution is improved to 1 m. In the processed high-resolution remote sensing image, 1802 agricultural parcels in the study area were sketched by hand as sample labels. Under the influence of topography and climate, the farmland in Hengyang is seriously fragmented, the agricultural parcels are scattered, of different sizes, have irregular shapes, and have a disorderly distribution, and farmland boundaries are fuzzy.

2.2. Data Preprocessing

Two regions were randomly selected in the Netherlands and Denmark datasets as the training set and test set of their respective images. The first 26 images in the iFLYTEK dataset were used as the training set, and the last 5 images were used as the test set. Most of the western regions in Hengyang dataset are used as the training set of the model, and a small part of the eastern regions are used as the test set of the model. In order to make the experimental data acceptable to the model, the image and label data were sliced. The remote sensing image is cut into 512 × 512 slices by a sliding window. According to experience, the overlap rate between adjacent parcels is 25%, and then the sliced data is cleaned to filter out the slices with less than 20% of the farmland area. For the Hengyang data, due to the small sample size, the data were enhanced, and the sample size was amplified by horizontal and vertical flipping, mixing, and color jitter. The relevant information of the obtained data is shown in Table 1. The number of training validation sample slices obtained from the Netherlands, Denmark, iFLYTEK, and Hengyang datasets was 2287, 3658, 3744, and 1152, respectively, and the number of test set samples obtained was 1428, 1356, 566, and 365. Then, the samples of the training validation set are divided into training samples and validation samples according to the ratio of 4:1. At the same time, the vector file of the parcel annotation is sorted into two labels, raster mask and COCO format, for model reasoning and accuracy evaluation. Finally, we obtained four datasets, which are mutually independent. The model was trained, validated, and tested on each dataset separately.

Table 1. Information about the Netherlands, Denmark, iFLYTEK, and Hengyang datasets.

Region	Satellite	Resolution	Bands	Date	Train/Val/Test
The Netherlands	Sentinel-2	10	4	1 May 2020	1830/457/1428
Denmark	Sentinel-2	10	3	8 May 2016	2927/731/1356
iFLYTEK	JiLin-1	0.75–1.1	4	-	2995/749/566
HengYang	GF-2	1	4	8 July 2023	922/230/365

2.3. Methods

In this paper, we propose a vector contour prediction network based on a hybrid backbone and multiscale edge feature extraction module (HEVNet) to extract agricultural vector parcels from remote sensing images. The network model is composed of the following components: a hybrid backbone, a multiscale edge feature extraction module, and a vector contour prediction module. The overall structure is shown in Figure 3. The hybrid backbone module combines two feature extraction models, Resnet 50 and Swin Transformer, to extract multiscale features from remote sensing images and generate feature pyramids. The multiscale edge feature extraction module predicts the edge pixels of the agricultural parcels by obtaining the extracted information, and returns the edge enhancement information. The vector prediction module uses the method of instance segmentation, regarding the farmland as a polygon, and generates the vector polygon boundary box of the farmland through the regression and discrimination of the position of the contour points, so as to solve the problems of internal cavity and the unclosed phenomenon of the parcel boundary in the previous pixel-based prediction.

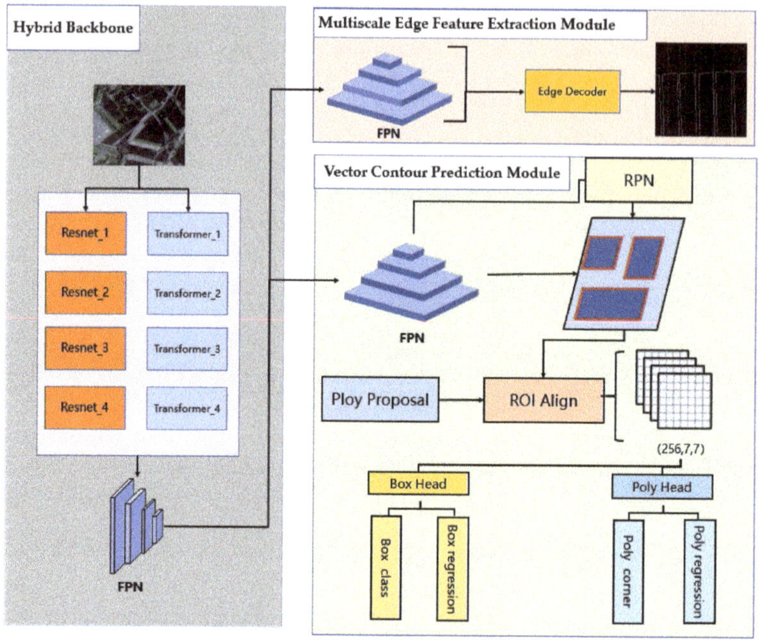

Figure 3. The overall structure of the model. The model consists of a hybrid backbone, multiscale edge feature extraction module, and vector contour prediction module. Hybrid backbone is used to extract image FPN. The multiscale edge feature extraction module predicts the edge of agricultural parcels at the pixel level based on the extracted features. The vector contour prediction module is responsible for predicting the boxes and poly points of agricultural parcels.

2.3.1. Hybrid Backbone

In the past, convolutional neural networks were usually used for the feature extraction of remote sensing images. Convolutional neural networks have a good effect on image local feature extraction, but due to the limitation of the convolutional kernel receptive field, it is difficult to recognize large-scale targets and obtain global information. In this regard, some scholars proposed the Transformer model with an attention mechanism as the core, and developed a Swin Transformer image feature extraction module similar to convolution feature extraction. The Swin Transformer feature extracts features efficiently through the attention mechanism of the window, breaks through the limitation of the convolution layer on the receptive field, and can use the self-attention mechanism to better capture the global context information. However, when dealing with fine-grained features, it may ignore local details, resulting in small local features that may be covered by global information.

According to the characteristics of convolutional networks with strong local information extraction and the Swin Transformer with strong global semantic information extraction, we use a feature extraction module that combines the convolutional network Resnet 50 and the Swin Transformer model, and propose a feature fusion module that can fuse the features of different scales extracted by the two in the channel and space. As shown in Figure 4, the two feature extraction modules extract features at different scales from remote sensing images, import the extracted image features at each level into the feature fusion module for feature fusion, and use the fused features as the input of the next level feature extraction module. The fused feature maps of different scales are used to form the feature pyramid of the remote sensing image.

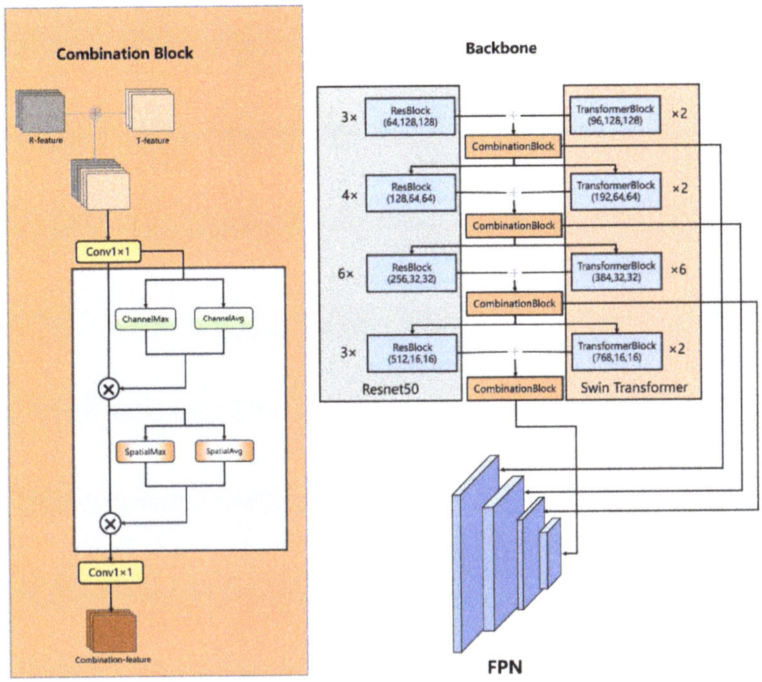

Figure 4. The hybrid backbone network is composed of Resnet and Transformer. On the right is the hybrid backbone structure. The features of different scales are fused and the feature pyramid (FPN) is output. On the left is the feature fusion module, showing the fusion of different scale features in the space and channel.

As shown in Figure 4, the feature fusion module references the previous CBAM attention mechanism and fuses image features from both spatial and channel perspectives [42]. Firstly, the feature images of different channel numbers are stitched, and the number of characteristic channels is adjusted through 1×1 convolution. Then, the channel and spatial position are weighted by means of average pooling, maximum pooling, global average pooling, and global maximum pooling, that is, spatial attention and channel attention. Finally, the fused feature layer is output through a 1×1 convolution layer.

2.3.2. Multiscale Edge Feature Extraction Module

The process of the continuous down sampling of image features will cause a loss of edge detail information. Therefore, we propose an edge prediction module to predict the edge pixels of agricultural parcels, and restrict the extraction of edge features by the hybrid backbone network. As shown in Figure 5, the edge decoder is connected to the four different scale features extracted from the hybrid backbone network. The decoder can obtain the probability prediction of the edges of agricultural parcels with different scales, and return the extracted edge features to the backbone network. Finally, the prediction results of each scale are superimposed and fused through the convolution network to obtain the final fused edge prediction map. Finally, the loss function of each scale edge prediction and the fused edge prediction loss function are calculated.

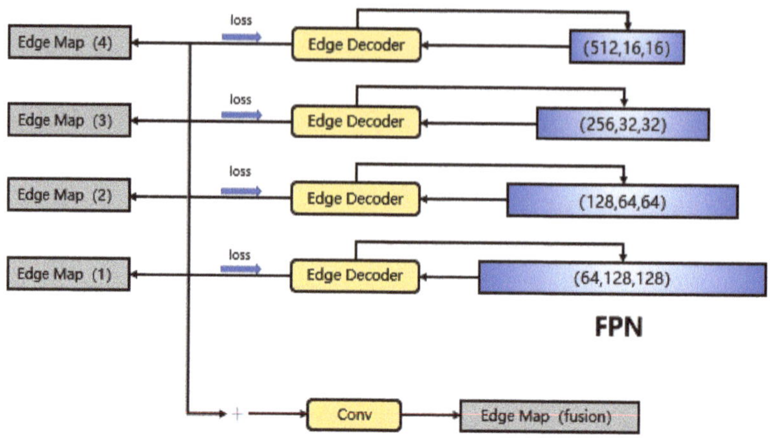

Figure 5. Different scale extraction of agricultural parcel edge module.

Edge decoder: In order to enhance the ability of the model to perceive the edge information in the feature graph, the edge perception module is added to integrate the gradient information in the feature graph. As shown in Figure 6, Sobel filters with four different directions are used for filtering. The edge of the agricultural parcels usually appear in the region with obvious difference in pixel values and fixed directions. Sobel filters can better capture the edge features of different directions in remote sensing images. The filtered image is stitched and fused through a 3 × 3 convolution network. The fused edge features are first returned to the backbone network, and then the fused edge feature image is reduced to one channel dimension through a 1 × 1 convolution operation. Finally, the image resolution is restored to the original image size using bilinear interpolation, and the prediction results of the agricultural parcels edges are obtained.

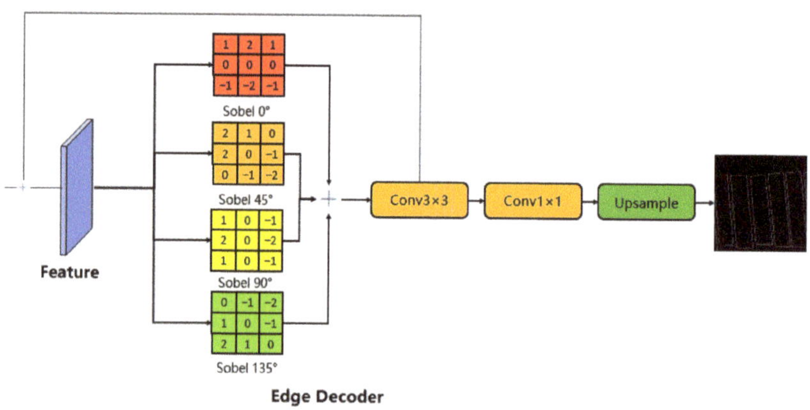

Figure 6. Agricultural parcel edge feature edge decoder.

Loss function: Referring to the HED edge detection loss function, we design the edge detection loss function in this paper [34]. For the prediction of agricultural parcel edges, the edge loss function is divided into the loss of the lateral output layer and the loss of the fusion layer. The loss of the lateral output layer is as follows:

$$L_{side} = \sum_{m=1}^{4} L_{side}^{m} \tag{1}$$

$$L_{side}^{m} = \begin{cases} -\beta \times \log(1 - \hat{y_E}) & y_E = 0 \\ -(1-\beta) \times \log \hat{y_E} & y_E = 1 \end{cases} \tag{2}$$

where L_{side}^{m} represents the loss of four different lateral output layers. y_E and $\hat{y_E}$ represent the sample label value and the predicted value, respectively. β is the weight parameter. Because the number of edge pixels and non-edge pixels in the edge detection task is extremely unbalanced, it is necessary to introduce a weight parameter to balance the weights of different categories, and its expression is as follows:

$$\beta = \frac{|Y_-|}{|Y_+ + Y_-|} \tag{3}$$

where Y_+ and Y_- represent the number of edge and non-edge pixels, respectively.

The loss of the fusion layer is calculated as follows:

$$L_{fusion} = \mathrm{Dist}(Y, \hat{Y}_{fusion}) \tag{4}$$

where Dist represents the cross-entropy loss function, and $\hat{Y}_{fusion}$ is the prediction result of fusing all side output layers through the activation function σ as follows:

$$\hat{Y}_{fusion} = \sigma(\sum_{m=1}^{4} A_{side}^{m}) \tag{5}$$

where A_{side}^{m} represents the result of each lateral output layer.

The final edge prediction loss function is the sum of the losses of the fusion layer and the lateral output layer as follows:

$$L_{Edge} = L_{fusion} + L_{side} \tag{6}$$

2.3.3. Vector Contour Prediction Module

In the past, the extraction of farmland blocks was essentially the pixel-level classification of remote sensing images, so the extracted farmland blocks would inevitably have the phenomenon of incomplete edges and empty parcels. In this regard, we propose a method to extract the parcel in the form of a vector, taking a single farmland area as an example, representing the contour of the target parcel with a polygon composed of the coordinate set of feature points, and extracting the agricultural parcel in an end-to-end manner. The contour prediction module proposed by us refers to the Fast R-CNN target detection model, and predicts the vector contour of the target object on the basis of generating the target detection box.

As shown in Figure 3, the contour prediction structure is similar to the Fast R-CNN structure [43]. Before vector contour prediction, target detection must be performed to generate a target detection box, and then the parcel contour is predicted in the target detection box. The remote sensing image is feature extracted through the previous hybrid backbone network, and edge features are enhanced through the multiscale edge feature extraction module. Then, the extracted feature map is put into the RPN (region proposal network) module to generate the extraction candidate box. According to the obtained candidate box, the features of the corresponding region are clipped in the feature map, and then the clipped part is imported into the ROI Align module. The ROI Align module is responsible for extracting the regional features in the candidate box and converting the regional features of different scales into the same scale size. The obtained regional features are transmitted to the full connection network and then transferred to the target box prediction module and contour prediction module. The target box prediction module generates the category and regression parameters of the target box, and the contour prediction module generates the confidence and regression parameters of the contour points.

The polygon representing the agricultural parcel is formed by a series of contour points. The prediction of agricultural parcel contour is essentially a task of identifying and regressing contour points. As shown in Figure 7, according to the target parcel boundary box obtained by target detection, points are generated by uniformly sampling along the surrounding boundaries of the target parcel boundary box as initial polygon contour points. According to the features in the target boundary box extracted by ROI Align, the offset of each initial contour point is obtained through the calculation of the convolution layer and the full connection layer, and the position of the contour points is adjusted. Each contour point is offset in the direction of the closest real contour point, and all contour points are regressed according to the offset between the original point and the real point, and then the redundant contour points are deleted through the confidence judgment of contour points, and finally a contour point set is obtained.

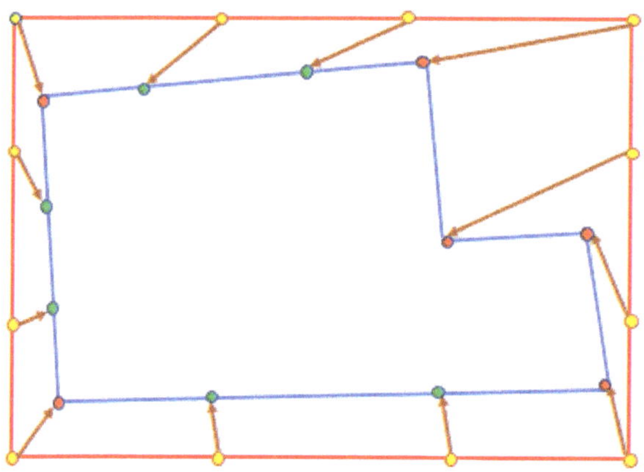

Figure 7. Schematic diagram of contour point position regression. The yellow dot is the prediction point, and the red dot and green dot are the corner and edge points of the real contour, respectively. The model first calculates the offset loss from the predicted point to the nearest real point, and then determines whether it is a corner point.

In this experiment, 64 contour points are preset for each polygon, and 64 points are also sampled in the annotation as the corresponding ones. According to the confidence of the contour points obtained, whether they are effective contour points is determined, and the vector polygons of agricultural parcel in remote sensing images can be obtained by connecting the effective contour points.

Loss function: For contour prediction, the loss function is divided into three parts, regional proposal network loss, detection box loss, and contour point loss. Each part of the loss is divided into classification loss and regression loss. The classification loss of RPN and the box is used to determine whether each anchor box contains targets, and the point classification loss is used to determine whether each point is a valid contour point. For the loss of classification and regression, the traditional Fast R-CNN loss function is used for calculation [43].The calculation formula of classified loss is

$$L_{cls} = -\frac{1}{N_{cls}}\sum_{i}\left[Y_p \log(\hat{Y}_p) + (1 - Y_p)\log(1 - \hat{Y}_p)\right] \quad (7)$$

where $\hat{Y_p}$ is the real label of the target frame (the foreground is 1 and the background is 0), Y_p is the foreground probability predicted by the model, and N_{cls} is the number of anchor frames participating in the training.

Regression loss is used to adjust the coordinates of the bounding box (or contour point) to make it closer to the true value. The calculation formula is as follows:

$$L_{reg} = \frac{1}{N_{reg}} \sum_i p_i^* \text{SmoothL}_1(t - t_i^*) \tag{8}$$

where p_i^* is the indicating function, which is 1 when the anchor box is a positive sample, t represents the predicted boundary box (contour point) coordinates, and t_i^* represents the real boundary box (contour point) coordinates. SmoothL_1 is used to calculate the offset between the predicted value and the real value.

The loss of vector contour prediction module is as follows:

$$L_{polygonal} = \lambda_1 L_{RPN\_cls} + \lambda_2 L_{RPN\_reg} + \lambda_3 L_{Box\_cls} + \lambda_4 L_{Box\_reg} + \lambda_5 L_{point\_cls} + \lambda_6 L_{point\_reg} \tag{9}$$

The classification loss and regression loss of RPN, box, and point are considered for contour prediction loss. λ is the weight of different losses. Referring to the settings of the weight of the loss function for instance segmentation in previous studies [38,39,43], and considering that this study focuses more on predicting contour points, we set the weight λ as 2, 2, 3, 3, 6, and 6 in this experiment.

The overall loss of this model: combined with the multiscale edge feature extraction module and the vector contour module, the overall loss function is constructed according to the proportion of 0.3 and 0.7.

$$\text{LOSS} = 0.7L_{polygonal} + 0.3L_{Edge} \tag{10}$$

2.4. Accuracy Evaluation

For the accuracy evaluation of the experimental results, in order to not only make a comparison with the traditional method, but also reflect the characteristics of vector extraction, we use two kinds of evaluation indexes for accuracy evaluation, the traditional semantic segmentation evaluation index and the evaluation index based on the contour object.

The semantic segmentation rating indicators include precision (P), recall (R), overall accuracy (ACC), intersection union ratio (IOU), and F1 score (F1). The calculation formula is as follows:

$$P = \frac{TP}{TP + FP} \tag{11}$$

$$R = \frac{TP}{TP + FN} \tag{12}$$

$$ACC = \frac{TP + TN}{TP + TN + FP + FN} \tag{13}$$

$$IoU = \frac{TP}{TP + FP + FN} \tag{14}$$

$$F1 = \frac{2 \times R \times P}{R + P} \tag{15}$$

where TP and TN are the areas correctly identified as agricultural parcels and non-agricultural parcels, while FN and FP are the number of pixels incorrectly identified as non-agricultural parcels and the area of agricultural parcels.

Object evaluation refers to taking the parcel as a single object to evaluate the geometric accuracy of the extracted results [38,44]. It includes over-classification error, under-classification error, and polygon similarity score (PoLiS).

The over-classification error and under-classification error [44] are used to indicate the excessive segmentation and insufficient segmentation of a single parcel object, as shown in the Figure 8. The smaller the error value, the better the segmentation effect.

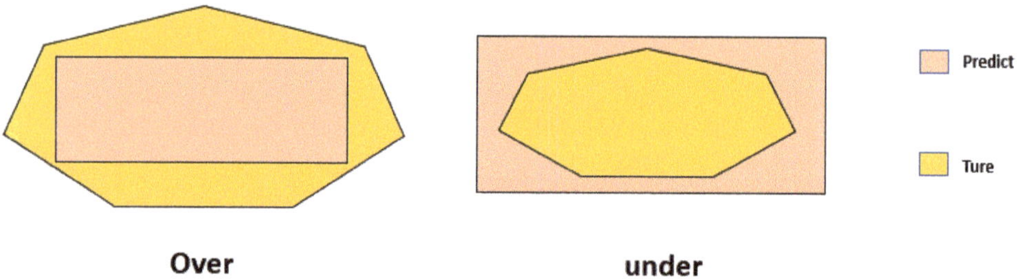

Figure 8. Schematic diagram of over-classification error and under-classification.

The calculation formula is as follows:

$$S_{over} = 1 - \frac{1}{n}\sum_{i}^{n}\frac{area(T_i \cap E_i)}{area(T_i)} \tag{16}$$

$$S_{under} = 1 - \frac{1}{n}\sum_{i}^{n}\frac{area(T_i \cap E_i)}{area(E_i)} \tag{17}$$

where T_i and E_i, respectively, represent the real value and predicted value of a single parcel. The over error and under error can be calculated by calculating the ratio of their overlapping area to their respective areas.

The polygon similarity score (PoLiS) is mainly used to measure the similarity between the extracted polygon contour and the real contour [38]. The smaller the value is, the closer the extraction result is to the real contour. The calculation formula is as follows:

$$PoLiS(X, Y) = \frac{1}{2M}\sum_{i}^{M}\min(x_i - y) + \frac{1}{2N}\sum_{i}^{N}\min(y_i - x) \tag{18}$$

where X represents the predicted fixed-point coordinate set, Y represents the real fixed-point coordinate set, and the formula represents the average distance between the predicted fixed-point coordinates and the nearest point of the real fixed-point coordinates.

3. Results

In this experiment, our proposed model and all the experiments involved are implemented in the pytorch 1.8 framework. The learning strategy is as follows: we use a stochastic gradient descent (SGD) optimizer with an initial learning rate of 0.001 and a weight attenuation of 0.00001, and the momentum is set to 0.9. The multisteplr method is used to adjust the learning rate within the set time interval. The batch size is configured as 8, and the model is trained for 100 rounds. All our experiments were conducted on an NVIDIA GeForce RTX 3090 GPU.

3.1. Ablation Experiment

In this experiment, we propose HEVNet, which includes a hybrid backbone module to extract remote sensing image features and a multiscale edge feature extraction module to constrain and enhance the edge features of agricultural parcels. In order to reflect the effectiveness of these two modules, we designed this ablation experiment. We use different backbone network models and delete the multiscale edge feature extraction module, and compare the agricultural parcel extraction results with the original model to verify the effect of the corresponding module on the extraction results.

Due to the large amount of data in the iFLYTEK dataset and the relatively regular agricultural parcels, it is conducive to vector contour extraction. Therefore, we verify the influence of different modules on the experimental results of the iFLYTEK dataset.

3.1.1. Influence of Different Backbones on the Results

The backbone network plays an important role in the process of extracting remote sensing image features. In this section, we use Resnet 50, Swin Transformer, and our proposed hybrid backbone of Resnet 50 and Swin Transformer feature fusion as the model backbone to extract agricultural fields in the iFLYTEK dataset. The results are shown in Table 2. From the evaluation index F_1 and IOU, it can be seen that the hybrid backbone network has the best result, and the F_1 score is 2.74% and 6.76% higher than the Resnet 50 and Swin Transformer backbones, respectively. The IOU value was 2.74% and 6.76% higher than the Resnet 50 and Swin Transformer trunks. From the perspective of object-based indicators, the over error and under score error of the hybrid backbone are less than Resnet 50, and the over error is slightly higher than that of the Swin Transformer, but the under score error has decreased significantly. This shows that the feature extraction performance of the Resnet 50 and Swin Transformer trunks is improved by the proposed hybrid trunk, especially for Swin Transformer.

Table 2. Evaluation results in iFLYTEK dataset using different trunks and whether to add multiscale edge feature extraction module. "Backbone" in the table represents different backbone types used; "Edge Predict" indicates whether the multiscale edge feature extraction module has been added to the model.

Data	Backbone	Edge Predict	P	R	ACC	F_1	IOU	S_{over}	S_{under}	PoLiS
iFLYTEK	Resnet50	√	84.54	90.62	85.25	87.47	75.28	39.24	16.51	1.47
	Swin Transformer	√	79.54	88.80	82.25	83.92	71.26	**38.94**	17.24	1.44
	Resnet50+ Swin Transformer		83.54	**92.80**	**88.25**	87.93	77.52	41.24	18.12	1.69
	Resnet50+ Swin Transformer	√	**86.54**	90.80	87.25	**88.62**	**78.02**	39.13	**14.12**	**1.38**

3.1.2. Influence of Multiscale Edge Feature Extraction Module on the Results

The edge feature extraction module plays a role in constraining and enhancing the edge information features of agricultural parcels at different scales in the model, and returns the extracted features to the backbone network. From the results obtained in Table 2, it can be seen that with the addition of the edge feature extraction module, the F_1 score and IOU index increased slightly, while the error in the object-based indicators decreased more significantly, in which the under score error and polis decreased by 4.00% and 0.31, respectively. It shows that the edge feature extraction module is helpful in improving the accuracy of parcel division and parcel boundary identification.

3.1.3. Visual Display of Extraction Results

Figure 9 shows the results of our model using different backbone networks. As can be seen from the figure, compared with our hybrid backbone, the Resnet 50 trunk model often has missing points, which cannot recognize the slender agricultural parcels, and for large farmland, it divides the area into several small blocks, which cannot guarantee the integrity of farmland extraction, as shown in the blue box in Figure 9. This is because the Resnet 50 trunk is affected by the receptive field and has an insufficient learning ability for a large range of irregular-shaped features. Compared with our hybrid trunk, Swin Transform will mistakenly identify wasteland and grassland as farmland and miss small agricultural parcels, as shown in the yellow box in Figure 9. This is because the Swin Transform trunk pays attention to the connection between the global vision and context, and ignores the texture features of agricultural parcels, resulting in the misjudgment of parcels.

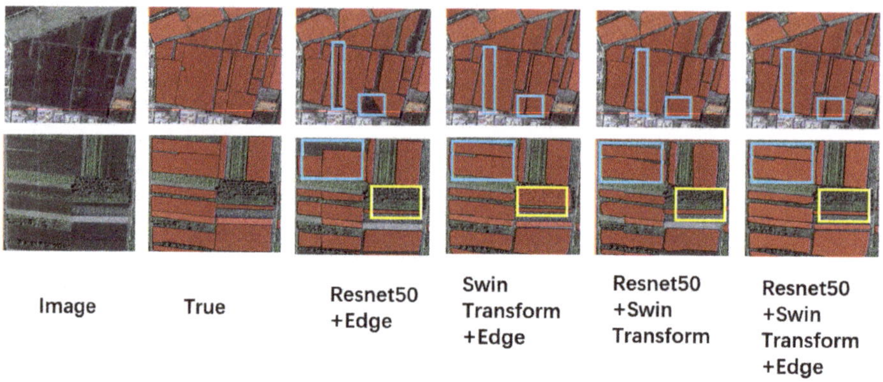

Figure 9. Results of extracting agricultural parcels in iFLYTEK dataset using different backbone and with/without the addition of the multiscale edge feature extraction module. The blue border indicates the parcel is incomplete, and the yellow border indicates the wrong parcel is extracted.

Figure 9 also shows the extraction results of adding the edge feature extraction module and not adding the model. As can be seen from the figure, compared with the model with the edge feature extraction module, the edge of the agricultural parcels extracted by the model without the added model is more tortuous (far from the edge position of the real parcel) and the shape of the parcel is irregular, so there is the phenomenon of missing the detection of small parcels. This is because the edge details of the image are lost in the process of down sampling, and the edge feature extraction module can constrain and integrate the edge features of the parcel at different scales, which improves the accuracy of the parcel edge.

3.2. Comparative Experiment

In order to verify the effectiveness of our proposed HEVNet, we selected the latest four semantic segmentation extraction models and one contour segmentation extraction model as the comparison method. The four semantic segmentation models are ResUNet-a [35], BisNet [37], SEANet [36], and DBBANet [45]. Among them, ResUNet-a is the first multitask learning network in the field of agricultural parcel extraction; the BisNet algorithm is based on the optimization and improvement of ResUNet-a; the SEANet algorithm integrates semantic edge perception into farmland extraction; and the DBBANet algorithm integrates spatial awareness and boundary awareness. Ploy r-CNN [38] is selected as the contour segmentation extraction model. The Ploy-r-CNN network was previously applied to extract the contour of buildings. Here we use it to extract agricultural parcels. The extraction result of the semantic segmentation model is in the form of a raster, and the contour segmentation result is in the form of a vector. In order to compare the accuracy between the vector extraction result and the raster extraction result, we convert the vector extraction result into a raster, and then use the semantic segmentation evaluation index to evaluate the accuracy.

Our model and the five control models were trained and tested on four different datasets. The parcels in the Netherlands and the iFLYTEK datasets are relatively regular, while the parcels in the Denmark and the Hengyang datasets are relatively fragmented, and the extraction effect may be different for different types of parcels.

3.2.1. Experimental Results and Analysis of Regular Parcels in Netherlands and iFLYTEK

The results of model testing on the Netherlands and the iFLYTEK dataset are shown in Table 3. Compared with the semantic segmentation model, the recall rate of our HEVNet model is relatively low, but the precision rate is significantly improved, which shows that the semantic segmentation model often misjudges non-agricultural parcels as agricultural parcels, while our HEVNet model has fewer problems in this regard. On the whole, the key evaluation indicators of our model are better than the latest semantic segmentation model, which is embodied in the F_1 score and IOU index. In the iFLYTEK dataset, the F_1 score of HEVNet was 2.09% higher than that of the DBBANet model, and the IOU index was 0.94% higher than that of the DBBANet model. In the Netherlands dataset, the F_1 score of HEVNet was 0.27% higher than that of the ResUNet-a model, and the IOU index was 1.32% higher than that of the DBBANet model. Compared with Ploy-r-CNN, our HEVNet has higher F_1 and IOU indexes by 2.33% and 1.70% in the iFLYTEK dataset and 1.82% and 3.00% in the Netherlands dataset. The F_1 score and IOU index are higher than the current optimal model.

Table 3. Evaluation results of different models in the regular parcels in the Netherlands and iFLYTEK datasets.

Data	Model	P	R	ACC	F_1	IOU
iFLYTEK	ResUNet-a	78.69	91.02	85.36	87.32	76.65
	BisNet	79.25	89.12	84.96	86.02	74.05
	SEANet	81.46	88.34	83.24	87.39	72.24
	DBBANet	80.25	**92.36**	**87.56**	88.34	77.08
	PloyR-cnn	83.56	88.62	87.35	86.01	76.32
	Ours	**86.54**	90.80	87.25	**90.43**	**78.02**
Netherlands	ResUNet-a	80.04	**92.77**	85.85	89.52	78.02
	BisNet	83.12	90.35	88.96	87.28	74.05
	SEANet	83.65	91.69	**90.24**	88.22	76.24
	DBBANet	82.02	90.45	89.52	86.50	80.03
	PloyR-cnn	86.36	89.65	87.25	87.97	78.35
	Ours	**88.35**	90.22	89.20	**89.79**	**81.35**

As shown in Figures 10 and 11, the visual images of the extraction results of agricultural parcels by our model and other models are displayed to better compare the difference between the vector contour extraction method and the raster extraction method. As can be seen from Figure 10, although the previous semantic segmentation based on raster extraction can also identify large areas of agricultural parcels, it is difficult to divide densely arranged parcels and cannot clearly show the boundaries of agricultural parcels. As a result, the boundaries of the parcel are blurred, the edges of the parcel are not closed (as shown in the red box in the figure), and the adjacent parcel is bonded. Neatly arranged parcels cannot be separated separately, and a whole large parcel often appears, which cannot reflect the real form of parcels. As can be seen in Figure 11, the raster extraction method can miss detecting parcels, resulting in empty parcels (as shown in the blue box), and wasteland can be wrongly detected as farmland (as shown in the yellow box). The HEVNet model does not have these problems. The extraction method based on contour regards each parcel as a separate object and describes the spatial position of the parcel by contour points, so there will be no interruption and lack of closure of the parcel boundary. Compared with Ploy-r-CNN, which is the same contour extraction model, our model has less missed detections and false detections. It can also be seen from the figure that the description of the edge of the parcel in our model is closer to the true value, and the spacing between adjacent parcels is more obvious, which can clearly express the boundary of adjacent parcels.

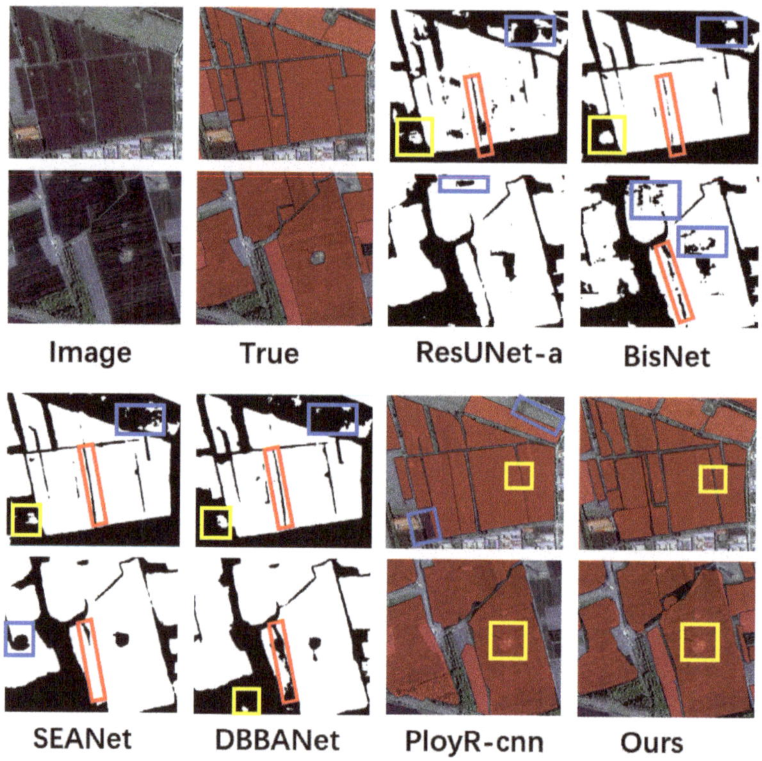

Figure 10. Prediction results of agricultural parcels in the iFLYTEK dataset.

3.2.2. Experimental Results and Analysis of Irregular Parcels in Denmark and Hengyang

The results of the model tests on the Denmark and Hengyang datasets are shown in Table 4. In the extraction of irregular parcels, our HEVNet model still has a high precision and low recall rate compared with the semantic segmentation model. In the Danish dataset, the HEVNet model achieved the highest precision, the highest accuracy, and the highest F_1 index, but in terms of IOU index, our model ranked second, which was 0.33% lower than the highest model, ResUNet-a. In the Hengyang dataset, our model obtained the highest precision, the highest accuracy, and the highest IOU index, which were 3.99%, 1.41%, and 1.2% higher than those of the second highest ResUNet-a model, respectively, but the F_1 index was lower than that of the ResUNet-a model, which was 3.3% lower. Compared with PloyR-cnn, the F_1 and IOU indexes of our model are 2.52% and 2.58% higher in the Denmark dataset, and 1.17% and 1.39% higher in the Netherlands dataset. This result shows that our model has an improved performance compared with the previous models in the process of extracting irregular parcels.

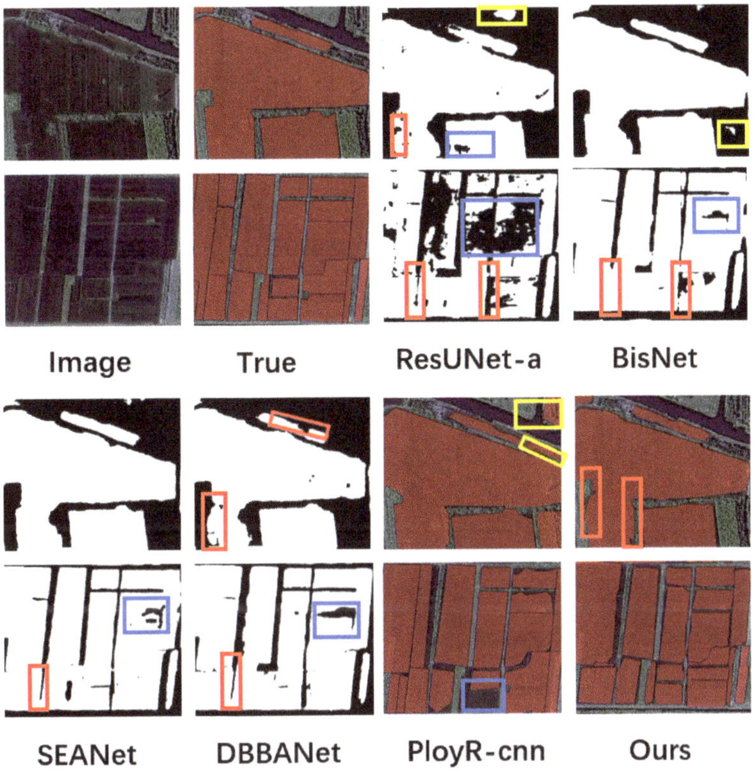

Figure 11. Prediction results of agricultural parcels in the Netherlands dataset.

Table 4. Evaluation results of different models in the irregular parcels of Denmark and Hengyang datasets.

Data	Model	P	R	ACC	F_1	IOU
Denmark	ResUNet-a	65.24	**81.35**	74.63	72.41	**68.25**
	BisNet	61.23	79.18	69.75	69.06	60.36
	SEANet	62.38	78.29	70.32	69.44	62.54
	DBBANet	63.29	78.36	72.14	70.02	61.76
	PloyR-cnn	67.03	73.52	75.12	70.12	65.34
	Ours	**69.76**	75.76	**77.25**	**72.64**	67.92
HengYang	ResUNet-a	68.36	**82.56**	79.54	**74.79**	65.15
	BisNet	59.85	78.96	68.96	68.09	62.45
	SEANet	61.96	75.45	72.75	68.04	59.36
	DBBANet	61.45	75.96	74.14	67.94	60.85
	PloyR-cnn	71.45	69.24	78.63	70.32	64.96
	Ours	**72.35**	70.65	**80.95**	71.49	**66.35**

Figures 12 and 13 show the extraction results of farmland in Denmark and Hengyang. From the results in Figure 12, in the Danish dataset, when extracting irregular and fragmented farmland based on raster extraction, there will be voids in the middle of the parcel (as shown in the blue box), and the edges between adjacent parcels cannot be separated (as shown in the red box), therefore, the complete agricultural parcel cannot be extracted.

Although the extraction method based on vectors also has deviations in the division of agricultural parcel edges, the extracted agricultural parcel is very complete and there is an obvious division between agricultural parcels. In the Hengyang dataset in Figure 13, the raster extraction method misjudges idle wasteland as farmland (shown in the yellow box), while the contour extraction model does not have such a phenomenon. In large and dense farmland, the vector-based model can adequately show the roads between farmland, but the raster extraction method cannot. Although there may be misjudgments in contour extraction, generally speaking, the results based on vector contour extraction are better than those based on raster extraction in terms of parcel integrity and edge thinning.

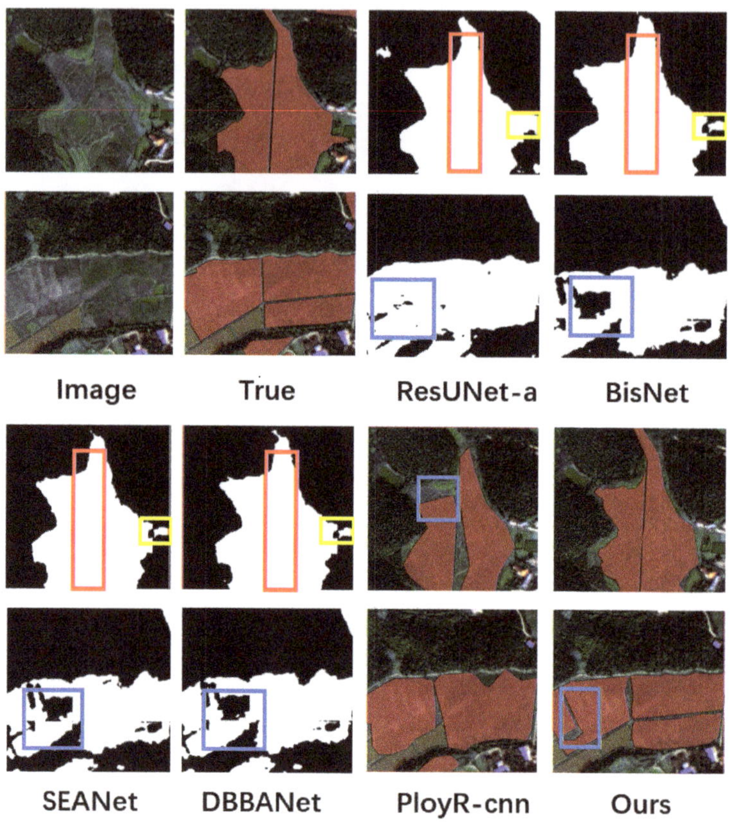

Figure 12. Prediction results of agricultural parcels in the Denmark dataset.

The extraction results of our model vary greatly in different agricultural environments. In the regular parcels of the Netherlands and the iFLYTEK datasets, our model's IOU index is 81.35% and 78.02%, and the F_1 scores are 89.79% and 90.43%. In the irregular parcel datasets of Denmark and Hengyang, the IOU index is 67.92% and 66.35%, and the F_1 scores are 72.64% and 71.49%. We can see that our model performs much better in extracting regular parcels than in extracting irregular parcels. From the comparison of the extraction results with existing models, our model has a more prominent advantage in extracting regular parcels, indicating that our model is more suitable for extracting regular agricultural parcels.

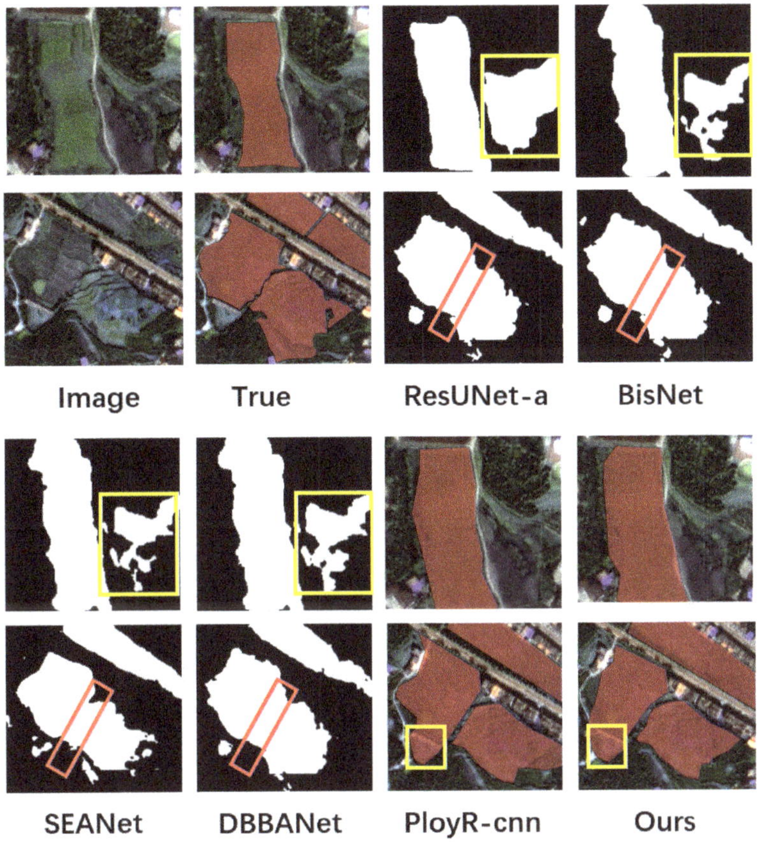

Figure 13. Prediction results of agricultural parcels in the Hengyang dataset.

4. Discussion

This study's key innovation was exceeding the previous prediction of agricultural parcels based on the semantic segmentation of raster data and the use of object-based vector contour prediction to extract agricultural parcels. At the same time, a hybrid backbone combining Resnet and Transform is proposed, which combines the features extracted from the two networks at different scales with the spatial and channel attention mechanisms, so that it cannot only contact the global context information, but also take into account the local details. Additionally, a multiscale edge feature extraction module is proposed to solve the problem of the fuzzy edges of agricultural parcels. By extracting and constraining the edges and shapes of agricultural parcels at different scales, the accuracy of the edge extraction of agricultural parcels is improved.

In terms of agricultural parcel extraction, our model has special advantages compared with previous raster-based extraction methods [35,36,45]. Previous methods aimed to improve the accuracy of agricultural parcel extraction by adding attention mechanisms [45], edge refinement [36], and multitask extraction [35]. However, due to the limitations of raster pixels, it was difficult to extract complete agricultural parcels. Our model extracts agricultural parcels in the form of objects, which is more efficient than raster-based methods. Compared with the raster-based approach, our model has a higher accuracy than the previous models in extracting agricultural parcels. The extracted parcels are more complete, the boundaries of the parcels are clearer, and the isolated segmentation is reduced, which shows a good effect in solving the hole phenomenon in the middle of large agricultural parcels. However, at the same time, our model has a low recall rate, which may result in missed detections for small agricultural parcels, and the recall rate of agricultural parcels is not as good as previous raster extraction methods.

Although the HEVNet model in this paper shows a good performance in extracting complete agricultural parcels, it still has some limitations. First of all, because the vector contour prediction model in this paper generates vector polygons by generating contour points, it can only predict closed agricultural parcels. For some farmland with houses and ponds inside, the agricultural parcels themselves have holes, so the model in this paper cannot predict their inner contours (such as Figure 10). The method of generating a vector image containing voids to represent the corresponding shape of the parcel is an improvement in the direction of this model.

Secondly, the contour prediction points generated by our model represent the polygon of agricultural parcels. The number of contour points is usually set quite high to cover most farmland plots, but this may cause contour point redundancy, which will cause the distortion of the predicted polygon of agricultural parcels [38] (such as Figure 11). The method for reducing the redundant contour points of simple-shaped agricultural parcels while extracting complex-shaped agricultural parcels remains to be optimized.

In addition, we find that the HEVNet model has a poor fitting effect on the shape of fragmented agricultural parcels, and there can be the problem of mistakenly dividing one parcel into multiple parcels (such as Figure 12). This may be because the edge features of scattered agricultural parcels are complex, and it is difficult for the multiscale edge feature extraction module to understand their features at a deeper level, thus similar texture features can be misjudged as agricultural parcel boundaries [26]. In the future, we can improve the fitting effect of agricultural parcels by optimizing edge feature extraction modules [36] and multitasking collaborative networks [35].

In general, although our model has some problems in extracting some special-shaped parcels, it has achieved remarkable results in solving the problems of agricultural parcel cavity, edge blur, and so on compared with the previous raster extraction model. In the future, we can conduct relevant research to optimize the extraction methods for the edges of small agricultural parcels and solve the problem of contour boundary extraction for special-shaped agricultural parcels, which can be used to extract fragmented agricultural parcels in special terrains.

5. Conclusions

This paper proposes a HEVNet model based on a hybrid backbone network and multiscale edge feature extraction, which is used to extract the vector polygon of agricultural parcels in remote sensing images. In this model, we propose a hybrid trunk module and a multiscale edge prediction module to assist in agricultural parcel contour prediction. By integrating features from the ResNet 50 and Transformer modules, the hybrid backbone module fully extracts the texture features of agricultural parcels, overcomes the limitations of traditional convolutional receptive fields, and improves the extraction of parcels of varying sizes. The multiscale edge feature extraction module identifies the edges of agricultural parcels on different scale features, and designs a multiscale edge loss function to restrict the extraction of edge features, reducing the loss caused by edge features in the down sampling process, so as to obtain agricultural parcels with clearer edges. In this aspect of mapping, we use the method of directly generating contour points to generate an agricultural parcel vector map, instead of the previous method of first using raster extraction and then using raster to vector mapping. We conducted experiments using datasets from Denmark, the Netherlands, iFLYTEK, and Hengyang, China. From the results, our model achieved IOU indices of 67.92%, 81.35%, 78.02%, and 66.35% on the four datasets, and F_1 scores of 72.64%,89.79%, 90.43%, and 71.49%, respectively. The evaluation metrics are mostly higher than those of the previous raster models (ResUNet-a, BisNet, SEANet, and DBBANet) and the Ployr-cnn model in terms of IOU and F_1 index. The effect extracted from regular parcels is obviously higher than that extracted from irregular parcels. From the visual map, the agricultural parcels extracted by our model are more complete and the boundary is more accurate and clear. Compared with raster-based extraction, vector-based extraction focuses more on the geometric information of agricultural parcels, effectively addressing the issues such as empty parcels, unclosed edges, and blurred boundaries that were present in previous extraction methods.

Author Contributions: Conceptualization, F.T. and L.W.; methodology, F.T.; validation, F.T. and L.W.; formal analysis, F.T. and S.L.; data curation, F.T. and S.L.; writing—original draft preparation, F.T. and S.L.; writing—review and editing, F.T. and L.W.; visualization, F.T.; project administration, L.W.; funding acquisition, L.W. All authors have read and agreed to the published version of the manuscript.

Funding: This research received no external funding.

Data Availability Statement: All code and data will be available from the authors upon request.

Acknowledgments: The authors wish to thank the anonymous reviewers for their constructive comments that helped improve the scholarly quality of the paper. This work was supported by the High-performance Computing Platform of China University of Geosciences Beijing.

Conflicts of Interest: The authors declare no conflicts of interest.

References

1. Huang, X.; Reba, M.; Coffin, A. Cropland mapping with L-band UAVSAR and development of NISAR products. *Remote Sens. Environ.* **2021**, *253*, 112180. [CrossRef]
2. Wang, L.; Wei, F.; Svenning, J.C. Accelerated cropland expansion into high integrity forests and protected areas globally in the 21st century. *Iscience* **2023**, *26*, 106450. [CrossRef]
3. Phalke, A.R.; Özdoğan, M.; Thenkabail, P.S. Mapping croplands of Europe, middle east, russia, and central asia using landsat, random forest, and google earth engine. *ISPRS J. Photogramm. Remote Sens.* **2020**, *167*, 104–122. [CrossRef]
4. Wu, J.; Peng, D. Tree-crown information extraction of farmland returned to forests using QuickBird image based on object-oriented approach. *Spectrosc. Spectr. Anal.* **2010**, *30*, 2533–2536.
5. Yin, H.; Brandão, A., Jr.; Buchner, J. Monitoring cropland abandonment with Landsat time series. *Remote Sens. Environ.* **2020**, *246*, 111873. [CrossRef]
6. Shafi, U.; Mumtaz, R.; García-Nieto, J. Precision agriculture techniques and practices: From considerations to applications. *Sensors* **2019**, *19*, 3796. [CrossRef]
7. Carfagna, E.; Gallego, F.J. Using remote sensing for agricultural statistics. *Int. Stat. Rev.* **2005**, *73*, 389–404. [CrossRef]
8. Yang, M.D.; Huang, K.S.; Kuo, Y.H. Spatial and spectral hybrid image classification for rice lodging assessment through UAV imagery. *Remote Sens.* **2017**, *9*, 583. [CrossRef]
9. Zhou, M.; Kuang, B.; Zhou, M. The spatial and temporal evolution of the coordination degree in regard to farmland transfer and cultivated land green utilization efficiency in China. *Int. J. Environ. Res. Public Health* **2022**, *19*, 10208. [CrossRef]
10. Li, Z.; He, W.; Li, J. Learning without exact guidance: Updating large-scale high-resolution land cover maps from low-resolution historical labels. In Proceedings of the IEEE/CVF Conference on Computer Vision and Pattern Recognition, Seattle, WA, USA, 17–21 June 2024; pp. 27717–27727.
11. Lv, Y.; Zhang, C.; Ma, J. Sustainability assessment of smallholder farmland systems: Healthy farmland system assessment framework. *Sustainability* **2019**, *11*, 4525. [CrossRef]
12. Belgiu, M.; Csillik, O. Sentinel-2 cropland mapping using pixel-based and object-based time-weighted dynamic time warping analysis. *Remote Sens. Environ.* **2018**, *204*, 509–523. [CrossRef]
13. Xu, L.; Yang, P.; Yu, J. Extraction of cropland field parcels with high resolution remote sensing using multi-task learning. *Eur. J. Remote Sens.* **2023**, *56*, 2181874. [CrossRef]
14. Chang, B.; Wang, J.; Luo, Y. Cultivated land extraction based on GF-1/WFV remote sensing in Shenwu irrigation area of Hetao Irrigation District. *Trans. Chin. Soc. Agric. Eng.* **2017**, *33*, 188–195.
15. Fritz, S.; See, L.; McCallum, I. Mapping global cropland and field size. *Glob. Change Biol.* **2015**, *21*, 1980–1992. [CrossRef] [PubMed]

16. Yang, H.; Sun, H.; Wang, K. Enhanced Farmland Extraction from Gaofen-2: Multi-Scale Segmentation, SVM Integration, and Multi-Temporal Analysis. *Agriculture* **2025**, *15*, 1073. [CrossRef]
17. Volpi, M.; Tuia, D. Deep multi-task learning for a geographically-regularized semantic segmentation of aerial images. *ISPRS J. Photogramm. Remote Sens.* **2018**, *144*, 48–60. [CrossRef]
18. Wagner, M.P.; Oppelt, N. Deep learning and adaptive graph-based growing contours for agricultural field extraction. *Remote Sens.* **2020**, *12*, 1990. [CrossRef]
19. Hong, R.; Park, J.; Jang, S. Development of a parcel-level land boundary extraction algorithm for aerial imagery of regularly arranged agricultural areas. *Remote Sens.* **2021**, *13*, 1167. [CrossRef]
20. Soille, P.J.; Ansoult, M.M. Automated basin delineation from digital elevation models using mathematical morphology. *Signal Process.* **1990**, *20*, 171–182. [CrossRef]
21. Zhao, X.; Zhu, L.; Wu, B. An improved mayfly algorithm based on Kapur entropy for multilevel thresholding color image segmentation. *J. Intell. Fuzzy Syst.* **2023**, *44*, 365–380. [CrossRef]
22. Xu, Z.; Liu, Y.; Gan, L. Rngdet: Road network graph detection by transformer in aerial images. *IEEE Trans. Geosci. Remote Sens.* **2022**, *60*, 4707612. [CrossRef]
23. Almeida, T.; Lourenço, B.; Santos, V. Road detection based on simultaneous deep learning approaches. *Robot. Auton. Syst.* **2020**, *133*, 103605. [CrossRef]
24. Marmanis, D.; Schindler, K.; Wegner, J.D. Classification with an edge: Improving semantic image segmentation with boundary detection. *ISPRS J. Photogramm. Remote Sens.* **2018**, *135*, 158–172. [CrossRef]
25. Mei, J.; Li, R.J.; Gao, W.; Cheng, M.M. CoANet: Connectivity attention network for road extraction from satellite imagery. *IEEE Trans. Image Process.* **2021**, *30*, 8540–8552. [CrossRef] [PubMed]
26. Liu, S.; Cao, S.; Lu, X. Lightweight Deep Learning Model, ConvNeXt-U: An Improved U-Net Network for Extracting Cropland in Complex Landscapes from Gaofen-2 Images. *Sensors* **2025**, *25*, 261. [CrossRef] [PubMed]
27. Mei, W.; Wang, H.; Fouhey, D. Using deep learning and very-high-resolution imagery to map smallholder field boundaries. *Remote Sens.* **2022**, *14*, 3046. [CrossRef]
28. Aung, H.L.; Uzkent, B.; Burke, M. Farm parcel delineation using spatio-temporal convolutional networks. In Proceedings of the IEEE/CVF Conference on Computer Vision and Pattern Recognition Workshops, Seattle, WA, USA, 14–19 June 2020; pp. 76–77.
29. Wang, M.; Wang, J.; Cui, Y. Agricultural field boundary delineation with satellite image segmentation for high-resolution crop mapping: A case study of rice paddy. *Agronomy* **2022**, *12*, 2342. [CrossRef]
30. Badrinarayanan, V.; Kendall, A.; Cipolla, R. Segnet: A deep convolutional encoder-decoder architecture for image segmentation. *IEEE Trans. Pattern Anal. Mach. Intell.* **2017**, *39*, 2481–2495. [CrossRef]
31. Zhong, B.; Wei, T.; Luo, X. Multi-swin mask transformer for instance segmentation of agricultural field extraction. *Remote Sens.* **2023**, *15*, 549. [CrossRef]
32. Xia, L.; Luo, J.; Sun, Y. Deep extraction of cropland parcels from very high-resolution remotely sensed imagery. In Proceedings of the 2018 7th International Conference on Agro-Geoinformatics (Agro-Geoinformatics), Hangzhou, China, 6–9 August 2018; IEEE: Piscataway, NJ, USA, 2018; pp. 1–5.
33. Liu, Y.; Cheng, M.M.; Hu, X. Richer convolutional features for edge detection. In Proceedings of the IEEE Conference on Computer Vision and Pattern Recognition, Honolulu, HI, USA, 21–26 July 2017; pp. 3000–3009.
34. Xie, S.; Tu, Z. Holistically-nested edge detection. In Proceedings of the IEEE International Conference on Computer Vision, Santiago, Chile, 7–13 December 2015; pp. 1395–1403.
35. Waldner, F.; Diakogiannis, F.I. Deep learning on edge: Extracting field boundaries from satellite images with a convolutional neural network. *Remote Sens. Environ.* **2020**, *245*, 111741. [CrossRef]
36. Li, M.; Long, J.; Stein, A. Using a semantic edge-aware multi-task neural network to delineate agricultural parcels from remote sensing images. *ISPRS J. Photogramm. Remote Sens.* **2023**, *200*, 24–40. [CrossRef]
37. Long, J.; Li, M.; Wang, X. Delineation of agricultural fields using multi-task BsiNet from high-resolution satellite images. *Int. J. Appl. Earth Obs. Geoinf.* **2022**, *112*, 102871. [CrossRef]
38. Jiao, W.; Persello, C.; Vosselman, G. PolyR-CNN: R-CNN for end-to-end polygonal building outline extraction. *ISPRS J. Photogramm. Remote Sens.* **2024**, *218*, 33–43. [CrossRef]
39. Xu, Y.; Zhu, Z.; Guo, M. Multiscale edge-guided network for accurate cultivated land parcel boundary extraction from remote sensing images. *IEEE Trans. Geosci. Remote Sens.* **2023**, *62*, 4501020. [CrossRef]
40. Xie, E.; Sun, P.; Song, X. Polarmask: Single shot instance segmentation with polar representation. In Proceedings of the IEEE/CVF Conference on Computer Vision and Pattern Recognition, Seattle, WA, USA, 13–19 June 2020; pp. 12193–12202.
41. Wei, F.; Sun, X.; Li, H. Point-set anchors for object detection, instance segmentation and pose estimation. In *Computer Vision–ECCV 2020, Proceedings of 16th European Conference, Part X 16 Glasgow, UK, 23–28 August 2020*; Springer International Publishing: Cham, Switzerland, 2020; pp. 527–544.
42. Woo, S.; Park, J.; Lee, J.-Y.; Kweon, I.S. CBAM: Convolutional Block Attention Module. *arXiv* **2018**, arXiv:1807.06521. [CrossRef]

43. Girshick, R. Fast r-cnn. In Proceedings of the IEEE International Conference on Computer Vision, Santiago, Chile, 7–13 December 2015; pp. 1440–1448.
44. Persello, C.; Bruzzone, L. A novel protocol for accuracy assessment in classification of very high resolution images. *IEEE Trans. Geosci. Remote Sens.* **2009**, *48*, 1232–1244. [CrossRef]
45. Li, J.; Wei, Y.; Wei, T.; Wei, H. A Comprehensive Deep-Learning Framework for Fine-Grained Farmland Mapping From High-Resolution Images. *IEEE Trans. Geosci. Remote Sens.* **2025**, *63*, 5601215. [CrossRef]

MDPI AG
Grosspeteranlage 5
4052 Basel
Switzerland
Tel.: +41 61 683 77 34

Remote Sensing Editorial Office
E-mail: remotesensing@mdpi.com
www.mdpi.com/journal/remotesensing

www.ingramcontent.com/pod-product-compliance
Lightning Source LLC
Chambersburg PA
CBHW061450210926
51850CB00008B/1667
* 9 7 8 3 7 2 5 8 5 8 4 3 9 *